ADVANCES IN IMAGING AND ELECTRON PHYSICS

VOLUME 105

Advances in

Imaging and Electron Physics

EDITED BY
PETER W. HAWKES
CEMES/Laboratoire d'Optique Electronique
du Centre National de la Recherche Scientifique
Toulouse, France

VOLUME 105

ACADEMIC PRESS
San Diego London Boston New York
Sydney Tokyo Toronto

This book is printed on acid-free paper. ∞

ACADEMIC PRESS
a division of Harcourt Brace & Company
525 B. Street, Suite 1900, San Diego, California 92101-4495, USA
http://www.apnet.com

Academic Press
24-28 Oval Road, London NW1 7DX, UK
http://www.hbuk.co.uk/ap/

International Standard Book Number: 0-12-014747-5

PRINTED IN THE UNITED STATES OF AMERICA
98 99 00 01 02 BB 9 8 7 6 5 4 3 2 1

CONTENTS

Near-Sensor Image Processing

ANDERS ÅSTRÖM AND ROBERT FORCHHEIMER

Digital Image Processing Technology for Scanning Electron Microscopy

EISAKU OHO

Design and Performance of Shadow-Mask Color Cathode Ray Tubes

E. Yamazaki

Electron Gun Systems for Color Cathode Ray Tubes

Hiroshi Suzuki

CONTRIBUTORS

Numbers in parentheses indicate the pages on which the author's contribution begins.

ANDERS ÅSTRÖM (1), Image Processing Laboratory, Department of Electrical Engineering, Linköping University, S-581 83 Linköping, Sweden

ROBERT FORCHHEIMER (1), Image Coding Group, Department of Electrical Engineering, Linköping University, S-581 83 Linköping, Sweden

EISAKU OHO (77), Department of Electrical Engineering, Kogakuin University, 1-24-2, Nishishinjuku, Shinjuku-ku, Tokyo, Japan 163-8677

HIROSHI SUZUKI (267), Matsushita Electric Industrial Co., Ltd., AVC Products Development Laboratory, AVC Company, 1-1 Matsushita-cho, Ibaraki, Osaka 567-0026, Japan

E. YAMAZAKI (141), LG Electronics Inc., Display Device Research Lab, 184 Kongdan-dong, Kumi, Kyoung-buk, 730-030 Korea

PREFACE

This volume contains four chapters, two on very different aspects of image processing and the other two on color cathode ray tubes. The authors of the first contribution, A. Åström and R. Forchheimer, both at the University of Linköping, have been exploring ways of improving the computer architectures used for image processing and have proposed a structure, inspired by the pixel-parallel bit-serial types contemplated long ago, that may well revolutionize the handling of large gray-level or color images. In their contribution, they present their ideas in considerable detail and explain the implications for the usual operations of image processing. The chapter also includes new experimental findings with the result that it is not just an extended survey of earlier material but contains original material as well.

This is followed by a survey by E. Oho from the Kogakuin University, who deals with a topic that I have planned to cover in these *Advances* for many years, and I am particularly glad that so knowledgeable an author as Professor Oho agreed to prepare it. The subject is digital image processing in the scanning electron microscope, where the possibility of image improvement, at first by analogue techniques and later digitally, owing to the sequential acquisition of the information used to generate the image, was realized almost as soon as the instrument became commercially available (in 1965). We find the first papers on the subject as early as 1968. All aspects of the topic are examined here, in the light of modern technology, and I suspect that this survey will be as heavily used as the first extended account of the scanning electron microscope, which appeared in these *Advances* in 1965, by the late Sir Charles Oatley, W. C. Nixon, and R. F. W. Pease.

In about 1929, the early gas-focused cathode-ray tubes were replaced by high-vacuum CRTs and, soon after, proposals began to appear in the patent literature for producing color images with the new tubes. However, none of these ideas seemed sufficiently practical to justify serious development work. In the late 1940s, with commercial black-and-white television well established and growing rapidly, the enormous commercial potential of color TV was recognized. Clearly, it could not become a reality until a satisfactory color tube was available, and so a massive research program was organized in late 1949 at the RCA Research Laboratories to solve this problem. This effort resulted in what is today known as the shadow-mask tube. Unfortunately, the new tubes were much more complex than black-and-white ones and they were also difficult to construct since the dimensional tolerances were far more stringent than any previously encountered. Not surprisingly, the brightness of the first color tubes, which appeared in 1950, was less than a tenth of that of existing black-and-white tubes and the resolution was inferior.

Encouraged by the improvements in these tubes during the next few years, the U.S. government adopted new broadcast standards in 1953. This allowed color and black-and-white TV signals to be broadcast on the existing black-and-white TV channels without any increase in bandwidth. Since no other color CRT that could function with the new color signals was available, this decision effectively ensured that the shadow-mask tube would be used for commercial television. During the next two decades, as successive improvements were made in these tubes, color broadcasting grew steadily in the USA and elsewhere.

The first comprehensive publication on color CRTs appeared in 1974 as a Supplement* to *Advances in Image Pickup and Display* (now amalgamated with these *Advances*). Since then, driven by intense competition among manufacturers, the performance of these tubes has greatly improved. Although the demands of television were paramount, the growing interest in color tubes for computer applications was also an important factor. For this application, however, difficult new performance goals had to be met, notably a substantial increase in resolution, a significant reduction in both flicker and image distortion, and more precise control of colors and their stability. Fortunately, considerable success was achieved in all these areas, thus opening a completely new mass market for color tubes.

The chapters by E. Yamazaki and H. Suzuki are intended to bring the reader up to date on the major developments in color tubes since 1974. The first, on the design and performance of shadow-mask color tubes, covers numerous aspects of the subject, such as improvements in phosphor screens, alternative combinations of shadow-mask and viewing screen structures, techniques for ensuring that the multiple electron beams converge to a common spot, the elimination of moiré patterns in scanned images, contrast enhancement, and problems of minimizing screen curvature.

The chapter by H. Suzuki, on electron gun systems for color CRTs, is primarily concerned with the basic problems of obtaining the high-current-density electron beams required for high performance color tubes. An important topic in this account is the development of asymmetric electron lenses with low aberrations. Other topics discussed are the use of quadrupole lenses to counteract the spot distortions resulting from magnetic deflection as well as the development of new cathodes with high emission current density to meet the exigencies of high-definition television.

**Color Picture Tubes*, by A. M. Morrell, H. B. Law, E. G. Ramberg, and E. W. Herold.

As always, I thank the authors of these surveys most sincerely for the time and effort they have devoted to their contributions and conclude with a list of papers planned for future volumes.

Peter W. Hawkes

Forthcoming Contributions

Mathematical models for natural images	L. Alvarez Leon and J.-M. Morel
Use of the hypermatrix	D. Antzoulatos
Modern map methods for particle optics	M. Berz and colleagues
Magneto-transport as a probe of electron dynamics in semiconductor quantum dots	J. Bird
Artificial intelligence and pattern recognition in microscope image processing	N. Bonnet
Distance transforms	G. Borgefors
Number-theoretic transforms and image processing	S. Boussakta and A. G. J. Holt
Microwave tubes in space	J. A. Dayton
Proton radiation and CCDs	J. Deen, T. Hardy, and R. Murowinski (vol. 106)
Fuzzy morphology	E. R. Dougherty and D. Sinha
Gabor filters and texture analysis	J. M. H. Du Buf
Liquid metal ion sources	R. G. Forbes
X-ray optics	E. Förster and F. N. Chukhovsky
The critical-voltage effect	A. Fox
Stack filtering	M. Gabbouj
The development of electron microscopy in Spain	M. I. Herrera and L. Brú
Contrast transfer and crystal images	K. Ishizuka
Conservation laws in electromagnetics	C. Jeffries
External optical feedback effects in semiconductor lasers	M. A. Karim and M. F. Alam
Numerical methods in particle optics	E. Kasper

Positron microscopy	G. Kögel
Spin-polarized SEM	K. Koike
Development and applications of a new deep-level transient spectroscopy method and new averaging techniques	P. V. Kolev and M. Jamal Deen
Sideband imaging	W. Krakow
Computer-aided design using Green's functions and finite elements: the case of inhomogeneous ferrite microstrip circulators	C. M. Krowne (vol. 106)
Memoir of J. B. Le Poole	A. van de Laak-Tijssen, E. Coets, and T. Mulvey
Well-composed sets	L. J. Latecki
Vector transformation	W. Li
Complex wavelets	J.-M. Lina, B. Goulard, and P. Turcotte
Discrete geometry in image processing	S. Marchand-Maillet (vol. 106)
Plasma displays	S. Mikoshiba and F. L. Curzon
Electronic tools in parapsychology	R. L. Morris
Z-contrast in the STEM and its applications	P. D. Nellist and S. J. Pennycook
Phase-space treatment of photon beams	G. Nemes
Electron image simulation	M. A. O'Keefe
Representation of image operators	B. Olstad
Aharonov-Bohm scattering	M. Omote and S. Sakoda
Fractional Fourier transforms	H. M. Ozaktas (vol. 106)
Geometric methods of treating energy transport phenomena	C. Passow
HDTV	E. Petajan
Scattering and recoil imaging and spectrometry	J. W. Rabalais
The wave-particle dualism	H. Rauch
Digital analysis of lattice images (DALI)	A. Rosenauer
Electron holography	D. Saldin
X-ray microscopy	G. Schmahl
Accelerator mass spectroscopy	J. P. F. Sellschop
Focus-deflection systems and their applications	T. Soma
Hexagonal sampling in image processing	R. Staunton
Recent developments in confocal microscopy	E. H. F. Stelzer (vol. 106)

Study of complex fluids by transmission electron microscopy	I. Talmon
New developments in ferroelectrics	J. Toulouse
Organic electroluminescence, materials, and devices	T. Tsutsui and Z. Dechun
Electron gun optics	Y. Uchikawa
Very high resolution electron microscopy	D. van Dyck
Mathematical morphology and scanned probe microscopy	J. S. Villarrubia
Morphology on graphs	L. Vincent
Representation theory and invariant neural networks	J. Wood
Magnetic force microscopy	C. D. Wright and E. W. Hill

ADVANCES IN IMAGING AND ELECTRON PHYSICS

VOLUME 105

ADVANCES IN IMAGING AND ELECTRON PHYSICS, VOL. 105

Near-Sensor Image Processing

ANDERS ÅSTRÖM

Image Processing Laboratory, Department of Electrical Engineering, Linköping University, S-581 83 Linköping, Sweden

and

ROBERT FORCHHEIMER

Image Coding Group, Department of Electrical Engineering, Linköping University, S-581 83 Linköping, Sweden

Parts of the material in this chapter are reused with the permission from IEEE.

ISBN 0-12-014747-5
ISSN 1076-5670/99 $30.00

I. Introduction

Most image processing applications are burdened with high volumes of data that have to be funneled through the image processing system. This is especially true in the early part of the system where the individual pixels are represented by full gray scale or color. Therefore, all efforts leading to a reduction of the pictorial data without too much loss of information are of interest. With the increased complexity that can be achieved by today's and tomorrow's silicon chips it is further tempting to revisit parallel architectures. Some of these architectures were suggested more than 30 years ago, namely, the pixel-parallel, bit-serial types, which will form the underlying theme of the current presentation (Reddaway, 1973).

We will introduce an aspect that was not taken into account in the earlier works, namely the added functionality gained by integrating also the photosensitive element into the distributed processing array. As it turns out, this is not only an issue of achieving a more compact device, such as a single chip image sensor and processor. If done properly, the added functionality will in fact alter the way image processing tasks will be approached and solved!

This presentation, an extension of material that was first published in Forchheimer and Åström (1994) and Åström *et al.* (1996), includes new experimental results and a more thorough discussion of applications. To understand the basic concept we will start by presenting a typical image processing sequence consisting of image acquisition, noise reduction and filtering, segmentation, feature extraction, and classification (see Fig. 1). The two left columns show examples of operations for each task, and the amount of input/output data for each operation. Here, N stands for the image size (square images assumed) and b for the number of bits in the input image. This type of image processing sequence is very common in both robot vision applications and industrial inspection tasks.

The amount of input data to the four first processing tasks are of order N^2, as shown in the rightmost column in Fig. 1. The first three stages consist of

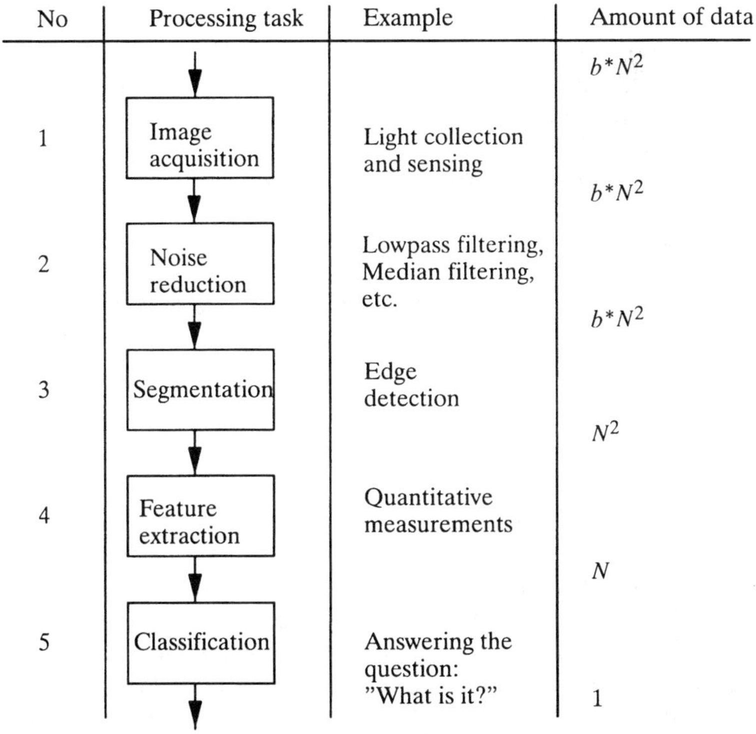

No	Processing task	Example	Amount of data
			b^*N^2
1	Image acquisition	Light collection and sensing	
			b^*N^2
2	Noise reduction	Lowpass filtering, Median filtering, etc.	
			b^*N^2
3	Segmentation	Edge detection	
			N^2
4	Feature extraction	Quantitative measurements	
			N
5	Classification	Answering the question: "What is it?"	
			1

FIGURE 1. Image processing sequence.

local operations while the fourth stage is a global feature measuring operation. As the input data for local neighborhood data operations are of similar type, this suggests that a physically integrated computer architecture is feasible and, as it all starts with an optical sensor, the ideal system for the four first stages would be a single solid-state component. The output from such a system would then be a feature vector to be forwarded to a classification system. Thus, what we are looking for is a system that accepts a time sequence of images $f(x, y, t)$, and for each image outputs a feature vector $\bar{g}(t)$, as shown in Fig. 2.

The architectures to be presented are derivations of the classical idea of the massively parallel single instruction multiple data (SIMD) processor (Batcher, 1980), (Fountain, 1987), and (Tucker and Robertson, 1988). However, when implemented on a single chip it will be seen that new constraints as well as new possibilities appear, to the extent that also the most basic image processing algorithms are affected. To decide what type of basic operations one would like to have available within such a component, it is necessary to

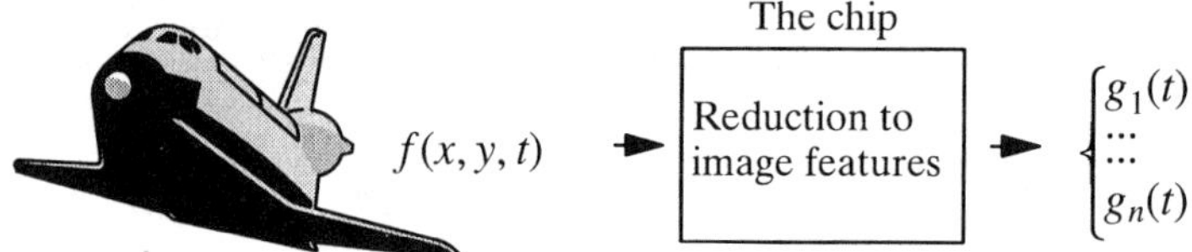

FIGURE 2. The goal for the image information reduction on a chip.

TABLE I
TRANSISTOR COUNT

Module	Transistors (A)	Transistors (B)	Transistors (C)
Sensor	1	1	1
ADC	110	90	10
ALU	100	60	60
Memory	110	50	50
Total	320	200	120

look at the cost, that is, the very large scale integration (VLSI) area occupied by the hardware supporting each operation. For the individual cell in a 2D array of sensor-processing elements (SPE), a minimal configuration would consist of a photodiode, an analog-to-digital-converter (ADC), a bit serial arithmetic and logic unit (ALU), and some memory for each pixel position. An important benchmark is the transistor count within each such SPE. Table I shows the transistor count for three different cases (A)–(C) containing sensor and processing units and with a memory of 16 bits per SPE.

In the case (A) we use "standard cell" technology. One possible realization of such an SPE is shown in Fig. 3. This design has a lot in common with PASIC (Forchheimer *et al.*, 1993) as it consists of one photosensor, one ADC, and a 1-bit bus connecting the ADC with three registers and a memory. Attached to the registers is a full adder logic, which enables us to perform all types of Boolean operations. Gates for 2D mesh connection are included. A 256×256 SPE array of this type would require 21 million transistors, which is far too much for present day technology.

The case (B) in Table I assumes an optimized VLSI design for this SPE, where dynamic circuitry has been used where appropriate. A 256×256 SPE array still requires 13 million transistors.

The last case (C) corresponds to the new concept of this chapter where the ADC has been removed and the memory per SPE has been reduced to 10 bits which leaves us with a moderate 6.5 million transistors for the 256×256 array of SPEs. We believe that this is a feasible number for a chip design in the near future.

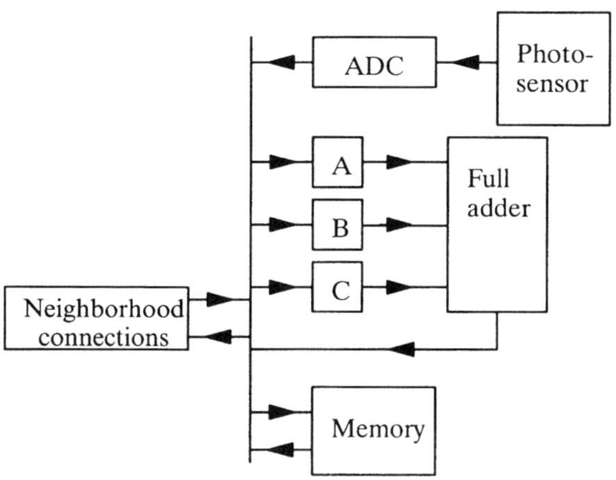

FIGURE 3. Logic scheme of a minimum configuration SPE based on traditional architecture.

Actually, in the sequel we will show that conventional A/D conversion is neither necessary nor particularly efficient. Also, by using nonconventional digital representation we will be able to save on memory registers. The concept that makes this possible is hereby coined Near Sensor Image Processing (NSIP).

To fully understand the NSIP concept it is essential to have a working model of the photo-electronic process that takes place within each SPE. For the discussion to be meaningful it is also necessary to define the way in which the analog sensor signal is converted into a measurable entity. These two topics are treated in section II. For the sake of simplicity we will start to describe NSIP for a one-dimensional SPE array that has been used in practice, the LAPP1100 (Forchheimer and Ödmark, 1983). Section III describes a number of simple NSIP operations and how they are implemented. It also describes how the special NSIP data representation is used throughout the system. Section IV describes adaptive thresholding while section V presents linear and nonlinear grayscale mapping techniques. Grayscale operations are discussed in sections VI and VII for grayscale morphology and linear convolutions, respectively. In section VIII we present a 2D SPE model and describe the consequences for architecture and programming. Section IX describes how global operations on the array are performed in 2D and section X how features are extracted in 2D using the COUNT net. Section XI describes an NSIP implementation and some resulting images from that system. Section XII gives an example to show that the NSIP concept could be implemented in special purpose sensors systems.

II. A Near-Sensor Image Processor

A. *The Photodiode Sensor*

The photoelectric effect is the basic physical principle used in all light sensors. Under certain conditions it is possible for a photon that hits a material to leave its energy to an electron. This causes a charge to build up in the sensor material. In a charge-coupled device (CCD) photoelement used in many TV cameras, the charge is collected in a well created by applying suitable voltages to a system of electrodes. During exposure the charge will grow until the well is full (overexposure) or until the charge is forced to move away.

The *photodiode sensor* utilizes the photoelectric effect differently, see Fig. 4. Here, the photocurrent is used to discharge a small capacitor, which initially has been charged to a nominal level. Thus, the remaining voltage over the capacitor will be an indicator of the accumulated light over the exposure time. It is essential to sample this voltage level before the capacitor is completely discharged, or overexposure (saturation) will result. Typically, the capacitor is obtained for free as a byproduct of the inversely biased diode (p/n-junction) in a semiconductor material. For silicon-based material the spectral response of a photodiode is rather broad; from infrared (1100 nm) to ultraviolet (300 nm).

There are several reasons to choose a photodiode element as the sensing device in machine vision applications. This sensor takes up more space than a CCD element but it is characterized by low noise, good uniformity, and somewhat higher sensitivity in the blue region. The most important property, however, is the possibility to read out information nondestructively through the use of a high-impedance amplifier/comparator. As will be seen, this property is essential for several of the things we want to accomplish.

Figure 4 shows schematically the photodiode sensor and its thresholding amplifier. By using a high gain differential amplifier, the output of the sensor will be a binary signal suitable for further digital processing. At first sight it may seem a considerable restriction to forward only binary values from the sensor. Disregarding this for the time being, however, let us note that the "infinite gain" operation taking place directly at the output of the sensor device is very robust to component variations and ensures that good response uniformity is maintained throughout an array of SPEs.

The relationship between light intensity I and voltage U over the photodiode during exposure can be approximated by the function:

$$U(t) = U_0 - \int_0^t kI(\tau)d\tau \qquad (1)$$

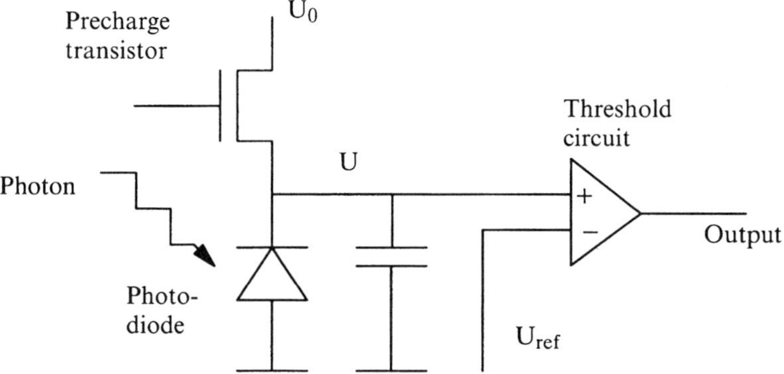

FIGURE 4. The photodiode element and sense amplifier.

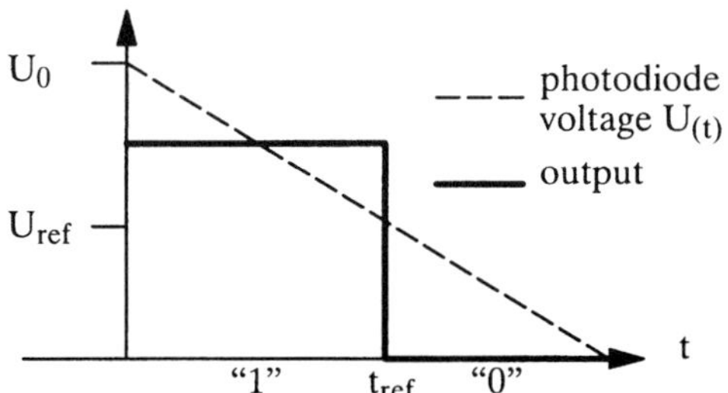

FIGURE 5. The voltage over the photodiode and the output from the comparator.

In the case in which the intensity $I(t)$ is constant over time t, we have

$$U(t) = U_0 - kIt \tag{2}$$

which is shown in Fig. 5. Here, t is the time interval from the precharging of the photodiode to the sampling moment, k is a constant that describes the light sensitivity of the photodiode, and U_0 is the precharge voltage. The relationship is valid as long as $U > 0$.

The time it takes for the voltage U to discharge to U_{ref} at a constant intensity I is

$$t_{\text{ref}} = \frac{U_0 - U_{\text{ref}}}{kI} \tag{3}$$

It is obvious that we can use the device in Fig. 4 to measure the light intensity in at least two ways. We may keep the exposure time constant and, after a complete integration time, we may vary the threshold level U_{ref}. The binary output reveals when a U_{ref} equal to the photodiode voltage is found. This corresponds to the action of an ordinary (CCD or photodiode) sensor system in which a fixed exposure time is used, after which the analog output is subjected to an analog to digital (A/D)-converter. The maximum dynamic range will be approximately 1000 : 1.

Alternatively we may keep the threshold at a constant value and measure the time interval t_{ref}, shown in Fig. 5. This is the NSIP approach. It yields a dynamic range in the order of 10^5:1 because valid exposure times may well range from 1 μs to 100 ms.

All NSIP algorithms are based on the observation of when the change of the binary output takes place. However, the output from a single photodiode, such as the one in Fig. 4, is not tested individually. Rather, all output data from an array of photodiodes are brought in for processing in parallel. In fact, the sensors are **interrogated** at even or uneven time intervals in the main NSIP loop during ongoing exposure. Hence, nondestructive read-out from the photodiode is essential to this concept.

Assume that the interrogation interval Δt is constant. Then, interrogation event number i where

$$i = 1, 2, \ldots, N$$

yields a linear relation to the interrogation points in time $t_1, t_2, \ldots t_N$

$$t_i = t_1 + (i - 1)\Delta t \tag{4}$$

that is shown in Fig. 6.

Suppose that a sensor element changes its output at time t_{ref}, where

$$t_i \leq t_{\text{ref}} < t_{i+1} \tag{5}$$

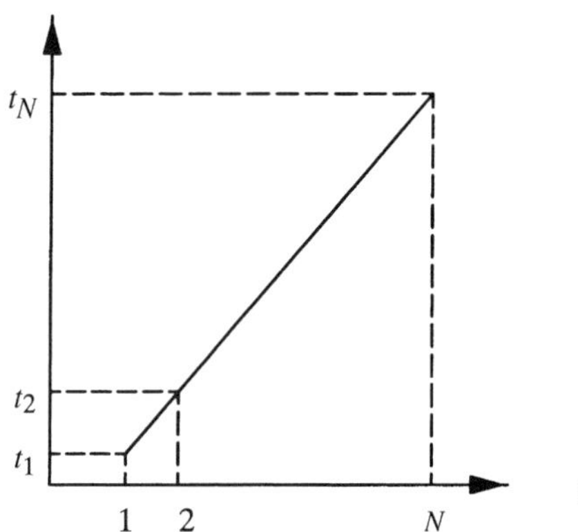

FIGURE 6. The relation between interrogation index and time.

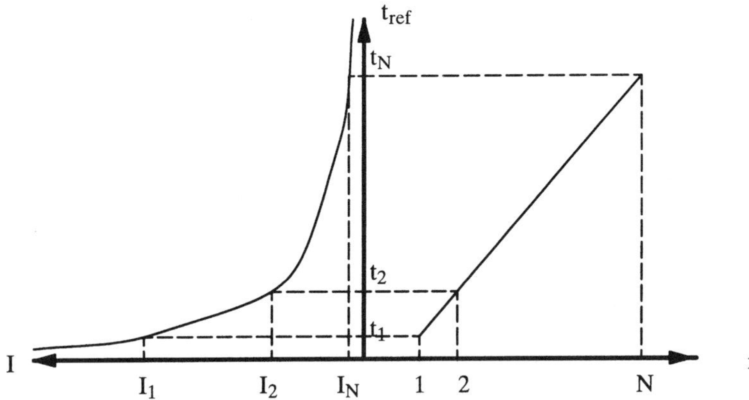

FIGURE 7. The relation between index, time, and intensity.

If the time interval Δt_i is small we assume

$$t_{\text{ref}} = t_i \doteq t(i) \tag{6}$$

For this sensor element we then have that

$$t(i) = t_1 + (i-1)\Delta t = \frac{U_0 - U_{\text{ref}}}{kI} \tag{7}$$

This gives us the intensity I as a function of the index i

$$I(i) = \frac{U_0 - U_{\text{ref}}}{k(t_1 + (i-1)\Delta t)} \tag{8}$$

Note that although the intensity is constant in time t for each sensor in Eq. (8), which is the I-value that causes switching at time t_{ref}, I is now a function of t_{ref} and, hereby, also of the index i as indicated in Fig. 7. As a consequence, the output quantity i, which is the only one the system can make use of, has an inverse linear dependence on the pixel intensity I according to Eq. (8).

B. N1D — A One-Dimensional NSIP Architecture

The architecture we will use in what follows to illustrate NSIP in one dimension, which we call N1D, is based on the LAPP1100 (Forchheimer and Ödmark, 1983). LAPP1100 is a traditional bit-serial single stream instruction multiple data (SIMD) parallel architecture found in a number of different machines, including MPP (Batcher, 1980), CLIP-7 (Fountain, 1987), and CM-2 (Tucker and Robertson, 1988).

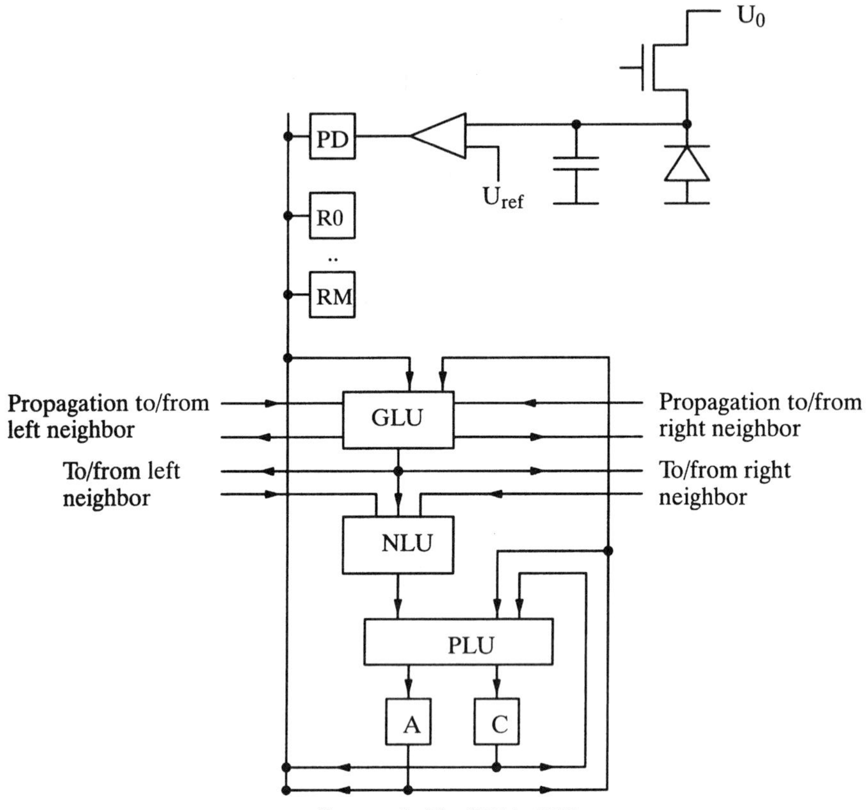

FIGURE 8. The SPE in N1D.

N1D is a 1D processor array with some extensions to handle image operations of a global nature. The processor is register-accumulator oriented and typical operations use one of the registers as the first operand and the accumulator as the second operand. The result is stored into the accumulator, thereby destroying its previous value. Multilevel (grayscale) operations are performed bitwise using the registers for intermediate storage (bit-serial operations). In the description that follows, in general, we set the array size to N and the number of registers to M. Figure 8 shows the architecture of one SPE.

The ALU is purely combinational and permits point operations (PLU), neighborhood operations of type 3×1 template matching (NLU), and global operations (to be explained). Operations belonging to different classes can be combined into one instruction. As an example, the instruction

$$\text{and } (010)\ r_5$$

will access register 5 and will then apply local template matching with (010) to each position. All matching positions are set to **1** and then AND-ed with the previous value of the accumulator. The result is that only those **1**s in the accumulator which correspond to isolated 1s in register 5 will remain as **1**s in the accumulator. This instruction is performed in one clock cycle. In place of r_5 one could write *pd* for the photodiode, *a* for the accumulator, or *c* for the carry register. The local template can be omitted, which is equivalent to using the template (x1x).

A typical GLU instruction is

lfill a

that sets all the accumulator bits to **1** if they are to the right of the leftmost **1**. The architecture of the GLU part will be discussed more in Section IX.

One instruction is particular to the photodiode array. This instruction is

initpd

that precharges the photodiode as described in the previous section. Sampling (read-out) of the binarized photodiode output is done whenever *pd* is used as an operand. Thus

ld pd

will store the current value in the accumulator. The accumulator content will then depend on the exposure time, which is given by the time elapsed between the "initpd" and the "ld pd" instructions. As already mentioned, the readout is nondestructive, which makes it possible to do several "ld pd" instructions during the exposure.

The N1D chip is equipped with an output status register called COUNT, which delivers a count of the number of PEs, which has a **1** in its accumulator registers.

The N1D chip communicates with a controlling microprocessor. The microprocessor issues instructions to the N1D and uses the status register to read out the result. A list of the instructions used in this paper follows. A Pascal-like notation will also be used throughout.

ld pd	Load accumulator with comparator output
ldi pd	Load accumulator with inverse value of comparator output
initpd	Precharge (initialize) the photodiode
lfill a	Set all accumulator bits to **1** that are to the right of the leftmost **1**
st r_0	Store the content of the accumulator in register 0
clr a	Set the accumulator to 0
or r_0	Or the accumulator with reg 0 and put the result in the accumulator

xor r_0 Exclusive-Or the accumulator with reg 0 and put the result in the accumulator

and r_0 And the accumulator with reg 0 and put the result in the accumulator

The following final comments are needed. The length of the programming examples to follow are deliberately kept to a minimum. Even so, NSIP programs in general often turn out to be very compact. It should be noted that this architecture is just one out of many possible architectures where the NSIP concept is applicable. We have chosen this LAPP-based architect because of its simplicity and because it has been tested and found to work well in practice. In the sequel most of the operations and algorithms will be described and exemplified using a linear array. However, one should bear in mind that the principal results are applicable also to 2D NSIP systems, which will be dealt with starting in Section VIII.

III. Near-Sensor Image Processing Algorithms

A. *General*

Practically all NSIP algorithms have the structure illustrated by the flowchart in Fig. 9. First, the photodiode in each SPE is initialized (=precharged). Then, a program loop is executed and for each pass the outputs from the comparators are interrogated and some logical operations are performed. When a certain condition is met the loop is terminated and a final feature extraction is performed.

In pseudo PASCAL, Fig. 9 corresponds to:

```
Initialize photodiode
REPEAT
   Interrogate the output and perform operations
UNTIL Condition is met
Perform feature extraction
```

The interrogation/execution loop, the foundation of the NSIP concept, is easily recognized in the routines that we will describe in what follows.

B. *Finding the Position of Maximum Intensity*

This is a very simple, but yet illustrative example of NSIP. After precharge, which takes place at time $t = 0$, we interrogate the array of the photodiodes repeatedly. For a 1D row of SPEs the initial status is shown in Fig. 10a, where

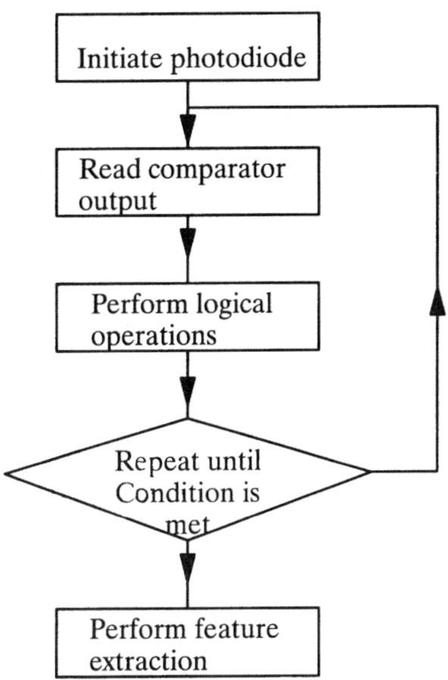

FIGURE 9. The flowchart for NSIP algorithms.

all the N diode voltages are shown along the x-axis. Then, the photodiodes will be discharged at a rate that is proportional to the incoming light. At a certain point in time $t > 0$ the voltage over the photodiode array may look like that found in Fig. 10b. During the discharge process we continuously interrogate the output from all comparators to see if the threshold has been crossed somewhere. When this occurs, the loop is terminated and we obtain the position of the pixel with the highest intensity by performing an LFILL operation on the accumulator followed by a COUNT (see Fig. 10c). COUNT outputs the number of set bits.

Using NSIP syntax this procedure can be described as:

```
initpd              ; Initialize the photodiode
REPEAT
  ldi pd            ; Load accumulator with inverse value of output
UNTIL COUNT < > 0
lfill a             ; Fill to the right
answer: = COUNT
```

The "initpd" statement precharges the voltage over the photodiodes in the array to U_0. The "ldi pd" instruction loads the inverted values of the photodiode comparator into the accumulator. Thus, zeroes will be loaded as

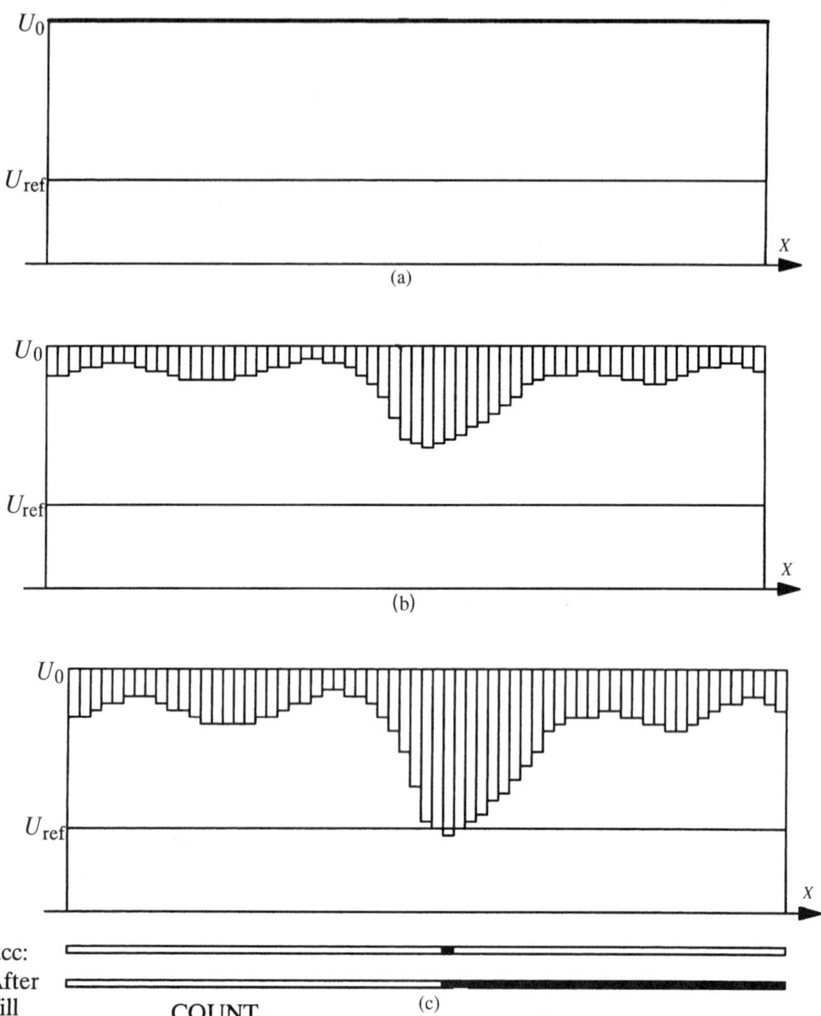

FIGURE 10. (a) Photodiode array status at $t = 0$; (b) photodiode array status at $t > 0$; (c) photodiode array status when brightest intensity reaches U_{ref}.

long as none of the diodes have reached the threshold level. The "UNTIL COUNT< >0" is the conditional branching operation that makes use of the global operation "COUNT". The "lfill" is a global instruction that sets all bits to the right of the leftmost "1". This will work as a tie-breaker in case several "1"s occur during the same sampling period but also, as a means to obtain the position of the leftmost "1" by reading the status variable COUNT. The status of the array after the termination of the loop is illustrated in Fig. 10c. At the bottom the accumulator content is shown at the instance of the first threshold passage and after the "lfill" operation.

This technique of finding the position of the maximum intensity is used in the sheet-of-light (SOL) application to be described in Section XII.

Finding the position of the **minimum** intensity requires the following changes to the preceding program:

```
initpd              ; Initialize the photodiode
REPEAT
  st r0             ; Store previous value of output
  ld pd             ; Load accumulator with noninverse value
UNTIL COUNT=0       ; Until all have passed
lfill r0            ; Fill previous readout to the right
answer: = COUNT
```

For comparison, in a conventional image processing system, the position of the maximum intensity is found in the following way. First, we will acquire the image by exposing the photosensitive sensor (CCD/photodiode/. . .) a given time. This exposure time should be set to ensure that the maximal value is not over- or underexposed. The analog values which we get from the sensor are then A/D-converted with an appropriate precision to ensure that we can distinguish between the largest and the second largest intensity. We then scan the sensor values, comparing each value with the maximum value found so far and exchange that value and its corresponding position when we find a larger value.

Note that in the traditional method we have to decide on an exposure time beforehand. In NSIP the exposure time depends on the light condition as it is exposed until at least one position accumulates enough light energy. Thus, tolerance to various light conditions has been traded for data-dependent non-fixed exposure time. To obtain such adaptation in traditional systems an outer control loop is needed to modify the exposure time and maintain a reasonable signal-to-noise (S/N) ratio. All this comes for free with NSIP.

In this NSIP example, the self-adaptivity to various light conditions is not limitless, however. Extremely low intensity levels would require extremely long and data-dependent operational times, which might be prohibited in a real-time application. To a certain extent long operational times can be shortened by decreasing $U_0 - U_{\text{ref}}$ but only at the cost of lower S/N ratio and the reduced possibility of distinguishing between high intensities. In fact, for any NSIP-application the global controllable system parameter $U_0 - U_{\text{ref}}$ should be optimized as a trade-off between speed and noise sensitivity.

C. Detecting Positive Gradients

This operation is equivalent to computation of the gradient (1D gradient = derivative) followed by thresholding with the threshold set to 0. The gradient kernel to be used is $\boxed{1}\boxed{-1}$.

The photodiodes are precharged at time $t = 0$ and then exposed by light. Then, the processor array executes a program that interrogates each pixel position to see if the voltage has passed the threshold while its left neighbor has not. If this is the case a register flag is set in that particular SPE. Figure 11a shows the status at a time $t > 0$ when there is one positive gradient which has just barely passed the threshold U_{ref}. Figure 11b shows the final result when all the voltages have passed the threshold and all the positive gradient pixels have been ORed together.

This operation is insensitive to various light intensities as it measures only the differences, and not the absolute value, of neighboring pixels. However, for very dark pixels the time-out for the total operation might prevent recording of their gradients. Also, if two neighbors have almost the same

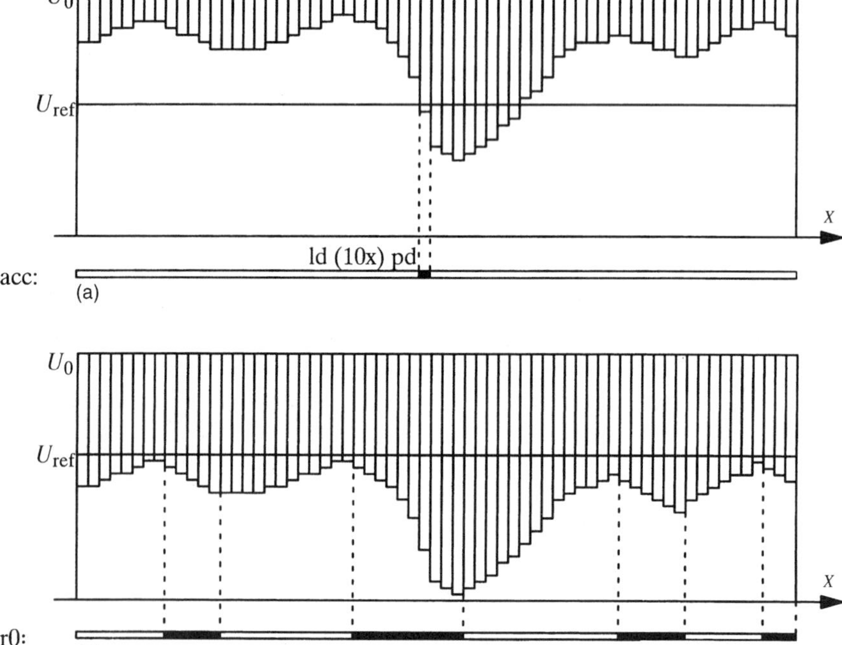

FIGURE 11. (a) A positive gradient is detected; (b) the result when all pixels are below the threshold voltage.

intensity they may change their binary outputs so close in time that their gradient escapes detection.

The following NSIP program will perform detection of positive gradients:

```
clr a
st r0
initpd
REPEAT
   ld (10x) a
   or r0
   st r0
   ld pd
UNTIL COUNT=0
```

The first two instructions clear the content of the accumulator and register 0. In the loop a "1" is accumulated in register 0 for all those pixels that have passed the threshold while their left neighbor has not. The "x" used in the template field in the ld instruction means "don't care." The loop continues until all photodiodes have passed the threshold (COUNT = 0).

D. Median and Rank-Order Filtering

Traditionally, median filtering is performed as a series of comparisons and multiplexing of grayscale values. Figure 12a shows how a three-way median operation can be performed with one max/min-, one max-, and one min-operation; the max/min-operation is shown in Fig. 12b.

In NSIP we use the fact that for each neighborhood of three, the second pixel to cross the threshold is the median pixel. We can generate the median value in the neighborhood in the same representation as the input, that is, a signal that is 0 before the median pixel has passed the U_{ref}, and 1 thereafter. If we assume that we use a 1×3 neighborhood, the Boolean function, which sets the output ($= r_2$) to one when the median pixel in the neighborhood passes, is

$$\begin{aligned} r_2 &= (1\mathrm{xx})\ r_1\ \&\ r_1 + r_1\ \&\ (\mathrm{xx}1)\ r_1 + (1\mathrm{xx})\ r_1\ \&\ (\mathrm{xx}1)\ r_1 \\ &= (11\mathrm{x})\ r_1 + (\mathrm{x}11)\ r_1 + (1\mathrm{x}1)\ r_1 \end{aligned} \tag{9}$$

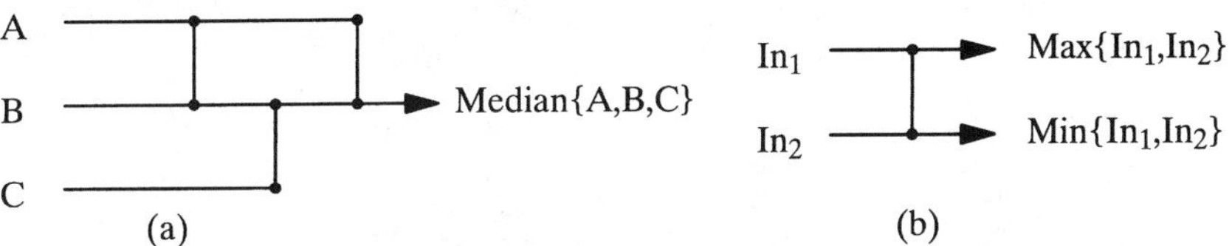

FIGURE 12. Three-input median operation.

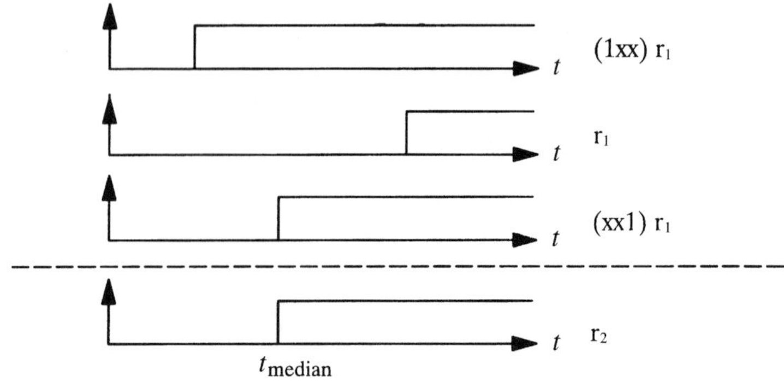

FIGURE 13. The median as a logic function of the three inputs.

Figure 13 shows the function of Eq. (9) graphically. The t-value when r_2 switches represents the median for the neighborhood as

$$I_{\text{median}} = \frac{U_0 - U_{\text{ref}}}{kt_{\text{median}}} \tag{10}$$

The NSIP program for this procedure is as follows:

```
initpd
ldi pd
REPEAT
   st r1
   ld (11x) r1          ;
   or (1x1) r1          ; from Eq. (9)
   or (x11) r1          ;
   st r2
   . .
   ldi pd
UNTIL COUNT=N
```

Here, r_1 holds the comparator output and r_2 the median. As will be shown in section F, this median signal can be used as input to other NSIP operations interwoven in the same program.

In the same manner we can obtain any rank-order function. For instance, a max function is implemented by

$$r_2 = (1\text{xx})\ r_1 + r_1 + (\text{xx}1)\ r_1 = \overline{(000)\ r_1} \tag{11}$$

and a corresponding min function by

$$r_2 = (1\text{xx})\ r_1\ \&\ r_1\ \&\ (\text{xx}1)\ r_1 = (111)\ r_1 \tag{12}$$

The right-hand sides of Eqs. (9), (11), and (12) are all positive Boolean functions. (A positive Boolean function may always be expressed with noninverted input variables.) Evaluation of such functions can be done with so-called *stack filters* (Wendt *et al.*, 1986), which require the input variables to be monotonically growing during evaluation. Thus, a binary variable may not switch back from value 1 to value 0 once it has been set. Obviously, this condition is fulfilled in our case as illustrated in Fig. 13 and, therefore, the NSIP implementation may be called stack filter. However, the signal representation illustrated in Fig. 13 is also equivalent with the **Umbra transform** (Sternberg, 1986) that will be discussed here.

E. Histogramming

The number of SPEs where the photodiode voltage has dropped below U_{ref} varies over time, for instance as shown in Fig. 14. This is the integral of the histogram of the image $t_{\mathrm{ref}}(\mathrm{x})$ that will be referred to as Inthist. To obtain the histogram we should differentiate the Inthist data.

Program for differentiating the inthist:

```
Old_Count: = 0
initpd
FOR i: = 1 TO n DO
   ldi pd
   hist[i]: = COUNT – Old_Count
   Old_Count: = COUNT
END–FOR
```

An alternative is to count only those SPEs for which the voltage passed the U_{ref} since the last interrogation. Obviously, this will give us the histogram directly.

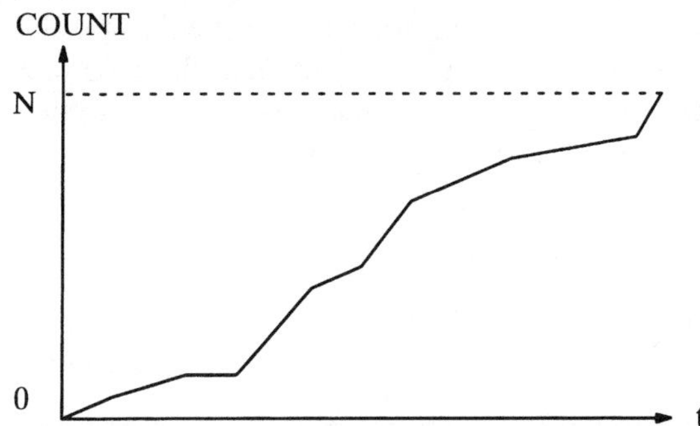

FIGURE 14. The content of COUNT as a function of time = Inthist.

Program for histogramming:

```
clr a
st r0
initpd
FOR i: = 1 TO n DO
   ldi pd
   xor r0                ; accumulator is 1 only when passing threshold
   hist[i]: = COUNT
   or r0
   st r0
END–FOR
```

The variable n corresponds to the number of bins, and hist[i] is the resulting histogram. In each SPE r_0 is used to signify that this SPE has already contributed to the histogram. The "xor r_0" instruction will set those pixels that contribute to the most recent bin.

It should be observed that histograms created by the forementioned programs are nonlinear in light intensity. This is due to the nonlinear relation (3) between the light intensity I and the time t_{ref}. Different schemes to obtain linear and nonlinear relations are discussed in Section V. It should also be noted that storage of the histogram data takes place in the control unit (host computer), not in the NSIP array.

F. Combining Operations

Image processing tasks often require a combination of several image operations. As an example, Fig. 15 shows a sequence of operations used to measure the area occupied by positive gradients. The image is first subjected to median filtering to suppress noise. We then detect where we have positive gradients. Finally, we count the number of these pixels to get the total area.

The median filter routine is obtained from Section D, Eq. (9), which delivers an output set to one as soon as the median pixel has passed the threshold. This is used by the positive gradient detector in Section C. The output from the combined NSIP routine is a count that represents the total area of positive gradients.

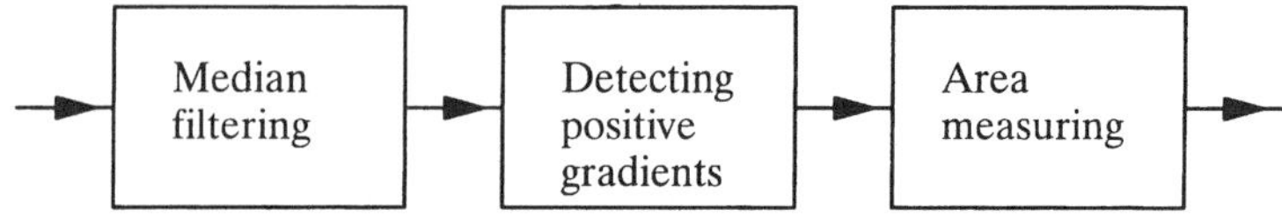

FIGURE 15. A three-stage image processing operation.

```
clr a
st r1
initpd
REPEAT
   ldi pd                          ; Input
   st r0
   ld (11x) r0
   or (1x1) r0
   or (x11) r0                     ; Median = ab + ac + bc
   st r0

   ld (01x) r0                     ; Positive gradient ?
   or r1
   st r1

   ld r0
UNTIL COUNT = n
ld r1
C := COUNT                         ; Area/length
```

The preceding example underscores some of the principles of NSIP. The first processing stage is the sensor and comparator, which delivers information about the t_{ref}-value in its special "thermometer" representation. This is input to the median filter which has the same representation as output that is fed to the positive gradient detector. The output from the gradient detector is not delivered in thermometer representation, however. Instead, the output is accumulated over time and when the interrogation loop is finished, a global count of the result can be achieved.

The general structure of a composite NSIP program is as follows:

```
Initializations
initpd
REPEAT
   ldi pd                          ; Input
   st r0
   r1 := Operation_1(r0)
   r2 := Operation_2(r1)
   ...
   rm := Operation_m(rm - 1)
   ld ri                           ; Termination control operand
UNTIL COUNT = n
ld rm
C := COUNT                         ; Feature extraction
```

All the operations in the inner loop are Boolean and the data transfers between the operations are implicit inside the loop.

IV. Adaptive Thresholding

A. Using the Maximum Intensity

Adaptive thresholding means that with a given threshold we want to find an exposure time so that the binarized image is optimal with respect to some parameter or quality. In the first case to be presented here, optimal thresholding is defined by the following procedure (see Fig. 16). Initially U_{ref} is set to 10% of U_0. When the first pixel in the image passes this threshold U_{ref} is changed to 55% of U_0. The binary image is then read out.

A generalization of the same idea would be to set the initial threshold to $A\%$ of U_0 and the read-out threshold to $(50 + \mathrm{A}/2)\%$. In an ideal noise-free situation the binarized result will not depend on A. Higher thresholds will automatically bring about shorter integration times. The advantages and disadvantages for high and low thresholds were discussed in Section B.

Program for performing the operation in Fig. 16:

```
ref: = 0.1 * U0
initpd
REPEAT
  ldi pd
UNTIL COUNT< >0
ref: = 0.55 * U0
ldi pd
```

The variable "ref" holds the value of the reference voltage. First, a low value is used to ensure that the highest intensity pixel exposes the photodiode array up to 90% of the saturation level. When this has happened the reference is set to the middle of the amplitude range and a new sample of the image line is read into the accumulator.

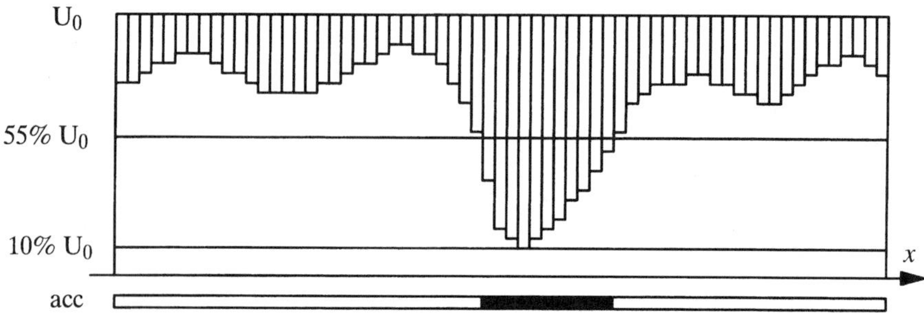

FIGURE 16. The result from the adaptive exposure.

B. *Pattern-Oriented Thresholding*

Another variation of adaptive thresholding is to interrogate the photodiode array until the pixels of an expected object or a specific pattern has passed the threshold and comes forward as a binary output. For example, if we want to detect the position of the object with the size S pixels in a single-object image, we can use the following simple binarizer.

```
initpd
REPEAT
  ldi pd
UNTIL COUNT > S
ld pd
lfill a                    ; Fill to the right
answer: = COUNT
```

In a more complex case we may have noise or unwanted light flashes in the image. Then several smaller objects (noise spikes) are to be expected. To avoid these we interrogate until an object larger than a predefined size shows up. The size of an object is obtained with the operations *ld (01x) lfill* and *mark*, which together reset all pixels except for the pixels in the leftmost object. The net effect of screening out small objects is a kind of median filtering.

```
initpd
REPEAT
    ldi pd
    st r0
    ledge r0               ; Get no of objects
    Max_size: = 0
    WHILE COUNT > 0
    ld (01x) lfill r0      ; Set leftmost pixel
    mark r0                ; Mark leftmost object

    Max_size: = max(Max_size,COUNT)

    xor r0                 ; Remove leftmost object from image
    st r0
    ledge r0               ; Get no of objects
  END–WHILE
UNTIL Max_size > constant
ld pd
```

Several other types of assumption and *a priori* information about the object can be utilized in the NSIP concept. It is interesting to note that a device that exploits a similar technique to achieve robust thresholding was described by

Lyon (1980). Here, a hard-wired logic network interacts with the photodiodes to achieve a particular sensitivity to the periodic pattern which is printed on a highly reflective board.

C. Histogram-Based Thresholding

In image processing it is very common to assume a bimodal histogram and to set the threshold at the min-point between the two maxima. Figure 17 shows a typical bimodal histogram and, ideally, the histogram contains two different Gaussian distributions that correspond to the background and the object. The min-point has a positive zero-crossing of the first derivative of the histogram.

In a well-behaved bimodal histogram we can easily compute the first derivative of the histogram by approximating the derivation with some differentiating operator. In the result *Hist'(I)* the point T is located where the sign has its positive zero crossing, that is, where the sign changes from minus to plus (see Fig. 18).

In the case of a discrete histogram the choice of the derivative estimator is critical. For the very noisy histogram shown in Fig. 19, the corresponding

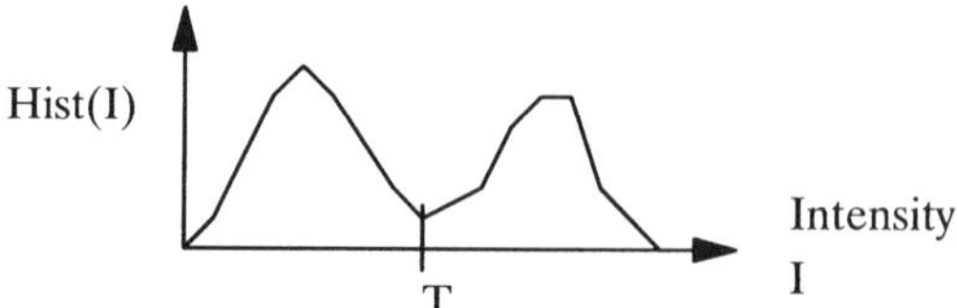

FIGURE 17. A histogram and its optimal thresholding.

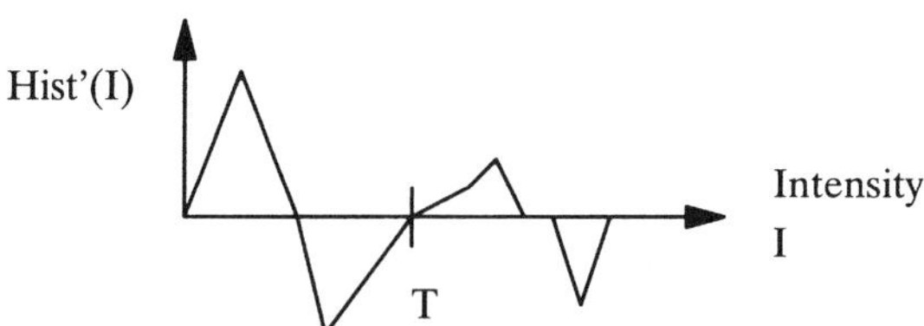

FIGURE 18. Zero-crossing of the first derivatives.

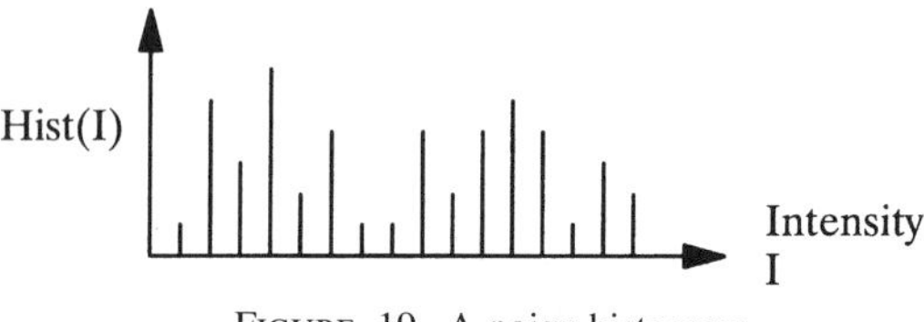

FIGURE 19. A noisy histogram.

convolution kernel must be wide and large, suppressing high frequencies to avoid false zero-crossings.

Large convolution kernels may create an excessive amount of computation. To avoid this we propose the following differentiating kernel $g_n(I)$:

$$g_n(I) = \underbrace{1\,1\,\ldots\,1}_{n} - \underbrace{1 - 1\,\ldots\,-1}_{n} = 1 - 1^{*}\underbrace{1\,1\,\ldots\,1^{*}}_{n} - \underbrace{1 - 1\,\ldots - 1}_{n} \tag{13}$$

We also propose to use the Inthist instead of Hist as input to the thresholding function. As was shown in Section III.E, the Inthist is as easy as Hist to compute. We then propose to compute

$$\begin{aligned} Hist'(I) &\doteq g_n{}^{*}\ Hist(x) = [1 - 1]^{*}\, g_n{}^{*}\ Inthist(I) \\ &= [\underbrace{1\,0\,0\ldots0}_{n} - 2\,0\,\underbrace{0\ldots0\,1}_{n}]^{*}\ Inthist(x) \end{aligned} \tag{14}$$

where we allow ourselves to employ the simple [1 – 1] kernel to estimate the second derivative in Eq. (14). To compute Hist'(I) in this manner in the host computer is trivial.

V. Variable Sampling Time

A. General

The result from the comparator in the SPE as a function of time is indicated by the thick line in Fig. 20. If we interrogate the comparator output we will first get a number of 1s followed by 0s, where the number of leading 1s is a measure of the intensity. This is how we made use of the intensity in the previous sections.

In the analog domain we observe the following. The voltage $U(t)$ over the photodiode as a function of time is a linear function shown as a dotted line in Fig. 20 and can be written as

$$U(t) = U_0 - kIt \tag{15}$$

As already explained in Section II.A, the time t_{ref} from the precharge of the photodiode until the voltage reaches the threshold level and thereby changes the sign of the comparator is inversely proportional to the illumination, so that

$$t_{\text{ref}} = \frac{U_0 - U_{\text{ref}}}{kI} = \frac{K}{I} \qquad \text{where } K = \frac{U_0 - U_{\text{ref}}}{k} \tag{16}$$

In the digital domain, the time variable t is mapped by the interrogating program onto a new quantity i, the interrogation cycle number. Obviously, by

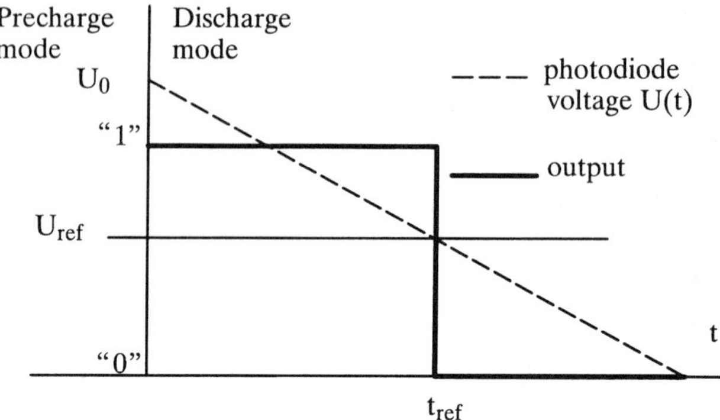

FIGURE 20. Output from an NSIP element.

introducing a **variable interrogation cycle time** we will obtain a relation between I and i quite different from the relation t_{ref} and I in Eq. (16).

As pointed out before, a typical NSIP program has the following structure:

```
REPEAT
  compute
UNTIL condition
```

Assume that the *compute* loop cycle time is t_d and that the discrete mapping between i and t is given by the time sampling set $(t_1, t_2, \ldots, t_i, \ldots, t_N)$. Given a fixed threshold level U_{ref}, to obtain a specific linear or a nonlinear relation between time t and the interrogation cycle i, we insert a variable extra delay in the loop. These delays are called δt_i and may be stored in a look-up table (LUT). They are defined as

$$\delta t_i = t_{i+1} - t_i - t_d = \Delta t_i - t_d \tag{17}$$

This will give us the following program structure:

```
wait(t1)
FOR i = 1 TO N − 1
  compute               ; execution time = td
  wait(δti)
END–FOR
```

Thus, to realize a certain relation between i and I in an NSIP system we need to know the time it takes to process one *compute* t_d, determine the number of quantization levels N which is also the number of times the interrogation loop will be executed, and, most importantly, determine the mapping functions $I = g(i)$ and $i = g^{-1}(I)$. These functions are given by the parameters N, t_d, t_1, and a set $(\delta t_1, \delta t_2, \ldots, \delta t_i, \ldots, \delta t_{N-1})$ stored in a LUT.

B. Linear Mapping

Figure 21 and Eq. (16) show the intensity as a function of the time it takes a photodiode voltage to reach the threshold. A linear mapping between time t and loop index i is to interrogate at N points in time, which are equidistant in intensity from I_N to I_1 so that

$$\Delta I \doteq \frac{I_1 - I_N}{N - 1} \tag{18}$$

The intensity I_i that changes the sign of its comparator at interrogation event i is given by the following linear function:

$$I_i = g(i) = I_1 f(i) = I_1 - (i-1)\Delta I = I_1 - (i-1)\frac{I_1 - I_N}{N-1} \tag{19}$$

For the case $N = 5$, Fig. 21 illustrates that we want to find sampling points t_1 to t_5. If the photodiode is precharged to U_0 at $t = 0$, the illumination intensity I_1 will cause the voltage to have discharged to the threshold voltage U_{ref} at $t = t_1$ so that

$$I_1 = \frac{U_0 - U_{\text{ref}}}{kt_1} = \frac{K}{t_1} \tag{20}$$

More generally, the intensity I_i that will make a photodiode reach the threshold at time t_i is

$$I_i = \frac{K}{t_i} \tag{21}$$

Using Eqs. (19) and (21) we get

$$t_i = (N-1)\frac{t_1 t_N}{t_1(i-1) + t_N(N-i)} \tag{22}$$

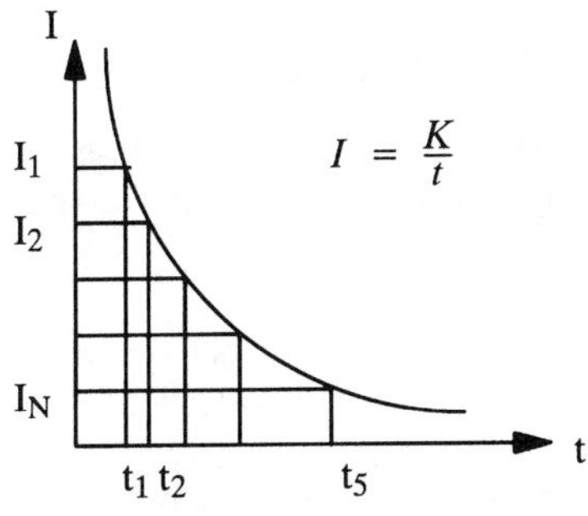

FIGURE 21. Equidistant sampling of I prolongs the time between two samples.

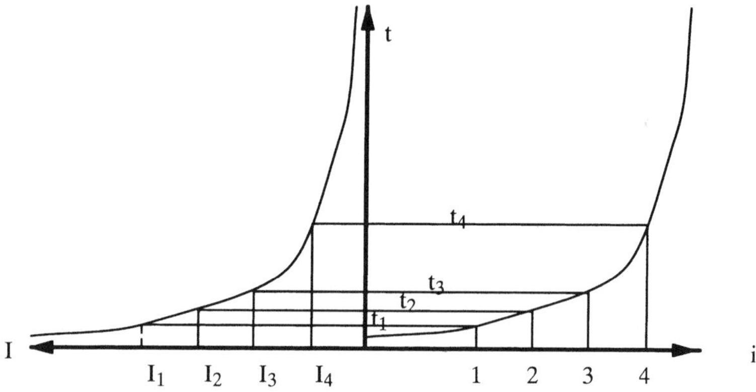

FIGURE 22. The relation between index, time, and intensity.

In the special case where $I_N = \Delta I$, which means that $I_1 = N^* I_N$, Eq. (22) simplifies to

$$t_i = t_1 \frac{N}{N + 1 - i} \tag{23}$$

Comparing Eq. (23) with Eq. (21) we see that a positive change in the variable i maps on t_i as a negative change in I_i. This is illustrated graphically in Fig. 22. In this case, according to Eq. (17), the difference between two time instances Δt_i, to be stored in a LUT as δ_i is

$$\Delta t_i = t_{i+1} - t_i = t_1 \frac{N}{(N - i + 1)(N - i)} \tag{24}$$

The iteration number that requires the fastest loop (shortest delay) is the first, which is evident from Figs. 21 and 22. This means that the value of Δt_1 to be selected must be equal to or larger than t_d, which is the time to complete the loop without any delays:

$$\Delta t_1 = t_2 - t_1 = \frac{t_1}{N - 1} \geq t_d \tag{25}$$

For the case of Eq. (23), the time t_N, which is the time for the lowest detectable intensity I_N to reach the threshold voltage, is

$$t_N = t_1 N \tag{26}$$

If the first sample should be able to detect the intensity I_1 at time t_1, the number of possible levels N according to Eq. (25) is bounded by

$$N \leq \frac{t_1}{t_d} + 1 = \frac{t_N}{N t_d} + 1 \Leftrightarrow N^2 - N \leq \frac{t_N}{t_d} \tag{27}$$

The quantity t_N is an important parameter because it typically corresponds to the maximal time to be allowed for the task.

To conclude, by making the rather natural choice $I_N = \Delta I$, we have only two adjustable parameters N and t_N. Another one is t_d and it is given by the actual NSIP algorithm. If either t_N or N is given the other one is constrained according to Eq. (27).

As an example, if we have a system with a minimum loop delay of $t_d = 10\,\mu\text{s}$ it is possible to make one of the following choices. We may set N to 100, for which Eq. (27) will give us $t_1 \geq 1\,\text{ms}$ and Eq. (26) $t_N \geq 100\,\text{ms}$. Alternatively, we may decide to make room for $N \leq 200$, which gives us $t_1 = 2\,\text{ms}$ and $t_N \geq 400\,\text{ms}$.

C. Nonlinear Mapping

We define a more general mapping from interrogation index i to intensity I_i as

$$I_i = I_1 f(i) \tag{28}$$

In the special case of linear mapping in Eq. (19), $f(i)$ is

$$f(i) = \frac{N - i + 1}{N} = 1 - \frac{i - 1}{N} \tag{29}$$

The constraints for $f(i)$ are that $f(1) = 1$ and $f(i + 1) < f(i)$, because I decreases and i increases over time. In the same manner as Eqs. (19) and (21) gave us Eq. (22), Eq. (28) will give us

$$t_i = \frac{K}{I_1} = \frac{t_1}{f(i)} \tag{30}$$

Assume that we would like to map I_i logarithmically onto i, that is,

$$I_i = I_1 f(i) = I_1 e^{-A(i-1)} \tag{31}$$

From Eq. (30) we get

$$t_i = t_1 e^{A(i-1)} \tag{32}$$

and

$$t_N = t_1 e^{A(N-1)} \tag{33}$$

that gives us an N value of

$$N = \frac{1}{A} \ln \frac{t_N}{t_1} + 1 \tag{34}$$

Again, the constraint is that the time difference between any two sampling events must be equal to or greater than t_d:

$$t_{i+1} - t_i = t_1 e^{Ai} - t_1 e^{A(i-1)} = t_1 e^{Ai}(1 - e^{-A}) \geq t_d \tag{35}$$

We notice that the difference is smallest for $i = 1$, due to the factor e^{Ai}. This means that

$$t_1(e^A - 1) \geq t_d \tag{36}$$

that can be viewed both as a constraint for t_1 and for A:

$$t_1 \geq \frac{t_d}{(e^A - 1)} \tag{37}$$

$$A \geq \ln\left(1 + \frac{t_d}{t_1}\right) \approx \frac{t_d}{t_1} \tag{38}$$

D. Data-Dependent Mapping

The interrogation program in Section V.B has the following structure:

```
initpd
initiate(time)
FOR i: = 1 TO N DO BEGIN
   REPEAT
   UNTIL time ≥ t1 * N/(N − i + 1)
   compute_operation
END
```

Thus, in Section V.B the delay in the REPEAT-loop is variable but predetermined and data-independent. In this section we will see some cases in which the delay is made dependent on the incoming sensor data.

Histogram equalization is a procedure that quantizes the pixel intensities so that the same number of pixels ends up in each graylevel bin. With conventional methods this is a rather complex operation. As was shown in Fig. 14 in Section III.E the count of the output from the comparators over time is the Inthist of the image. This means that we should be able to obtain equalization by simply observing when the appropriate number of pixels have passed the threshold. Given an image size of M and a bin number of N, this is an NSIP program for positive gradient detection, as described in Section C, performed on a histogram equalized image:

```
FOR i = 1 TO N
   REPEAT
      ldi pd
   UNTIL COUNT>i*M/N
   ld (01x) a                    ; ldi pd : ld (01x) a = ld (10x) pd
END–FOR
```

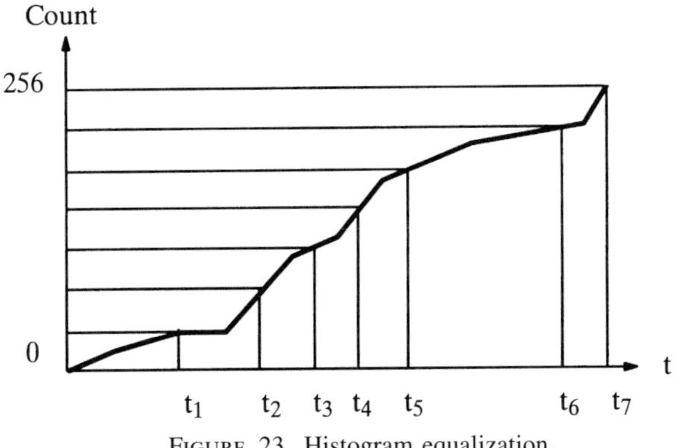

FIGURE 23. Histogram equalization.

Note that the complete equalized image never exists in the NSIP program. Instead, the outer loop uses one quantization level at a time of the equalized image to execute (in this example) positive gradient detection.

Figure 23 shows graphically an equalization procedure with $N = 7$. The count bins on the vertical axis will all contain the same number of pixels. This is accomplished by sampling (read out) at times $(t_1, \ldots, t_7)$ when the appropriate number of pixels have passed the threshold. By changing the UNTIL condition in the REPEAT-loop of the previous program we can modify the implicit flat equalized histogram to something of our own choice. In Fig. 24 is shown the Inthist for a perfect equalized case (a) and a case that allocates more dynamic range to lower intensities. An NSIP program for this modified equalization without the gradient detection can be written as

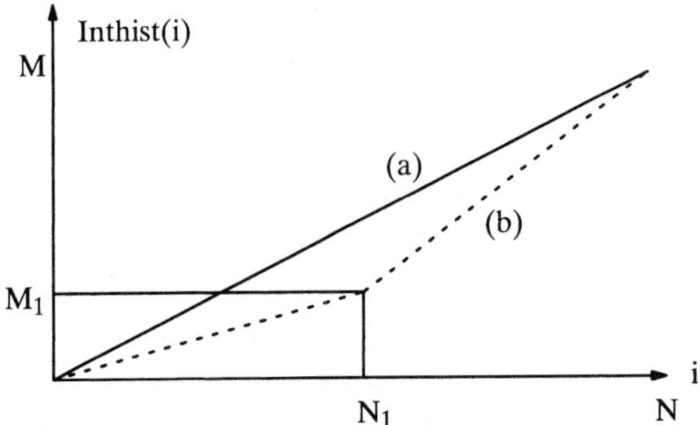

FIGURE 24. Different specifications of the Inthist.

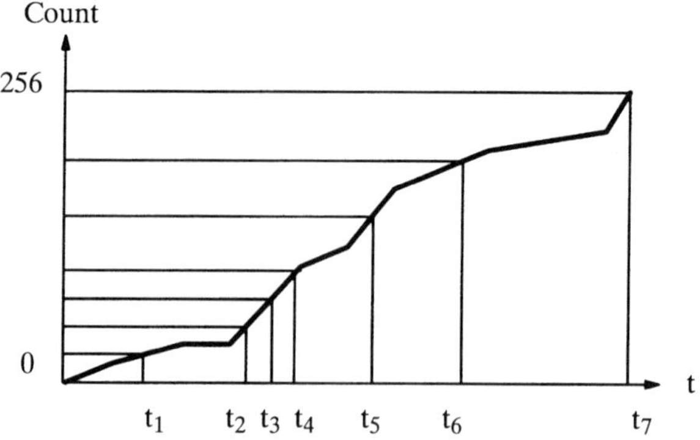

FIGURE 25. Modified histogram equalization.

```
FOR i = 1 TO N1
  REPEAT
    ldi pd
  UNTIL COUNT>i*M1/N1
  compute
END–FOR
FOR i = N1 + 1 TO N
  REPEAT
    ldi pd
  UNTIL COUNT > M1 + (i – N1)*(M – M1)/(N – N1)
  compute
END–FOR
```

The graphic description of this operation is shown in Fig. 25. In fact, the Inthist could be modified to any monotonically growing function $f(i)$ using the following program:

```
FOR i = 1 TO N
  REPEAT
    ldi pd
  UNTIL COUNT > f(i)
  compute
END–FOR
```

An even more general approach to use the COUNT operation for data-dependent mapping of the grayscale of the input image is to wait, within each iteration, until a certain condition is satisfied. In the case of the histogram equalization, this condition is simply that the right number of pixels have passed the threshold. An example of a more sophisticated but totally arbitrary case is to leave the inner REPEAT-loop when five new left edge pixels have passed the threshold. This can be written as

```
FOR i: = 1 TO N DO BEGIN
   sum: = 0
   REPEAT
      st pd,r0
      ld (01x) r0
      sum: = sum + COUNT
   UNTIL sum > 5
   ld r0
   compute_operation
END
```

In a general case, the number that must satisfy the condition could be a function $f(i)$. In this case the program may look like:

```
FOR i: = 1 TO N DO BEGIN
   sum: = 0
   REPEAT
      st pd,r0
      compute_condition
      sum: = f(sum,COUNT)
   UNTIL sum > f(i)
   ld r0
   compute_operation
END
```

The interesting thing about this type of mappings in the NSIP concept is that the complete mapping table must not be computed before we can output data. For instance, in the histogram equalization example we do not generate any histogram, but rather it is done implicitly in the program.

VI. Grayscale Morphology

Grayscale morphology is an extension of binary morphology (Sternberg, 1986). In binary morphology various operators are applied to a binary image. Typical operators are expansion, erosion, and thinning. In grayscale morphology these operators are generalized to operate on grayscale images where the image is viewed as a terrain map and the pixel value is interpreted as height. Such representation of the pixel value is called the **Umbra** transform. An Umbra transform $\mathcal{U}$ of a function $f(x)$ is defined as

$$\mathcal{U}(f(x)) = u(x, y) = \begin{cases} 1 & y \geqq f(x) \\ 0 & y \geqq f(x) \end{cases} \tag{39}$$

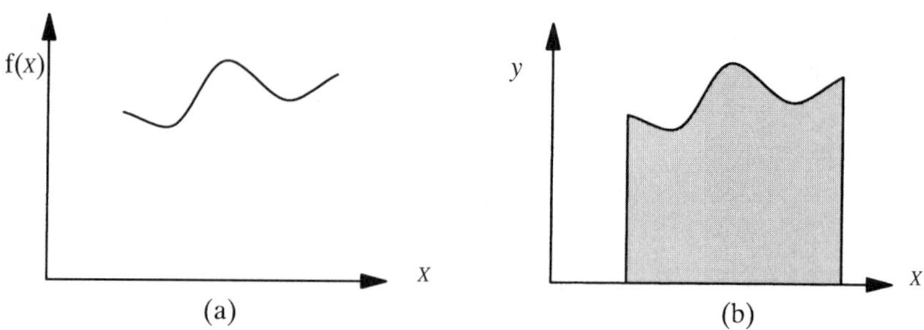

FIGURE 26. The Umbra transform.

that is illustrated in Fig. 26 where (a) represents the original function and (b) the $u(x, y)$ function. Note that $u(x, y)$ is a binary 2D function. In a similar fashion, the Umbra transform of a 2D function $f(x, y)$ defines a binary 3D function $u(x, y, z)$.

NSIP is particularly well adapted for grayscale morphology. The reason is that the output from the individual pixel comparators as a function of time corresponds to the Umbra representation of an image. Figure 27a shows a coarsely digitized 1D grayscale signal, where the Y-axis represents light intensity. The Umbra representation of this signal is shown in Fig. 27b. At time $t = 0$, no pixels have passed the threshold. When $t = 1$ the comparator with the pixel corresponding to the highest intensity in Fig. 27a is set. At $t = 2$ another comparator is set, and at $t = 3$ three new comparators are set, and so on. As pointed out in Section III.D this is the same as the stack or thermometer representation (Wendt *et al.*, 1986). In the following sections we will present some grayscale morphological operations and show how they are implemented in a NSIP system.

As described in Section III.D we can perform any type of rank-order filtering, for instance, max or min, using NSIP techniques. These operations

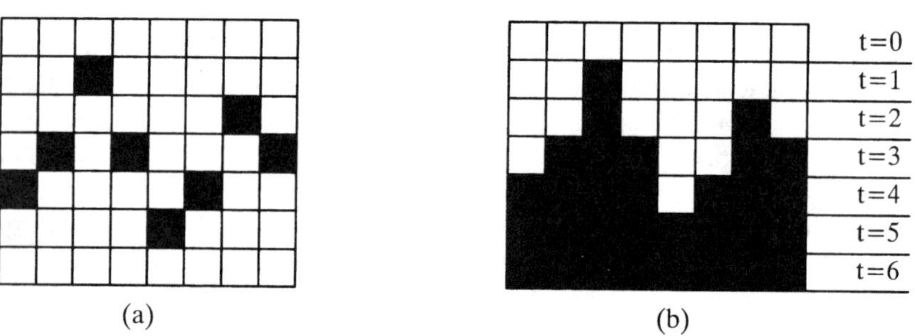

FIGURE 27. The comparator output viewed as an Umbra transform.

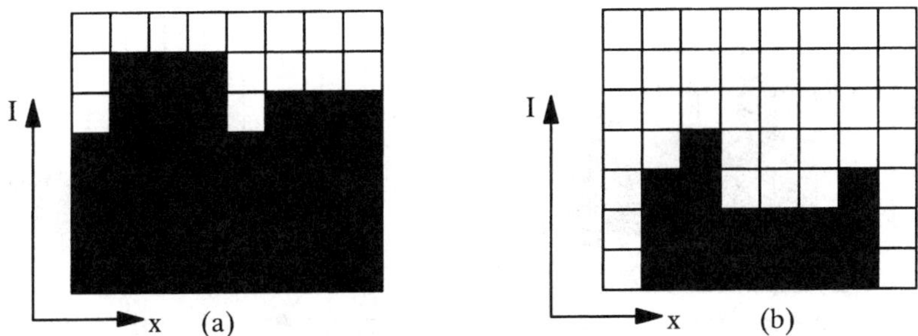

FIGURE 28. Dilation and erosion.

FIGURE 29. A structure element.

applied on an Umbra transformed image are grayscale *dilation* and *erosion* respectively.

Figures 28a and b show the result after dilation and erosion, using the image input of Fig. 27a and the structure element shown in Fig. 29. The structure element works in the following way. For dilation, the structure element will set the pixel to **1** if the center pixel is already set (white circle) or if either one or both of the two neighbors is set (black circles). For erosion, the structure element will keep the pixels set (in the white circle position) if both of the two neighbors also are set (black circles), otherwise it will be changed to **0**.

A program for grayscale dilation will look like:

```
REPEAT
   ldi pd                    ; invert pd-value
   st r0
   ldi (000) r0              ; dilation operation
   st r1
   ..
   ld r0
UNTIL COUNT=N
```

The operation ldi (000) r_0 performs a dilation because

$$a + b + c = \overline{\bar{a}\&\bar{b}\&\bar{c}} \tag{40}$$

In the same way as for binary images *closing* and *opening* are combinations of erosion and dilation. Figures 30a and b show the result of closing and opening applied to the signal in Fig. 27a.

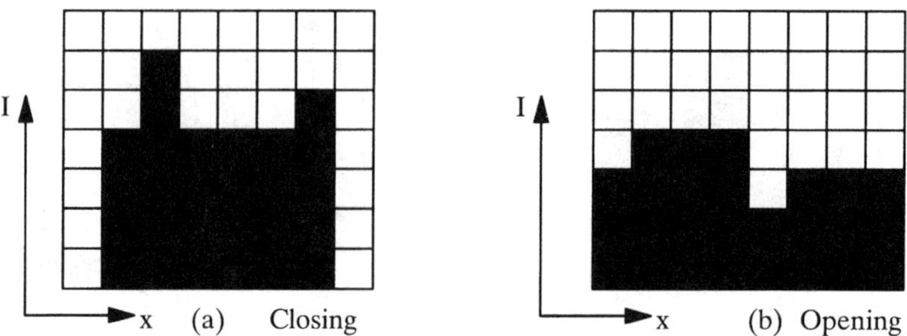

FIGURE 30. Closing and opening.

The program for grayscale closing will be:

```
REPEAT
   ldi pd
   st r0
   ldi (000) r0            ; dilation
   ld (111) a              ; erosion
   st r1
   ..
   ld r0
UNTIL COUNT=N
```

The input data to be operated upon by the structuring element of Fig. 29 are all coming from the same interrogation event. Figure 31 illustrates three different structuring elements which utilize input data from three successive cycles i, $i+1$, and $i+2$.

If we erode the image of Fig. 27 with the three operators in Fig. 31, the result is as shown in Fig. 32, where the dark area represents the result and the shaded pixels the eroded area.

An approximation of the Rolling ball operator (Sternberg, 1986) is shown in Fig. 33. The following NSIP program using this element for dilation consists of two parts. The first part computes the output and the second part rearranges register data for the next interrogation and computation cycle.

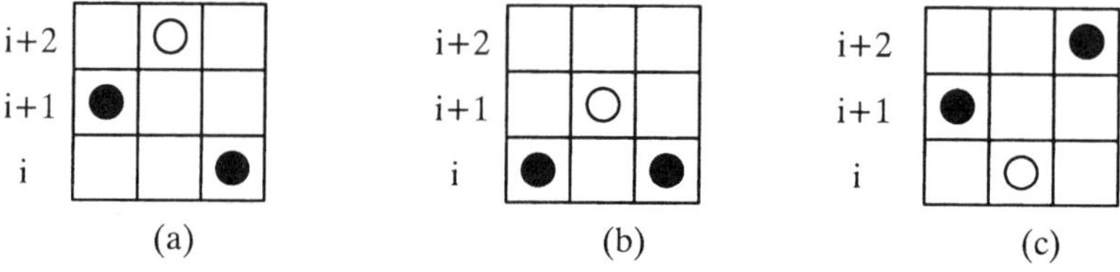

FIGURE 31. Structure elements both in the x and time dimension.

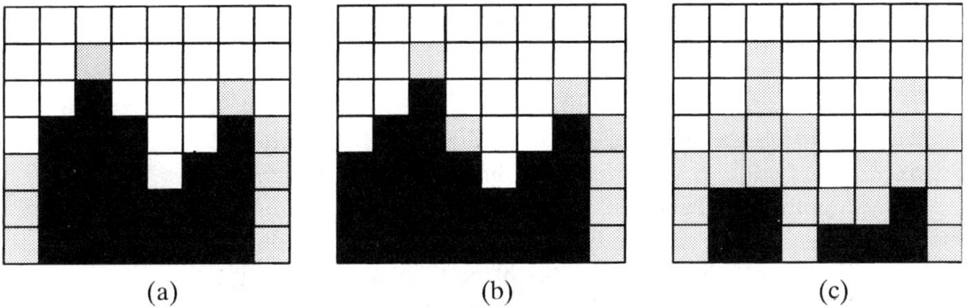

FIGURE 32. The effect of dilation and erosion using the structure elements in Fig. 31.

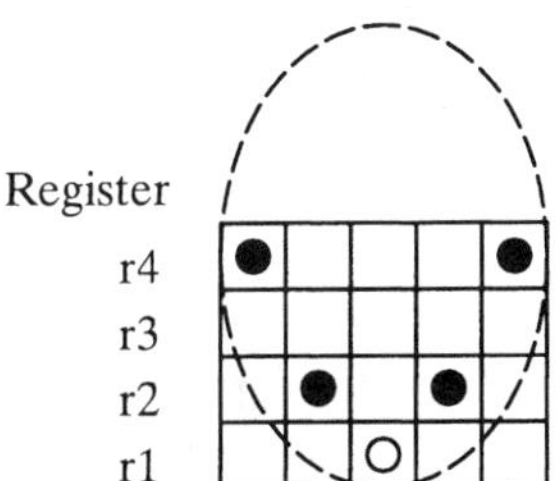

FIGURE 33. A Rolling ball operator.

The vertical size of the operator determines the number of registers to be employed because this size indicates how many cycles back in time the output requires.

```
clear register r2..r4
REPEAT
   ldi pd
   st r1
   or (1xx) r4          ; Compute dilation
   or (xx1) r4
   ori (0x0) r2
   st r0                ; Dilation output
   ..
   ld r3                ; Shift rows
   st r4
   ld r2
   st r3
   ld r1
   st r2
   ld r1                ; Load input
UNTIL COUNT=N
```

VII. Linear Convolution

In this section we will utilize grayscale pixels where the grayscale is represented by the interrogation index i. As described in Section II.A we do not automatically get a linear relation between the time t and the intensity I, but using the strategies described in Section V we can make the relation linear. For the sake of simplicity, however, we omit the mapping procedures here and assume that they can be introduced wherever wanted.

The time, or rather the interrogation index, elapsed from the precharging of a photodiode till the diode passes the threshold is measured as follows (also shown in the program that follows). A stack of registers is used as a counter and the incrementing bit from the photodiode is placed in the accumulator. If the counter is incremented more than 256 times the counter would normally wrap around back to zero. This is prevented by a few extra lines of code and an overflow bit r_8 which is set when the counter reaches 256. The number of instructions will average approximately three per intensity level if Gray-code representation is used.

Program for obtaining graylevel pixels in natural binary code:

```
clear register r0..r8
initpd
ld pd
REPEAT
   increment(r0..r7)
   ld c                  ; Generate overflow ...
   or r8
   st r8                 ; stored in r8
   ld pd
UNTIL COUNT=0
```

The function *increment( )* will treat $r_0 \dots r_7$ as an 8-bit number and perform the appropriate binary operations in order to perform an increment. The result in $r_0 \dots r_7$ will have an inverted grayscale because a high intensity corresponds to a low interrogation index and vice versa. To produce a noninverted result, we can either decrement a fixed value (in this case 255 with underflow check), or we can change *ld pd* to *ldi pd*, which means that the counter counts from passing the threshold until exit from the loop.

In Section III.C we described a positive gradient detection routine where the left edges in the input image were marked. We will now modify this routine to measure the steepness of the gradient. As long as the center pixel has not passed the threshold the counter is idle, which corresponds to the situation in Fig. 34a. When the pixel passes the threshold corresponding to

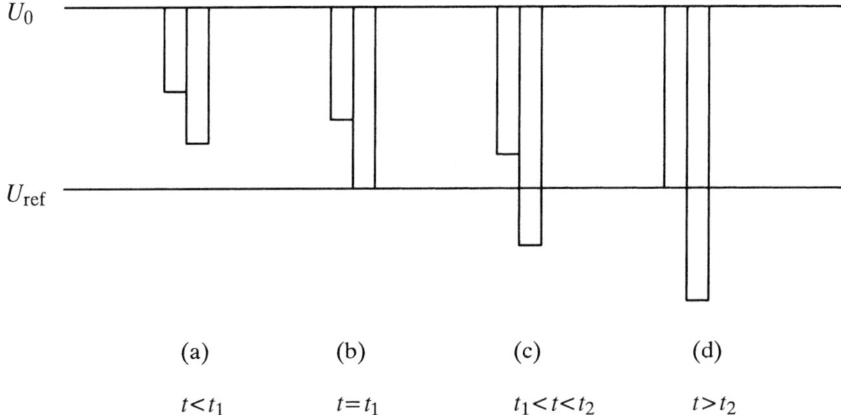

FIGURE 34. Four different time instances during discharge of the photodiode.

(b), we continue to increment as long as the left neighbor has not passed the threshold (c).

Program for computation of positive gradients.

```
clear register r0..r4
initpd
ld pd
REPEAT
   ld (10x) a
   increment(r0..r3)
   r4 = overflow check(carry,r4)
   ld pd
UNTIL COUNT = 0
```

Compared to the previous program we use a smaller number of bits in the counter ($r_0 - r_4$) because the dynamic range is expected to be less for the gradient than for the absolute values. An overflow check is inserted.

By adding the instruction

or (01x) pd

before *increment*($r_0 .. r_3$) the counter will be incremented as long as there is a left <u>or</u> a right edge. Hence, this modified routine computes the absolute value of the gradient.

Figure 35 shows a 1D example of a Laplacian convolution kernel. The straightforward way to compute this convolution is to do the following in the interrogation loop. Increment the stack of registers by 1 if the left neighbor has passed the threshold. Increment this value by 1 if the right neighbor has

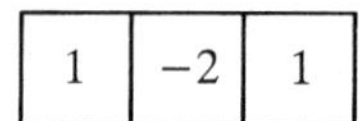

FIGURE 35. A 1D Laplacian operator.

passed the threshold. Decrement this value by 2 if the center pixel has passed the threshold.

The program becomes:

```
clear register r0 .. r7
initpd
ld pd
REPEAT
   ldi (1xx) pd
   increment(r0 .. r7)
   ldi (xx1) pd
   increment(r0 .. r7)
   ldi pd
   decrement(r1 .. r7)                ; −2 = −1 shifted one step
   ld pd
UNTIL COUNT = 0
```

In most cases it is faster to do the convolution by performing a delta computation ahead of the incrementation. As long as none of the pixels in a neighborhood has passed the threshold no increment should take place, that is, $\Delta := 0$. If one of the neighbors has passed the threshold but not the center pixel or the other pixel, then $\Delta := 1$. For similar obvious reasons it is understood that the total incrementation to be done in the loop can take the values $\Delta := -2, -1, 0, 1, 2$. Thus, the delta value can be represented by 3 bits. Figure 36 shows four different time instances and the corresponding delta values.

Program for Laplacian using Δ-increment:

```
initpd
ldi pd
REPEAT
   st r0
   generate delta value in r1 .. r3 given the comparator value in r0
   add(r1 .. r3,r4 .. r11)
   ldi pd
UNTIL COUNT = 256
```

The function *add*( ) adds the 3-bit number $r_1 \,..\, r_3$ to the 8-bit number $r_4 \,..\, r_{11}$ and store the results in $r_4 \,..\, r_{11}$. The word-length of the final value has been set to 8 bits. The input data are stored in r_0, the delta value in $r_1 \,..\, r_3$, and the final value in $r_4 \,..\, r_{11}$.

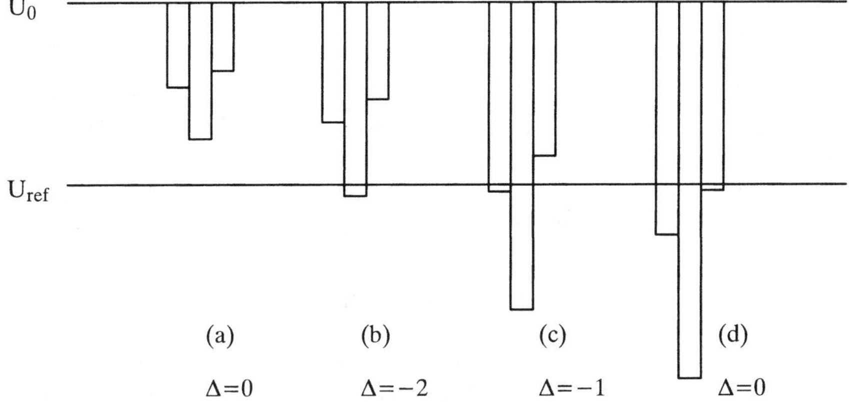

FIGURE 36. An illustration of the delta-increment method.

The number of bits b_1 for the delta value is determined from the coefficients in the kernel. Given the arbitrary n-sized convolution kernel, the formula for the number of bits b_1 is

$$b_1 = \left\lceil {}^2\log\left(\sum_i |c_i|\right)\right\rceil + 1 \tag{41}$$

To see when the delta-increment method is more efficient than the straightforward method, we assume the following parameters:

i) An addition takes **3b** cycles.
ii) Each of the coefficients can be described by $\mathbf{b_2}$ bits.
iii) There are **n** coefficients.
iv) The final result contains $\mathbf{b_3}$ bits.

Then, with the straightforward method, the cycle count is

$$S_{S\text{-}F} = 3nb_3 \tag{42}$$

For the delta-increment method the count yields

$$S_\Delta = 3(n-1)(\log n + b_2) + 3b_3 \tag{43}$$

The difference

$$S_{S\text{-}F} - S_\Delta = 3(n-1)b_3 - 3(n-1)(\log n + b_2) \tag{44}$$

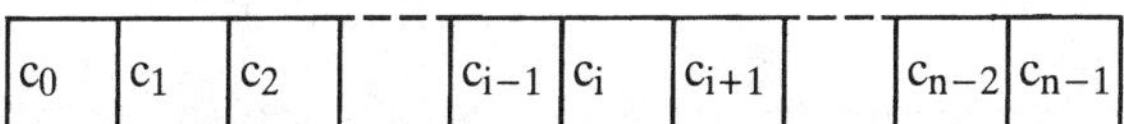

FIGURE 37. An arbitrary 1D convolution kernel.

is equal to or greater than zero when

$$b_3 \geq \log n + b_2 \tag{45}$$

From Eq. (41) we know that $b_1 = b_2 + \log n$. Thus, as the wordlength of the final result must be larger than the Δ-value, the delta-increment method is always faster.

VIII. A 2D NSIP Model

A. *Programming Model*

The 2D NSIP system to be described here, which we will refer to as N2D, is largely a generalization of the N1D in Section II.B. A floorplan for a conceptual N2D chip is shown in Fig. 38. The SPEs are distributed along the 2D array in a square grid. As in the 1D case the N2D SPE consists of a photosensor, an analog comparator, a digital processor, and memory in the form of binary registers.

The photosensor (photodiode) and the comparator work in the same way as in the 1D system. However, the rest of the VLSI implementation is different as the area available for each SPE has to be smaller. Therefore, the processing power is likely to be smaller and we may have to accept that some of the single-cycle instructions in N2D have to be performed in a number of cycles in the 2D case.

Figure 39 shows the logic function of the SPE, which is a 2D generalization of Fig. 8. With a few exceptions, the instructions for the point logic unit (PLU) will be the same. The neighborhood logic unit (NLU) receives data from its four nearest neighbors and the output from the global logic unit (GLU). Therefore, the NLU mask, which controls the operation, has five positions. An NLU-mask in an instruction such as

ld (abcde) pd

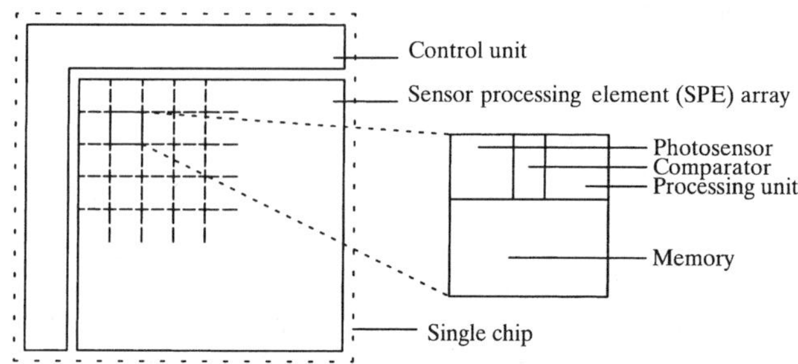

FIGURE 38. A floor plan of a 2D NSIP architecture.

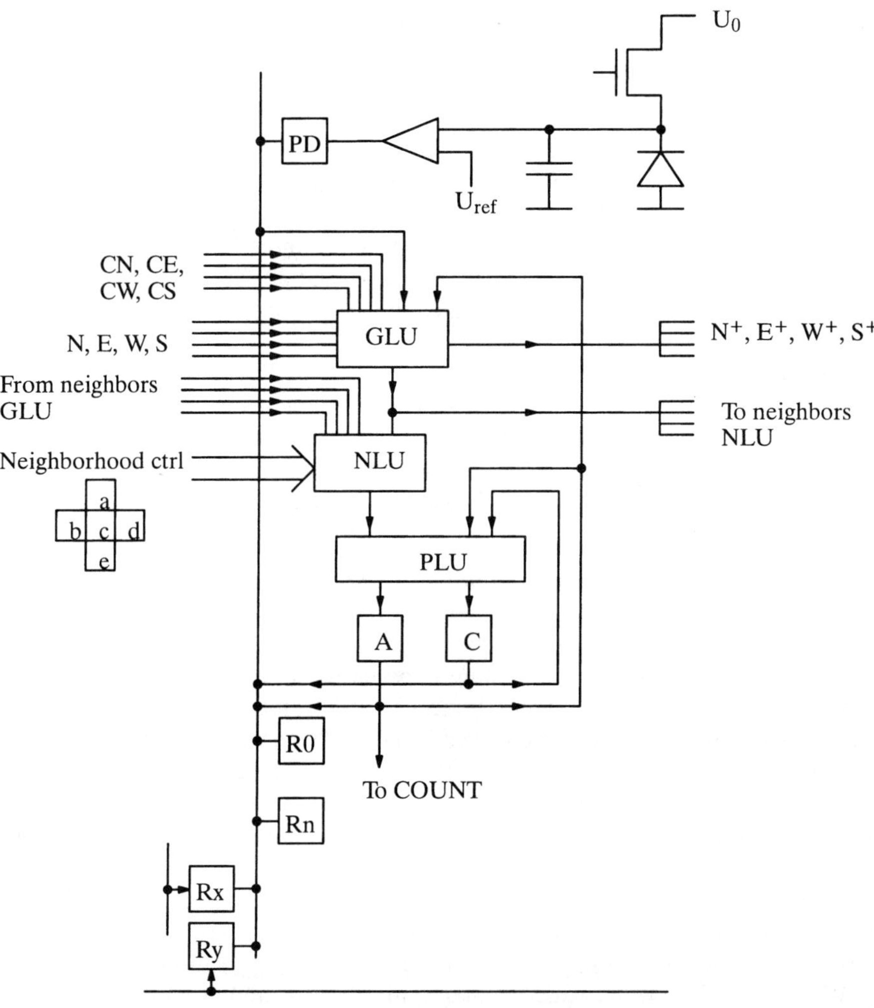

FIGURE 39. The programmer's model of one SPE.

corresponds to the neighborhood shown in Fig. 39. Each parameter (a,b,c,d,e) takes one of the values (0, 1, X (= don't care)).

This means that the instruction

ld (x10xx) pd

will perform the same operation as

ld (10x)

in the 1D case.

The 2D GLU shown in Fig. 39 performs global operation with the bus and the accumulator as input operands. The output is the result after the global operation, the bus, or the accumulator. Unlike in the N1D, the GLU neighbor connections are controlled in a way that makes it possible to select certain propagation directions across the array. The architecture differs a great deal from the 1D case and is described in more detail in Section IX.

The memory size will not be specified here although it is reasonable to assume a maximum number of bits between 16 and 32. The memory includes two special registers Rx and Ry, which are connected to a horizontal and a vertical bus, respectively.

It should be emphasized that the diagram of Fig. 39 is strictly a programmer's model. What the programmer sees as one single-step operation could be executed as a multistep sequence in a vastly different manner. Several of the connections and functional units in Fig. 39 can actually be implemented by the same piece of hardware as will be shown in Section XI.A.

B. Generation of Coordinates

As shown in Fig. 39, each SPE is connected to one vertical and one horizontal bus. A similar 2D data broadcasting network exists in the digital assembly program (DAP) array processor (Reddaway, 1971). It would be quite feasible to connect the bus-connections available at the array edge to a broad-bandwidth data channel for all types of communications to and from the array. Note that if we want to load data into a specific row of the array we can use the vertical buses and do the selection with a proper control word on the horizontal bus.

We propose a somewhat more limited use of the buses and the Rx, Ry registers. In Fig. 40 a 16×16 array with 16 horizontal and 16 vertical buses is shown. At the left and upper edge of the array are two read-only-memories (ROM), four bits wide, which contain only the row and column number in ordinary binary code. Clearly, if these numbers were transported into the array, the combined result is that each SPE would have access to its own coordinates. Thus, as soon as our application contains instructions or algorithms that depend on the position inside the array, we may expect to be able to make use of this feature. Actually, as will be shown, in many cases it is not necessary to make the full (x, y)-coordinate available to the SPEs.

In the following programs, the bit-planes in the ROMs to be connected to the horizontal and vertical buses are set with the instructions

```
setrx <bit-plane x>
setry <bit-plane y>
```

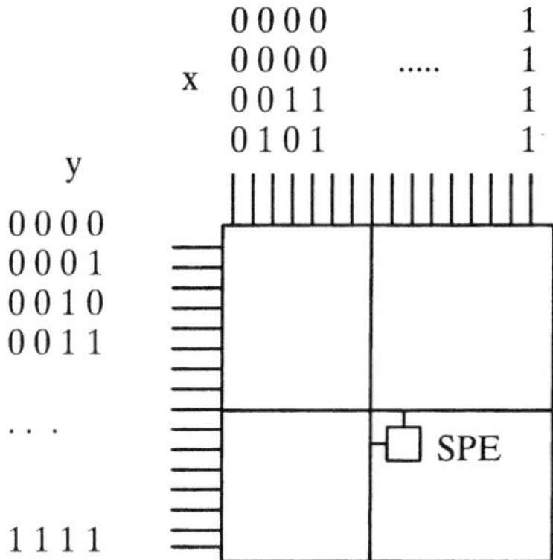

FIGURE 40. Position constants.

The content of the buses will be available in memory cells *rx* and *ry* in the SPE.

In the first example we want to select a region-of-interest (which consists of the lower right-hand corner shown in Fig. 41). Assuming that $(x, y) = (0, 0)$ in the upper left-hand corner, this is equivalent to stating that only SPEs with the MSB set in **both** x and y should be set. The following program produces such a mask array:

```
setrx n − 1          ; MSB in x
setry n − 1          ; MSB in y
ld rx
and ry
st a,r0              ; Store mask in r0
```

By using more sophisticated formulas than those previously used, it is possible to achieve regions-of-interest as shown in Fig. 42.

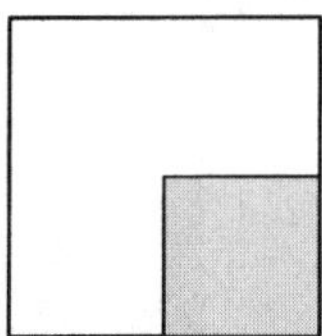

FIGURE 41. Region of interest. SPEs in shaded area are selected.

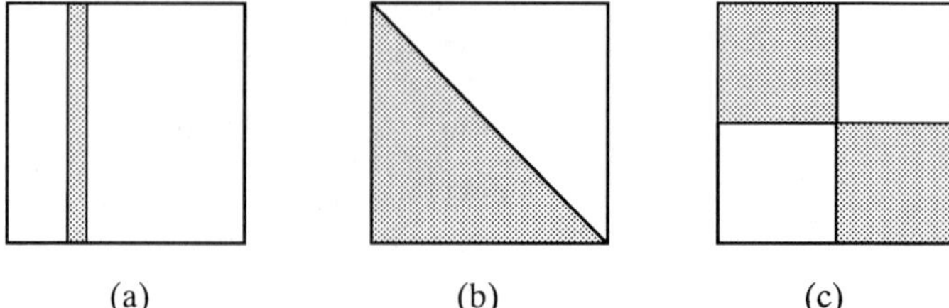

FIGURE 42. Different regions-of-interest patterns.

The vertical line in Fig. 42a is achieved with the following program.

```
clr a
FOR i = 0 TO n − 1
   setrx i
   IF Pattern[i] = 1 THEN
      and rx
   ELSE
      andi rx
   END–IF
END–FOR
st a,r0                        ; Store mask in r0
```

Here, the Pattern[i] is a bit vector that defines the position of the column. For instance, if we want to use column 34, the vector Pattern will look like [0, 1, 0, 0, 1, 0, 0, 0].

Figure 42b shows a region where the y-position is greater than the x-position of each SPE. This is achieved in the following program by subtracting the y-position from the x-position. If the final carry bit (= borrow) is set, the result $x - y$ is negative and the SPE belongs to the region-of-interest.

```
clr a
st r1
xori a                                   ; Carry = 1
st r1
FOR i = 0 TO n − 1
   setrx i
   setry i
   r0 = rx&¬ry+rx&r1+¬ry&r1              ; Carry
   r1 = r0                               ; New carry
END–FOR
                                         ; Mask in r0
```

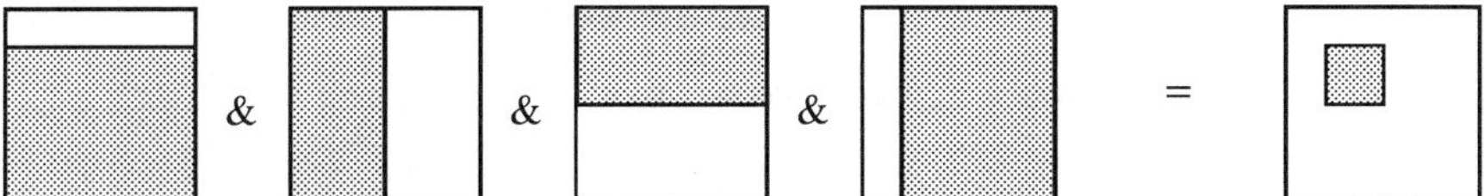

FIGURE 43. Constructing a region-of-interest as an arbitrary rectangle.

Figure 42c is a union (= OR) between two regions. The following program computes the mask for each individual region and then ORs them together:

```
setrx n − 1          ; MSB in x
setry n − 1          ; MSB in y
ld rx
and ry
st a,r0              ; lower right-hand corner in r0
setrx n − 1          ; MSB in x
setry n − 1          ; MSB in y
ldi rx
andi ry              ; upper left-hand corner in accumulator
or r0                ; upper left-hand OR lower right-hand
st a,r0              ; Store mask in r0
```

In the same manner, the region in Fig. 43 is obtained by ANDing a series of simple regions.

Obviously, the more irregular the region-of-interest is, the longer is the program required to produce it. In the worst case, we might have to define and generate the region as a set of individual 1-pixel regions that are ORed together. The complexity of such a program is $\mathbf{O}(\mathbf{N}^2 \mathbf{2} \log \mathbf{N})$ because there are $\mathbf{O}(\mathbf{N}^2)$ pixels, each of which requires $\mathbf{2} \log \mathbf{N}$ accesses from the ROMs. Fortunately, masks to be used in practice are much simpler to produce.

IX. GLOBAL OPERATIONS

A. *MARK and FILL*

Before we discuss 2D global operations, let us reconsider the 1D case.

The instructions performed by the global logic unit, the GLU-instructions, have proved to be a very powerful tool both in real applications for LAPP1100, and in our NSIP studies. For instance, the pattern adaptive thresholding in Section IV.B used the GLU instructions *mark* and *fill* to compute the size of the object in the image, and the max detector in Section III.B used the instruction *lfill* to determine the position of the maximum intensity before the feature extraction.

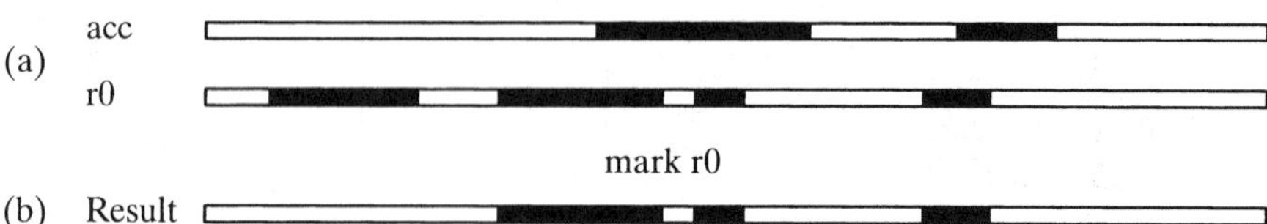

FIGURE 44. The MARK operation.

The basic GLU-instruction in LAPP is *mark*. Other GLU instructions can be derived from *mark*. The mark operation keeps those objects in the operand that overlap in any position any object in the accumulator. The result is placed in the accumulator. For example, given an accumulator and r_0 content as in Fig. 44a, the instruction *mark* r_0 will result in an accumulator content as in Fig. 44b.

The circuitry of one processor slice, SPE, that performs these operations is shown in Fig. 45. Here, Mask and Image represent two register, which in the N1D case is the accumulator and the operand bus, respectively. First the Mask (accumulator) is applied and then the Image (bus). Positions, that is, those SPEs where both the Mask and Image are set, will give a "1" at the Result output. This combination will also generate a propagation to the left and to the right, thus setting all the Result bits in those SPEs that are connected to a set Result bit. The Result bit is loaded to the accumulator but not until the next clock cycle.

It might seem possible to reduce the logic of Fig. 45 to Fig. 46. However, with a common wire for left and right propagation, latching and oscillations may appear. These effects should also be avoided in the 2D case.

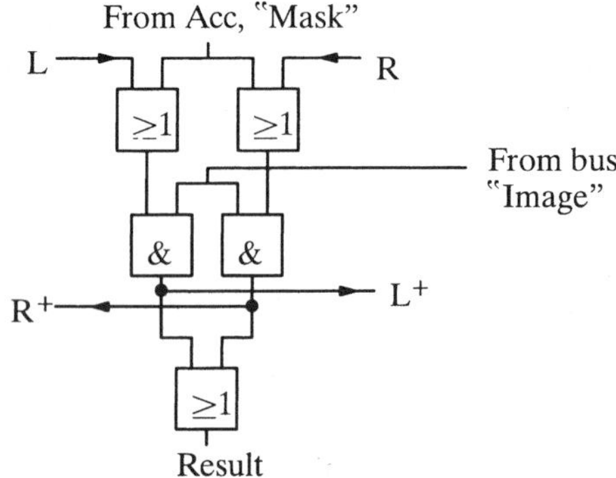

FIGURE 45. The GLU-net in LAPP1100.

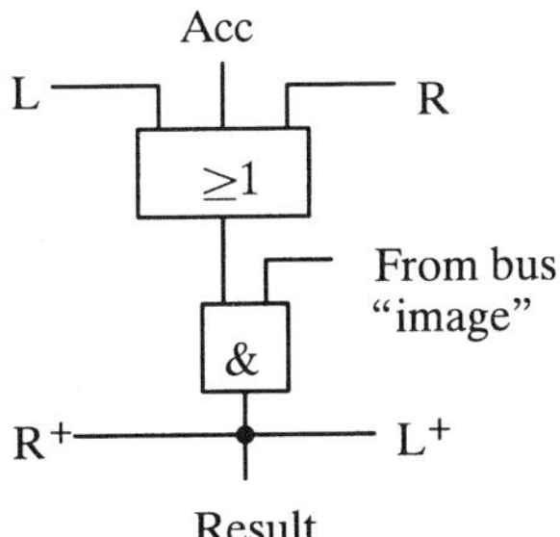

FIGURE 46. Reduced GLU-net in LAPP1100.

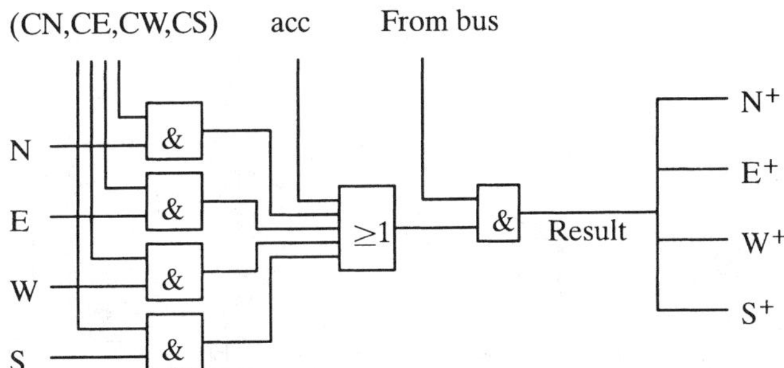

FIGURE 47. A 2D MARK-net with input controls.

The reduced GLU-model of Fig. 46 can be generalized to two dimensions (see Fig. 47). The four nearest neighbors are used. The connections to these are controlled by four control signals (CN,CE,CW,CS). Thus, while we distribute data to neighbors from one point as in Fig. 46 we safeguard ourselves with extra gating. The potential problems with 2D bidirectional propagation across the array will be discussed in Section IX.B.

The gates for direction control can either be placed at the input, as in Fig. 47, or at the output, as in Fig. 48. Actually, these two nets are the same. This is illustrated in Fig. 49 where Fig. 49a corresponds to the net in Fig. 47 and Fig. 49b to the net in Fig. 48. As the control bits are addressed individually (represented by the small boxes on the arrows) it does not matter if the direction control is performed on the incoming or the outgoing signal.

To perform the 2D operation *mark*, all the bits in the direction control vector (***CN***,***CE***,***CW***,***CS***) are set to 1, the mask bit is placed in the Mask register (which we will assume to be the accumulator) and the image bit is taken from the bus (which is typically a register). This will initiate a 2D

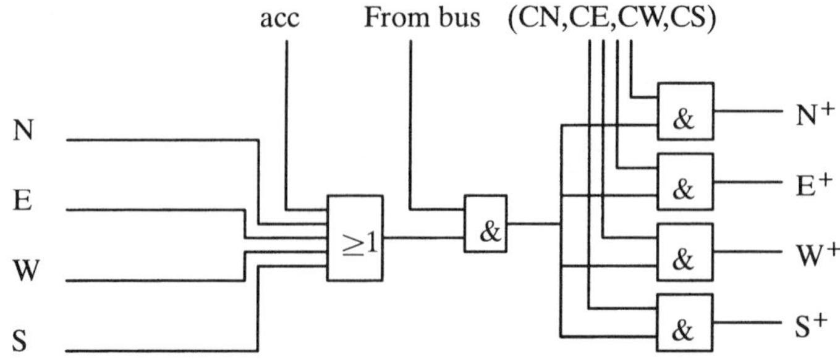

FIGURE 48. A 2D MARK-net with output controls.

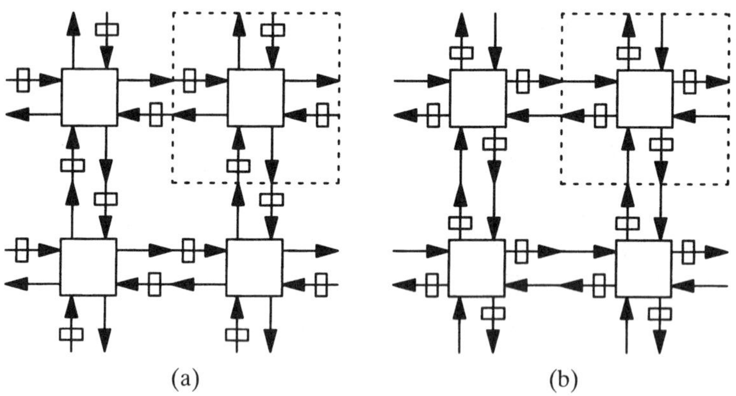

FIGURE 49. The equivalence of input and output control.

propagation equivalent to Fig. 44 and, when it is finished, the Result bit can be stored in the accumulator.

In the same manner we can obtain *fill* by simply bringing in the inverted value from the bus. As in N1D a fill operation preserves holes rather than objects as determined by the mask (see Fig. 50). Both **MARK** and **FILL** operations may use an instruction specific mask rather than the accumulator content. In the 1D case we had three choices (lfill, rfill, and lrfill) for these fill operations. Conceptually, it is convenient to think of the fourth choice as the one defined by the image in the accumulator. In the 2D case we have 15 border *fill* alternatives as listed in Table II. They are all controlled by the vector (***CN***,***CS***,***CE***,***CW***) and they turn out to have some nontrivial effects compared to the 1D case. Two rather powerful operations can be carried out using the functions in Table II and will be described here. The first one expands each object to the smallest possible rectangle that fully covers the

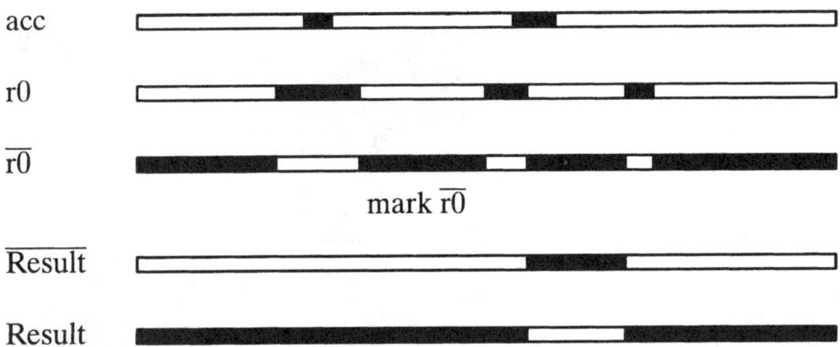

FIGURE 50. The FILL operation.

TABLE II
BORDER FILL

No	CN	CS	CE	CW	Name
0	0	0	0	0	
1	0	0	0	1	rfill
2	0	0	1	0	lfill
3	0	0	1	1	lrfill
4	0	1	0	0	tfill
5	0	1	0	1	trfill
6	0	1	1	0	tlfill
7	0	1	1	1	tlrfill
8	1	0	0	0	bfill
9	1	0	0	1	brfill
10	1	0	1	0	blfill
11	1	0	1	1	blrfill
12	1	1	0	0	tbfill
13	1	1	0	1	tbrfill
14	1	1	1	0	tblfill
15	1	1	1	1	tblrfill (hole_fill)

object. The second one removes such holes, that are enclosed by an object and do not touch the image border.

Figure 51 shows the operational steps required to expand an object into a circumscribing rectangle. Here, Fig. 51a is the original image. First, we perform operation 9 in Table II which is a *brfill*, and set the shaded area in Fig. 51b. Similarly, the shaded areas in Figs. 51c, d, and e are set by operations 6, 10, and 5 in Table II. Each of these operations will produce a corner of the final rectangle. When these images are ORed together we obtain the final result in Fig. 51f, which is the smallest possible rectangle that covers the object. Normally, this operation works regardless of the number of objects

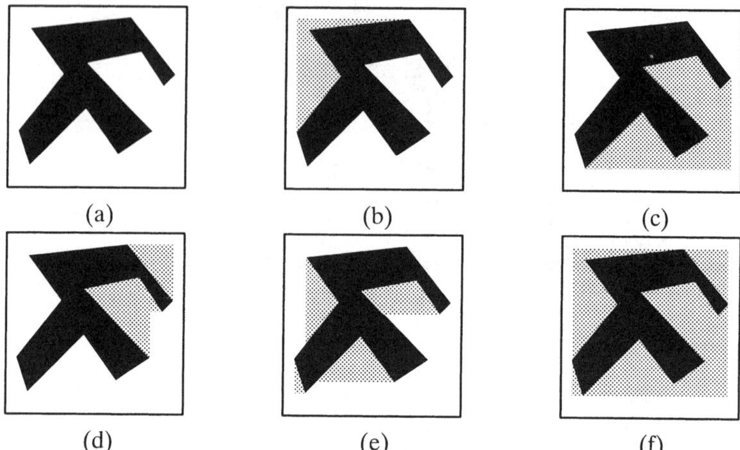

FIGURE 51. Smallest possible rectangle.

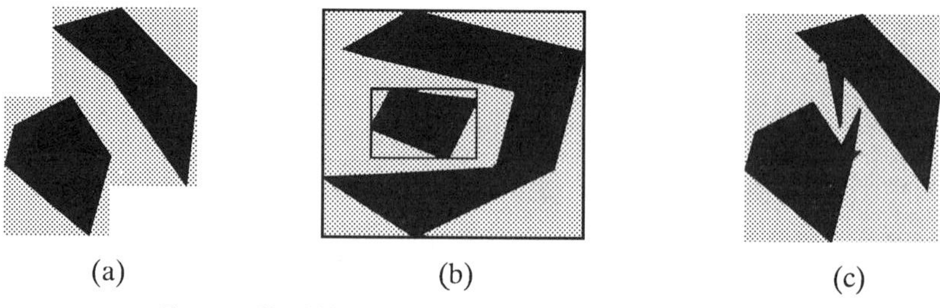

FIGURE 52. Objects that intersect each other's rectangles.

in the image. However, in some cases the rectangles of two objects are overlapping (see Fig. 52a), and in yet other cases one rectangle might cover another rectangle completely (see Fig. 52b). In the worst case, two objects are connected in a way that they will be regarded as one object by the operator (see Fig. 52c).

To remove holes in objects we first use the *tblrfill* operation. As shown in Fig. 53 this operation sets the background pixels that are connected to the image border.

A very useful operation is **thresholding with hysteresis**. A 2D NSIP system with the GLU-function mark is well suited for this task. Two binary images are produced for the same scene using two different thresholds. The first one is set so low (i.e., high U_{ref}) that all parts of the object come through

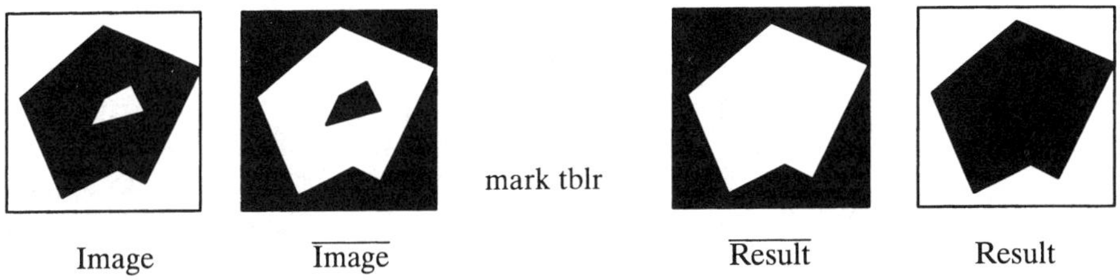

FIGURE 53. Hole filling.

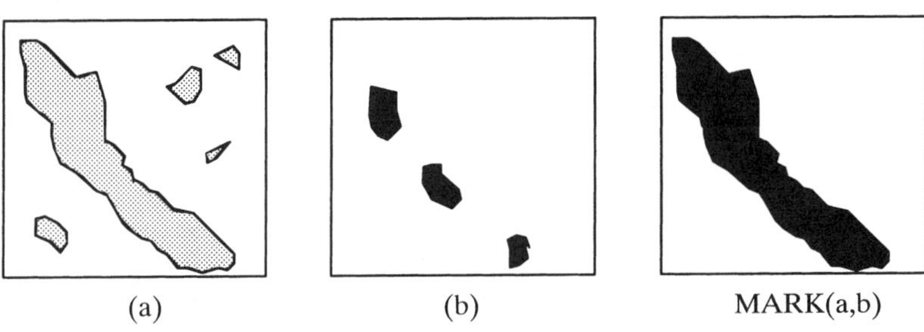

FIGURE 54. Thresholding with hysteresis.

along with some unwanted noise as illustrated by Fig. 54a. The second image is taken with a high threshold (i.e., low U_{ref}) so that only the brightest parts of the intensity objects will come through. This is the binary image of Fig. 54b. The final result is obtained with the mark-operation.

Program for thresholding with hysteresis.

```
ref: = 0.5 * U0          ; High Uref
ldi pd
st a, r0
ref: = 0.1 * U0          ; Low Uref
ldi pd
mark r0
```

Pure unmasked propagation is of course also possible with GLU. It is implemented as a special case of the mark-operation by simply setting all accumulator bits to 1 along the array. Once a "seed" has been injected into the global net it will continue to propagate according to what is allowed by the control vector (***CN*,*CE*,*CW*,*CS***). The possibilities are given by Table III and the case when the input image contains a single 1-pixel is shown in Fig. 55.

In Fig. 55, the numbers below each square correspond to an entry in Table III. The single initial 1, the seed, is placed as in plot number 0.

TABLE III
PROP OPERATIONS

No	CN	CS	CE	CW	Direction	Name
0	0	0	0	0	No propagation	
1	0	0	0	1	W	wprop
2	0	0	1	0	E	eprop
3	0	0	1	1	E-W (Horizontal)	hprop
4	0	1	0	0	S	sprop
5	0	1	0	1	SW	swprop
6	0	1	1	0	SE	seprop
7	0	1	1	1	~N	hsprop
8	1	0	0	0	N	nprop
9	1	0	0	1	NW	nwprop
10	1	0	1	0	NE	neprop
11	1	0	1	1	~S	hnprop
12	1	1	0	0	N-S (Vertical)	vprop
13	1	1	0	1	~E	vwprop
14	1	1	1	0	~W	veprop
15	1	1	1	1	All	vhprop

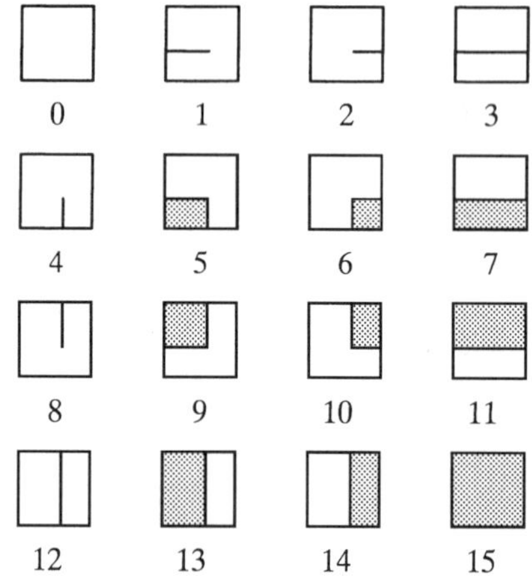

FIGURE 55. PROP propagations.

B. Safe and Unsafe Propagation

Consider the 1D GLU-net shown in Fig. 56. The operation to be performed is LRPROP. The two processors shown receive only zeroes from the left and right and from the mask and should, therefore, deliver zero at their outputs.

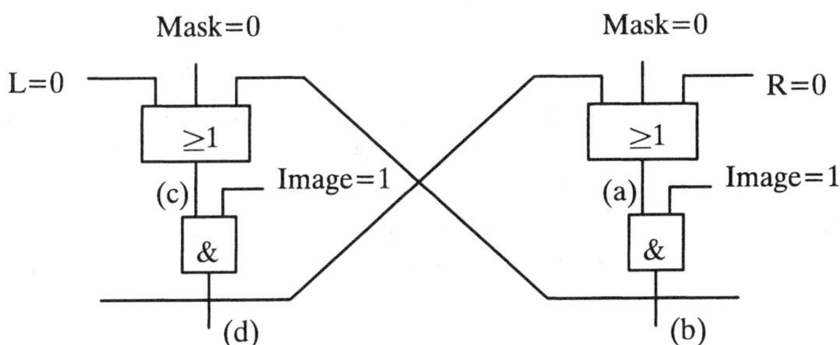

FIGURE 56. Stable error.

TABLE IV
SAFE GLU OPERATIONS

No	CN	CS	CE	CW	Mark	Fill
1	0	0	0	1	rmark	rfill
2	0	0	1	0	lmark	lfill
4	0	1	0	0	tmark	tfill
5	0	1	0	1	trmark	trfill
6	0	1	1	0	tlmark	tlfill
8	1	0	0	0	bmark	bfill
9	1	0	0	1	brmark	brfill
10	1	0	1	0	blmark	blfill

However, because of clock-signals and external noise sources we can never rule out the possibility of short false spikes, say, at point (a). Because of the double-directed propagation this spike may perpetuate to (b), (c), and (d). Then, the two SPEs latch into a false stable state with both outputs equal to one. Even worse, this false condition will be propagated throughout the array.

The underlying physical reason for this effect is the positive feedback between the two SPEs. No such feedback can be allowed. For the 2D case the implication is that we allow only the propagation directions for mark and fill as shown in Table IV.

It should be noted that the **propagation** operations in Table III are different in purpose and function. Here, all Mask bits in Fig. 56 are set to 1 and the result is not at all dependent on this operand. Instead, these propagations just expand the object unconstrained up to the image border. Clearly, the propagation distance in this case is limited to 2N SPEs, the worst case being propagation from one image corner to the diagonally opposite one.

The propagation that takes place in **MARK**- and **FILL**-operations is constrained to the input image. For a given image size of $\mathbf{N*N}$ pixels, the longest and most difficult object to propagate though is the Meander curve shown in Fig. 57. The length of this single one-pixel wide object is $\mathbf{N^2/2}$. Thus, even if we could allow the propagation to take place asynchronously within one single instruction as in the pure propagation instructions, we would have to allow this instruction an execution time of $\mathbf{O(N^2)}$. But as we can use only the restricted directions for **MARK** and **FILL** we also have to divide the total procedure into an iterative sequence of mark/fill steps as shown by the following program.

```
REPEAT
  ld r0
  st r1                 ; Store previous row
  ld r2                 ; Mask bit
  trmark r0
  st r0
  ld r2                 ; Mask bit
  tlmark r0
  st r0
  ld r2                 ; Mask bit
  brmark r0
  st r0
  ld r2                 ; Mask bit
  blmark r0
  st r0
  xor r1
UNTIL COUNT = 0
```

Suppose that the control bits (CN,CE,CW,CS) are stored in register positions within each SPE. Because we may also bring in the SPE-position number, we would then be able to configure arbitrary propagation paths across the array. For instance, by using the last bit in the row (y) and column (x) number and XORing these bits, we get a checkerboard pattern image C as shown in Fig. 58. Then, if CN: = C, CE: = 1, CW: = 0, and CS: = C, the

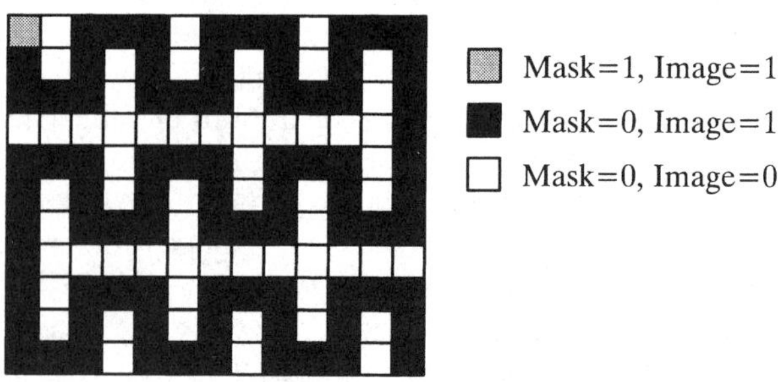

FIGURE 57. The Meander curve.

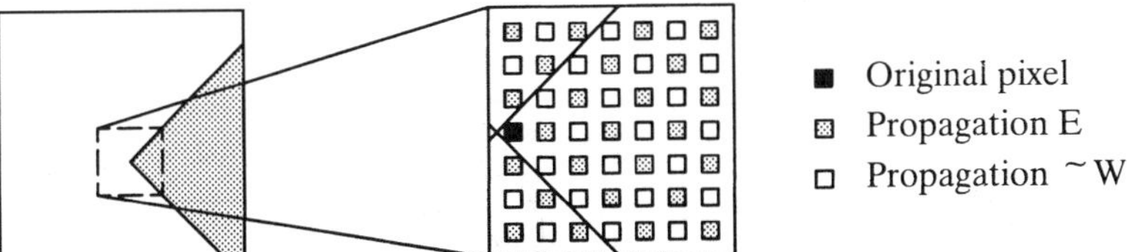

FIGURE 58. NE propagation tilted 45°.

FIGURE 59. Diagonal line propagation.

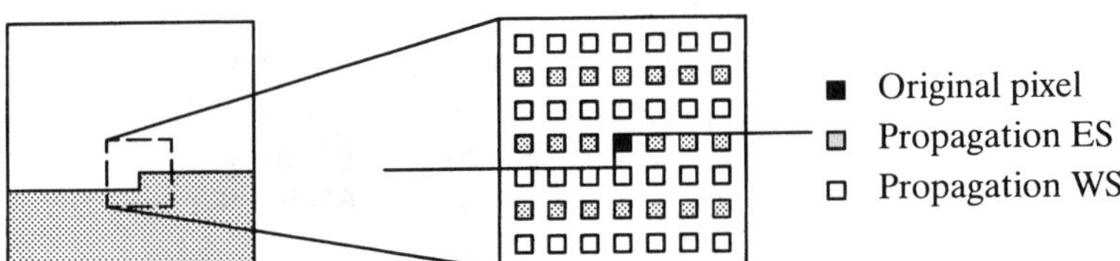

FIGURE 60. Almost 180° of propagation.

result will be a propagation within a 90° cone directed eastwards. The global propagation goes both north and south but is still safe because the latch condition of Fig. 56 does not exist.

It is even simple to obtain only diagonal propagation from the checkerboard pattern by ***CN***: = C, ***CE***: = $\overline{\mathrm{C}}$, ***CW***: = 0, and ***CS***: = 0, as shown in Fig. 59.

As a final example, let the LSB position in row number be the image B as in Fig. 60. Then, an almost 180° propagation cone angle can be obtained by ***CN***: = 0, ***CE***: = B, ***CW***: = $\overline{\mathrm{B}}$, and ***CS***: = 1.

X. FEATURE EXTRACTION

A. *Finding a Position*

In NSIP acquiring and processing of an image is performed at the same time. To perform a certain processing task may then require global data from the array at two stages in the process. The first stage is when global data is used for termination of the interrogation loop. The second stage is for delivering features of the final result.

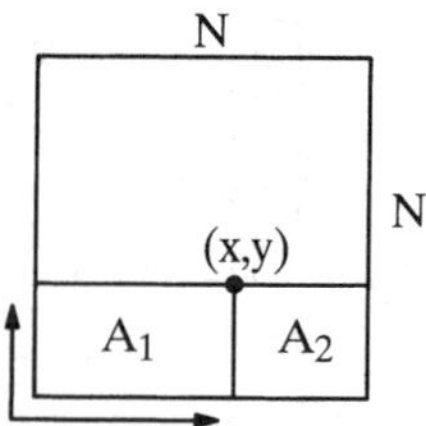

FIGURE 61. Position acquired using COUNT.

In the 2D-NSIP chip proposed here, the only available global data is COUNT, which is constantly available and delivered to the host and its program control after every instruction. COUNT is a number that tells how many SPEs there are with a value of **1** in their accumulators. As we have already seen in numerous examples, this is a typical termination parameter for an NSIP loop. However, as will be shown, the COUNT operation is also extremely useful for feature extraction. In the examples to follow, we assume that the NSIP loop has been terminated and a result remains in the form of one or several images where each one occupies one bit-plane of memory in the SPE array.

Many of the features to be extracted from an image are answers to questions "How many?" or "Where?". Obviously, features that answers the question *How many* are to be extracted by the COUNT operation. As was shown in the preceding, the *Where* question in the 1D LAPP was answered by a combination of PROP and COUNT operations. In the proposed 2D architecture the global propagation operations in four possible directions serve the same purpose.

To extract the position of a single 1-pixel is fairly simple. As can be understood from Fig. 61 we may extract values for areas A_1 and A_2 with the following program, where the GLU operations are taken from Table III and Fig. 55:

ld swprop r_0
$A_1 := COUNT$
ld seprop r_0
$A_2 := COUNT$

In the host computer we may then compute

$$y = \frac{A_1 + A_2}{N} \tag{46}$$

$$x = \frac{A_1}{y} = N\frac{A_1}{A_1 + A_2} \tag{47}$$

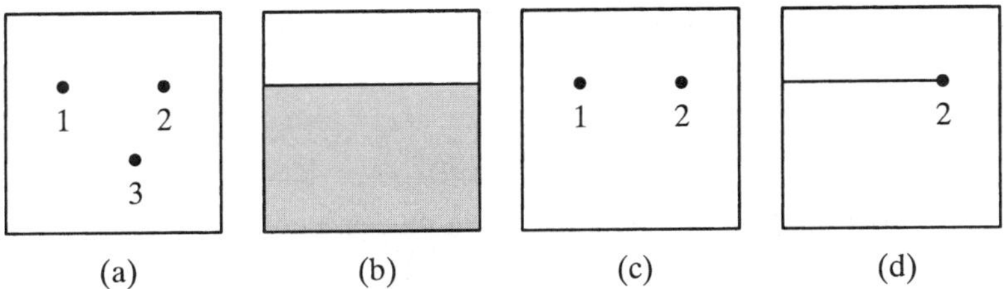

FIGURE 62. Select rightmost uppermost pixel.

We refer to a situation illustrated in Fig. 62 with three points, of which 1 and 2 have the same y-coordinate. In the first step, we perform a *hsprop*, which produces the result in Fig. 62b. A count gives us the y-coordinate for 1 and 2. A neighborhood condition (0x1xx) selects the upper edge of this result, which is ANDed with the original to obtain 62c. We may then propagate west and COUNT to obtain the x-coordinate for 2. The result is shown in Fig. 62d. A neighborhood condition (xx10x) selects the rightmost pixel of the line and by XOR with the original image we may obtain an image without point 2. From here we may repeat the procedure and bring out the coordinates for the remaining points. The following program performs this procedure:

```
hsprop r0
y: = COUNT / N
ld (0x1xx) a
and r0
wprop a
x: = COUNT
ld (xx10x) a
xor r0
....
```

B. Moments

Moments are often used in image processing to characterize and recognize binary objects and shapes.

Let P_{xy} be the value of pixel at position (x, y). Then, the zeroth-order moment is

$$m_{00} = \sum_{y=0}^{N-1} \sum_{x=0}^{N-1} P_{xy} \tag{48}$$

and it is computed as follows in a 2D NSIP system, where we assume that the binary image is stored in r_0.

ld r_0
$m_{00}\colon = COUNT$

The first-order moments in the x-direction m_{10}, are computed as

$$m_{10} = \sum_{y=0}^{N-1}\sum_{x=0}^{N-1} xP_{xy} \tag{49}$$

Because

$$x = \sum_{i=0}^{n-1} x_i 2^i \tag{50}$$

Equation (49) can be written as

$$m_{10} = \sum_{y=0}^{N-1}\sum_{x=0}^{N-1} P_{xy}\sum_{i=0}^{n-1} x_i 2^i = \sum_{i=0}^{n-1} 2^i \sum_{y=0}^{N-1}\sum_{x=0}^{N-1} x_i P_{xy} \tag{51}$$

We identify the inner product as an AND between the image P_{xy} and a position-dependent binary constant. The two summations are equivalent to a COUNT readout. Finally, the COUNT values are to be weighted by a factor 2^i and summed, which takes place in the host.

Program to compute m_{10}:

```
m10: = 0;
FOR i: = 0 TO n − 1 DO BEGIN
   setrx i;
   ld r0
   and rx
   m10: = m10 + COUNT * 2^i          ; m10 is accumulated and stored in the host
END
```

The moments of second order m_{11}, m_{20}, and m_{02} can be derived in the same manner as for the first-order moments. For instance, the formula for m_{20}

$$m_{20} = \sum_{y=0}^{N-1}\sum_{x=0}^{N-1} x^2 P_{xy} \tag{52}$$

can be rewritten using binary weights as

$$m_{20} = \sum_{y=0}^{N-1}\sum_{x=0}^{N-1} P_{xy}\sum_{i=0}^{n-1} x_i 2^i \sum_{j=0}^{n-1} x_j 2^j = \sum_{i=0}^{n-1}\sum_{j=0}^{n-1} 2^{i+j} \sum_{y=0}^{N-1}\sum_{x=0}^{N-1} x_i x_j P_{xy} \tag{53}$$

Again, the inner product and summation can be identified as an AND between the image and position-dependent constants x_i and x_j, followed by a COUNT readout. The value we get for each summation is then weighted with a factor 2^{k+1}.

Program to compute m_{20}:

```
m20: = 0;
FOR i: = 0 TO n – 1 DO BEGIN
   setrx i;
   ld r0
   and rx
   st r1
   FOR j: = 0 TO n – 1 DO BEGIN
      setrx i;
      ld r1
      and rx
      m20: = m20 + COUNT * 2^(i+j)
   END
END
```

The second-order moment m_{11} can be derived in almost the same manner.

$$\begin{aligned} m_{11} &= \sum_{y=0}^{N-1}\sum_{x=0}^{N-1} xyP_{xy} = \sum_{y=0}^{N-1}\sum_{x=0}^{N-1} P_{xy} \sum_{i=0}^{N-1} x_i 2^i \sum_{j=0}^{n-1} y_j 2^j \\ &= \sum_{i=0}^{n-1}\sum_{j=0}^{n-1} 2^{i+j} \sum_{y=0}^{N-1}\sum_{x=0}^{N-1} x_i y_j P_{xy} \end{aligned} \tag{54}$$

Moments calculations for graylevel images require an extra inner bit loop. If

$$P_{xy} = \sum_{k=0}^{b-1} P_{xyk} 2^k \tag{55}$$

m_{10} can be computed as

$$m_{10} = \sum_{y=0}^{N-1}\sum_{x=0}^{N-1} P_{xy} \sum_{i=0}^{n-1} x_i 2^i = \sum_{i=0}^{n-1}\sum_{k=0}^{b-1} 2^{i+k} \sum_{y=0}^{N-1}\sum_{x=0}^{N-1} x_i P_{xy,k} \tag{56}$$

With 8-bit graylevel pixels stored in $r_0 - r_7$ (MSB in r_7), the following program computes Eq. (56).

```
m20: = 0;
FOR i: = 0 TO n − 1 DO BEGIN
   setx i;
   FOR j: = 0 TO 7 DO BEGIN
      ld r0 + i
      and rx
      m20: = m20 + COUNT * 2^(i+j)
   END
END
```

C. Shape Factor

A shape factor for a binary object very often represents the compactness of the object where the circle is defined to be the most compact object. One such shape factor can be computed from the distance transform of the object (Danielsson, 1978). If the object is defined as in Fig. 63a the distance transform in city-block metric gives the result shown in Fig. 63b.

The average distance to the edge is obtained as Eq. (57) and the final shape factor as Eq. (58).

$$\bar{d} = \frac{\sum_A \left[d(x, y) - \frac{1}{2} \right]}{A} = \frac{\sum_A d(x, y)}{A} - \frac{1}{2} \tag{57}$$

$$Shape = \frac{1}{9\pi} \frac{A}{\bar{d}^2} \tag{58}$$

To implement this shape factor in a 2D NSIP system we do not have to generate a distance map and then calculate the $\bar{d}$. Instead, we calculate the contributions to $\bar{d}$ iteratively while the distance map is generated. The iteration starts by setting the object pixels, which are connected to the background, that is, the edge pixels. The number of such pixels is then collected using COUNT and the sum is weighted by the factor 1 (in the host).

(a)

	1	1	1	1	1
1	2	2	2	2	1
1	2	3	3	2	1
1	2	3	3	2	1
1	1	2	2	2	1
		1	1	1	

(b)

FIGURE 63. Distance transform of a binary object.

All the object pixels, which now have been counted, are set to 0, that is, turned into background pixels. We again select all pixels belonging to the reduced object, which are connected to the background, count them and add the count to the previous sum weighted by a factor of 2. This procedure is repeated until we have reached the center of the object, that is, when the object is annihilated.

The following program shows this procedure. The image is in r_0.

```
A = COUNT                          ; Area of the object
d = 0
dist = 0
REPEAT
   ldi (11111) r0                  ; Select border pixels
   and r0                          ;
   tmp = COUNT
   d: = d + dist * tmp             ; Add one layer to the total sum
   xor r0                          ; Remove border pixels
   st r0                           ;
   dist = dist + 1
UNTIL tmp = 0
d = d/A − 0.5
Shape_factor: = A/(9*pi*d*d)
```

XI. AN NSIP SYSTEM

A. SPE Design

An NSIP system can be implemented in many different ways. The common problem will always be to minimize the number of transistors within each SPE. In Eklund *et al.* (1996) an NSIP array of 32 × 32 SPEs, including the 2D GLU, was implemented on a single chip and simulations were made on a 128 × 128 array. Figure 64 shows one SPE from that design. Each SPE has a local bus, a global communication unit, an external I/O unit (GLU and NLU), five registers and a diode comparator. ALU functions are performed directly on the write bus. Some instructions can then be executed very efficiently, for example, reading from two different memory cells and writing to a third (R0: = R1 + R3). External communication with the processor matrix is done via two global buses. The X-bus is connected to every pixel in a row. It is used for global functions, for example, to find if any pixel is set. The Y-bus is used for high I/O-data rate, for example, to input a mask pattern. The COUNT operation is performed using this bus. The neighborhood data exchanges use the same physical wires as the GLU. The unit that performs operations on

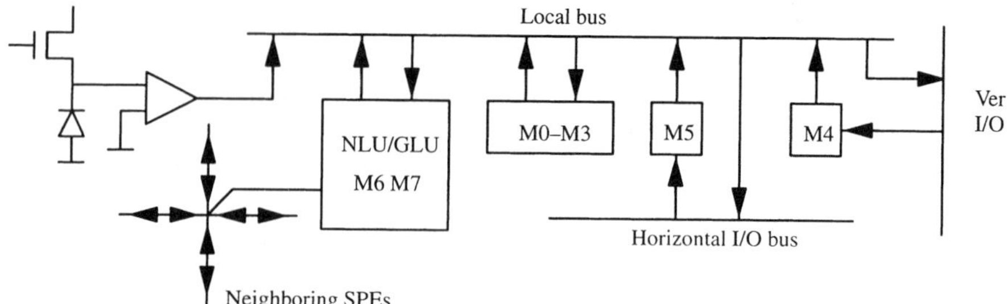

FIGURE 64. One SPE.

neighborhood data are called NLU (Forchheimer and Ödmark, 1983). NLU instructions are performed synchronously in two steps, load and propagate. As already described, the GLU operation uses the same hardware but performs an asynchronous propagation over the entire matrix.

This SPE design could be built using approximately 100 transistors (Eklund *et al.*, 1996). That design was verified to run at 100 MHz. However, as all operations are local, except for the GLU, which is asynchronous, it is likely that we can benefit from the enormous development in complementary metal oxide semiconductor (CMOS) technology.

Based on the conservative clock frequency of 100 MHz or higher, it is possible to estimate the performance of a practical NSIP system. For this purpose we use two different programs. The first one is the program that calculates the maximum intensity pixel, which corresponds to Section III.B. It contains only 10 instructions within the NSIP loop. The other program is 10 times more complicated and consists of a median filter, grayscale morphology filter, and gradient detectors, as described in the previous sections. Table V shows the relation between the frame rate and the number of intensity levels, that is, the number of loops for each frame.

TABLE V
FRAME RATE VERSUS INTENSITY RESOLUTION

	Program complexity	⇒ No. of intensity levels
Frame rate (Hz)	10 cycles/loop	100 cycles/loop
100	100,000	10,000
1000	10,000	1000
10,000	1000	100
100,000	100	10

For simple programs in real application, the number of intensity levels required might be 1000, which corresponds to 10 bits. This limits the frame rate to 10,000 Hz. For more complicated programs we can either keep the frame rate and reduce the resolution, or we can keep the resolution and lower the frame rate.

B. Application Examples

The 32×32 NSIP chip was build into a camera system and controlled by a PC. The PC is responsible for sending the instructions to the chip as well as performing the NSIP program loops. All the following images have been acquired and processed on an NSIP chip. Normally, the output from the chip should not be an image. However, for demonstration purposes, the results from all our examples will be images.

Figure 65a shows a grayscale image that has been acquired by grouping, in software, the register in each SPE to a counter. This result is a 6-bit gray level image. Figure 65 shows a thresholded image obtained using the adaptive threshold in Section IV.A.

Figure 66 shows the highest intensity pixel in the image. Here, the algorithm in Section III.B has been used. Figure 66 shows the functionality of the 2D GLU, which selects the region in Fig. 65b connected to the brightest pixel, which is detected in Fig. 66a. This operation is called thresholding with hysteresis and is explained in Section IX.A.

Figure 67 performs the same counting procedure as in Fig. 65a. However, before each counting step we calculate the smallest circumscribing rectangle,

(a)

(b)

FIGURE 65. (a) Grayscale image; (b) thresholded image.

(a) (b)

FIGURE 66. (a) Maximum intensity pixel; (b) combining Fig. 65b and Fig. 66a.

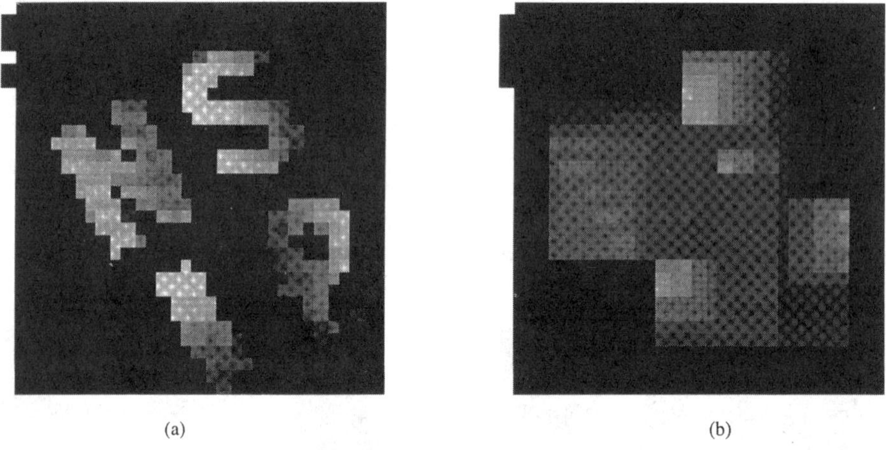

(a) (b)

FIGURE 67. (a) Original image; (b) smallest circumscribing rectangle.

which means that all SPEs within that area will increment their counters (i.e., we perform the operation for each bit plane in the image).

Figure 68b shows the smallest circumscribing rectangle operation per bit plane for the standard image in Fig. 68b.

By comparing the output from two neighboring SPEs we can calculate the gradient of the image. Figure 69b shows the horizontal gradient of the image in Fig. 69a. The idea is the same as in Section III.C; however, we have here a

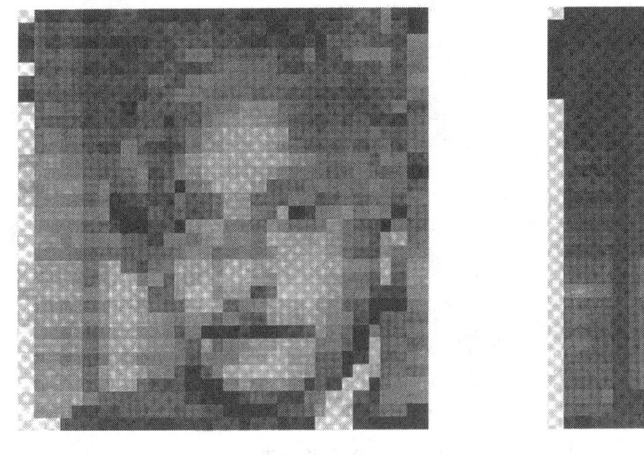

(a) (b)

FIGURE 68. (a) Original image; (b) smallest circumscribing rectangle.

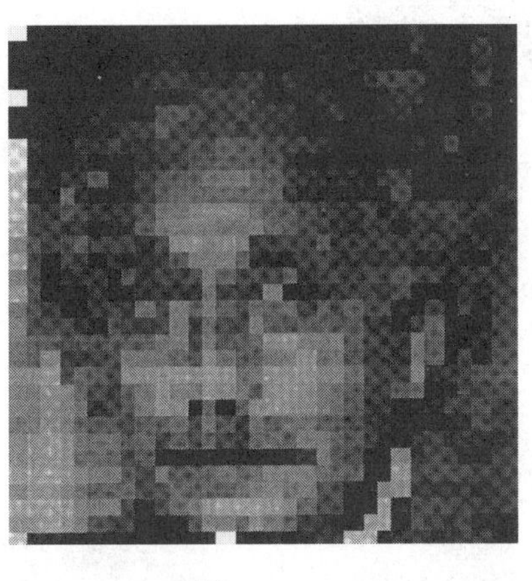

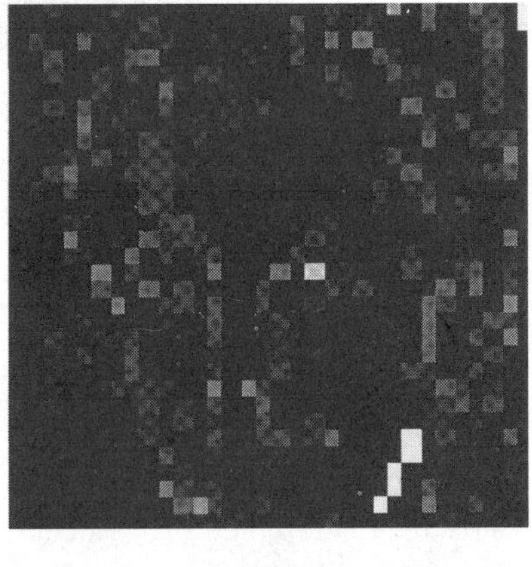

(a) (b)

FIGURE 69. (a) Original image; (b) horizontal gradient.

counter that counts the number of cycles that fulfills the condition for a positive gradient.

It is also possible to perform temporal filtering on the chip. The software counter in each SPE is then used in the following way. During the first frame the counter is incremented and during the second image the counter is decremented. If the content in the counter after these two operations is not zero we have had some motions in the scene. Figure 70 shows the effect when the image in Fig. 69a is moved through the scene.

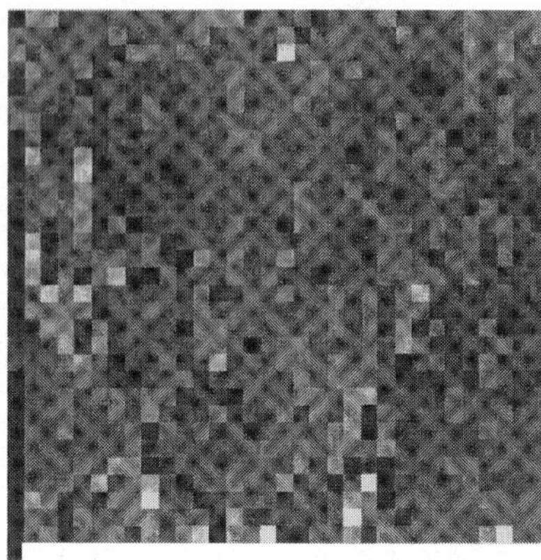

FIGURE 70. Temporal filtering (horizontal move of image in Fig. 69a).

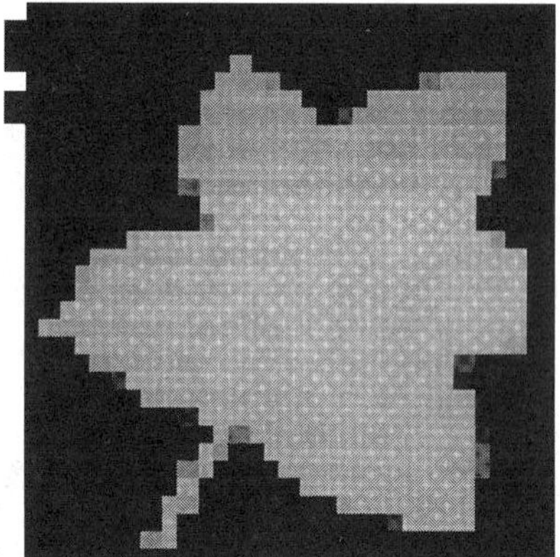

FIGURE 71. Original image.

As already described, binary operation such as erosion and dilation can easily be implemented in this concept. Figure 71 shows a grayscale image. After a threshold, that is, a readout at a certain time from the comparator, we can perform thinning operations of the image. Figure 72a shows the result of an iterative thinning to point operation in a four-connectivity neighborhood. Figure 72b shows an iterative thinning to skeleton operation. In Fig. 72 the resulting images overlay the original thresholded image.

An inspection task for NSIP could be to find etching errors on PCBs. Figure 73a shows the input image where there are two errors. We start by looking for vertical errors by repeatedly eroding the borders from the south and north, in

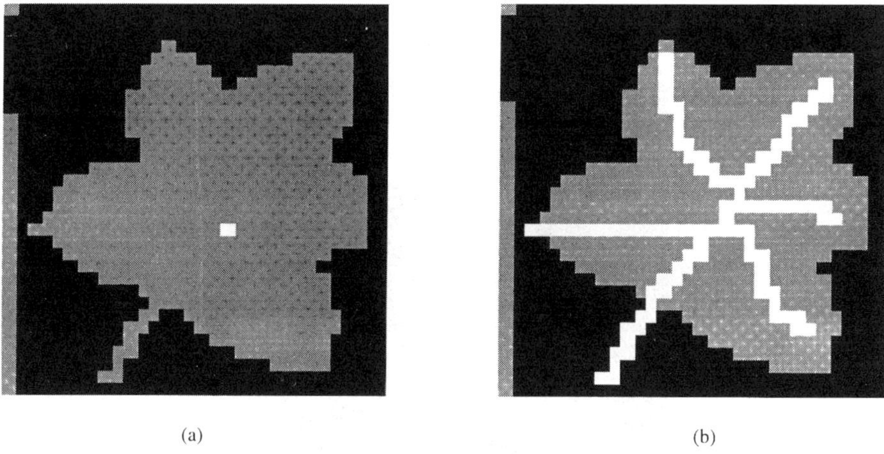

(a) (b)

FIGURE 72. (a) Thinning to point; (b) thinning to skeleton.

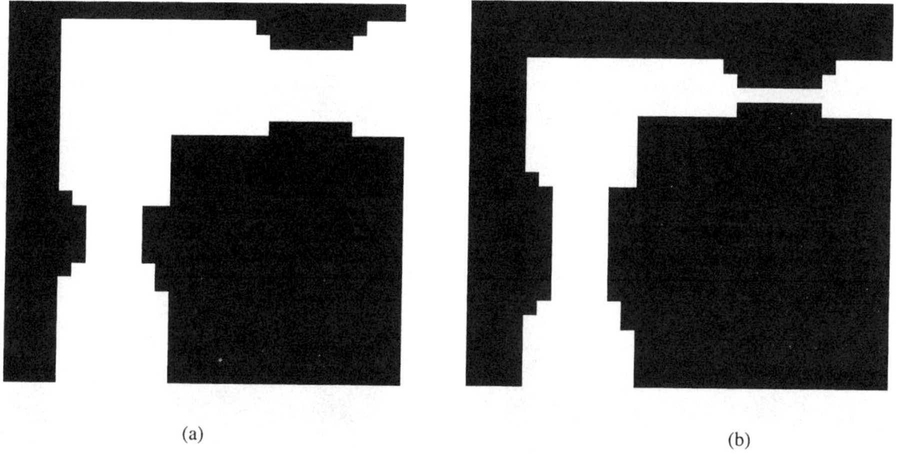

(a) (b)

FIGURE 73. (a) Threshold input image; (b) vertical eroded image.

this case four times. If during any of these phases we get a very narrow line, as in Fig. 73b, a mask bit will be set. This bit will be the input to the GLU net, together with the original image, and a North-South propagation will be performed. The same operation in the horizontal direction is then performed. The final result is shown in Fig. 74.

Section V.D described how histogram equalization and other histogram related mappings can be achieved. Figure 75 shows the effect of histogram

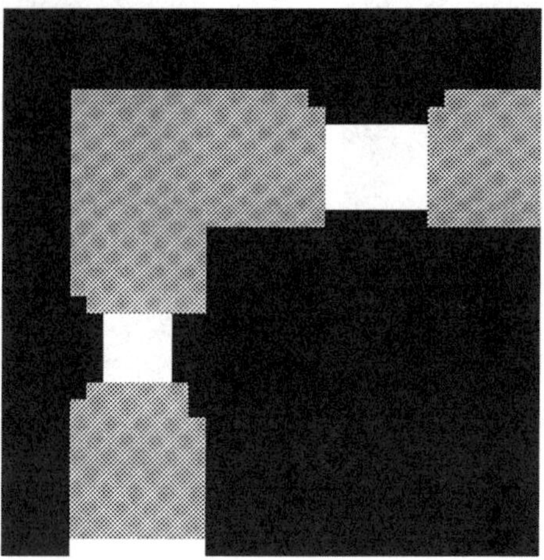

FIGURE 74. Marked erroneous areas.

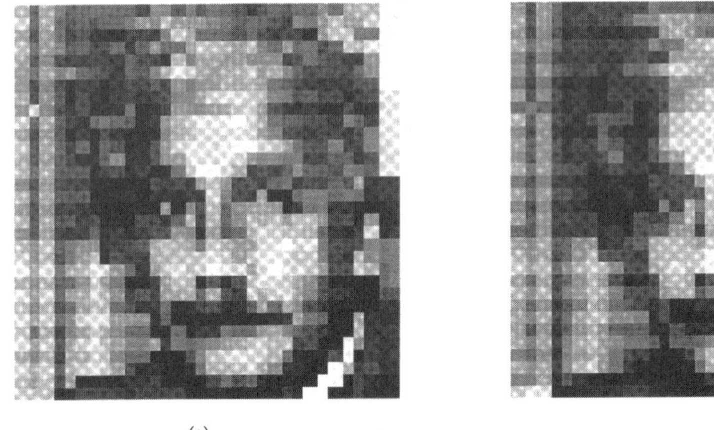

(a) (b)

FIGURE 75. (a) Iris = 1.4; (b) iris = 4.

equalization. In Fig. 75a the image is acquired using the histogram equalization program described in Section V.D. The iris of the lens is set to 1.4. Still running the same program we adjust the iris to 4 (i.e., the amount of light that reaches the sensor during a certain time has been reduced by a factor of eight). However, the resulting image will be the same, as shown in Fig. 75b, because the image is histogram equalized in real-time.

XII. Sheet-of-Light Range Images

A. *Implementation in NSIP*

SOL range imaging is a well-known technique where the distance to each point along a line in a scene is measured. The camera and the line projector are spatially separated (see Fig. 76), which means that the impact of the line on each column of the 2D sensor is a measurement of the distance to that point in the scene.

A scene illuminated with a SOL may result in an illumination of the sensor array as shown in Fig. 77. For each x-coordinate the y-coordinate is a measure of the distance from the laser to the object. A program that detects the maximum intensities for each column can be implemented in the general NSIP architecture by the following program:

```
clr a
st r1                    ; inhibit column
st r0
REPEAT
   ldi pd                ; Read from photodiode
   xor r1                ;
   or r0                 ;
   st r0                 ; Conditional storage
   vprop a               ; Has any pixel passed the threshold?
   st r1
UNTIL COUNT=N
sprop r0
st a ,r0
```

The vertical propagation instruction *vprop* ensures that the first SPE in a column to pass the intensity threshold blocks the possibility to set r_1 in all other SPEs. When the program has terminated, the image in Fig. 77 will result

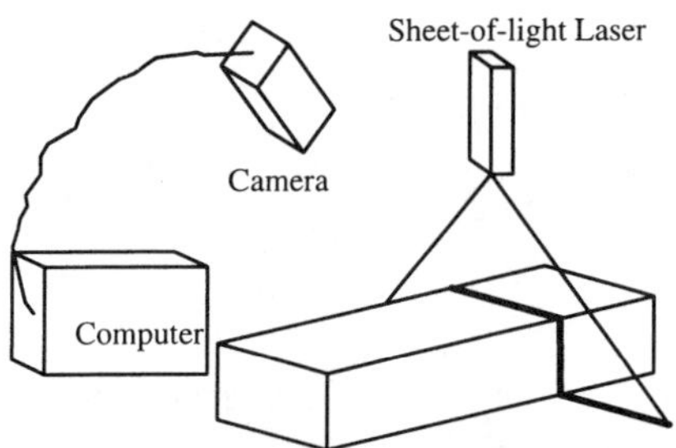

FIGURE 76. SOL range system.

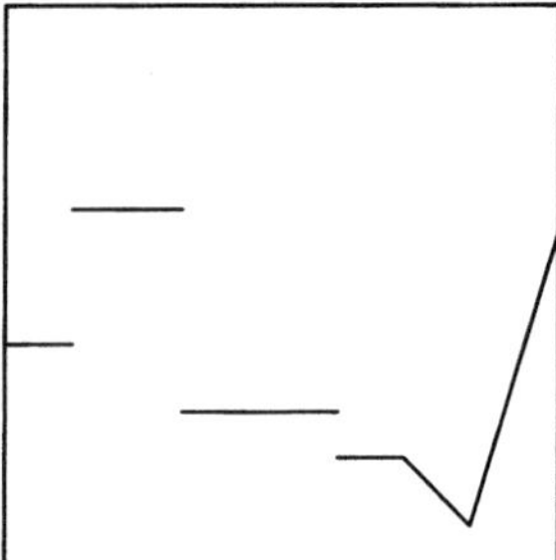

FIGURE 77. A scene illuminated with a SOL as acquired by the sensor.

FIGURE 78. The contents of R0 after the loop has terminated.

in Fig. 78. Here, the shaded area represents set bits in r_0. The result can then be read out, column by column, using COUNT as is shown in the following program:

```
setx 0
ldi rx
FOR i = 1 TO n − 1
   setx i
   andi rx
END–FOR
st a,r1                  ; set r1 in leftmost column to 1
FOR i = 0 TO N
   ld r0
   and r1
   Range = COUNT
   ld (x1xxx) r1         ; shift readout to next column
   st r1
END–FOR
```

The max detection program executes as long as needed, or until a certain time-out arrives from the host. In any case, this part of the total procedure will not set any limit on speed as long as the NSIP loop can be executed fast enough. Very likely, it is the propagation instructions that will dominate.

B. Special NSIP Architecture for Sheet-of-Light

If we need a high resolution it might be the case that a full 2D NSIP chip meeting our demand is not likely to be designed and manufactured in the immediate future. With this in mind, however, the basic idea of NSIP may be implemented for some special purpose application where fewer transistors per SPE are needed. One such application is the SOL camera sensor. A proposal for this application is shown in Figs. 79 and 80.

Note that in Fig. 79 the jagged line runs in the vertical direction. To understand this function, see Fig. 80, which shows the design of an individual SPE. In each row all SPEs have a common inhibit bus and they are precharged at the same time. As soon as the first SPE passes the threshold, the inhibit bus goes low because of the (a) transistor. This inhibits all other SPEs in the same row and only the inhibiting SPE is able to introduce a new state at the inverter input (b). The data is transferred to the readout register when the exposure time is over and then read, one row at a time, to the position unit. This unit performs a 1D lfill operation followed by a COUNT, which gives us the position of the maximum intensity. The intensity, or rather the time to inhibit, can be registered for each column. Saved in a shift register these data may be shipped out from the chip together with the position for each row. Ideally,

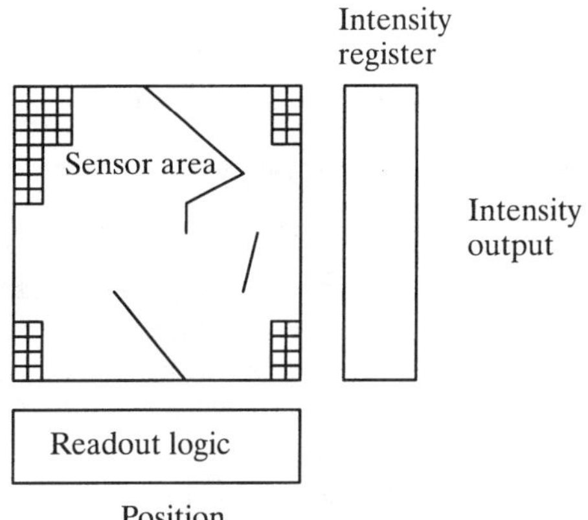

FIGURE 79. Special architecture for SOL range imaging, using the NSIP technique.

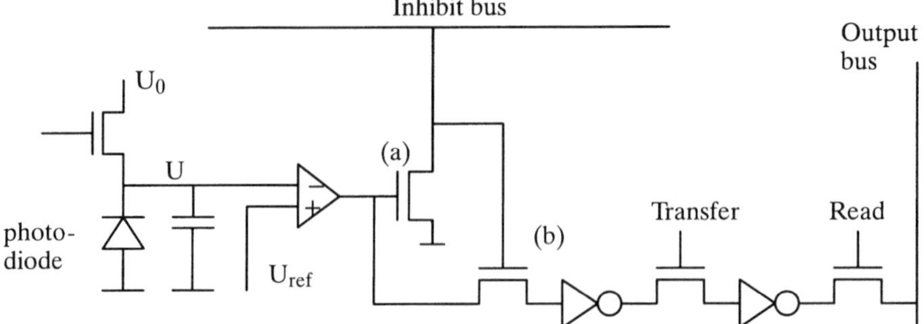

FIGURE 80. An SPE in the special architecture.

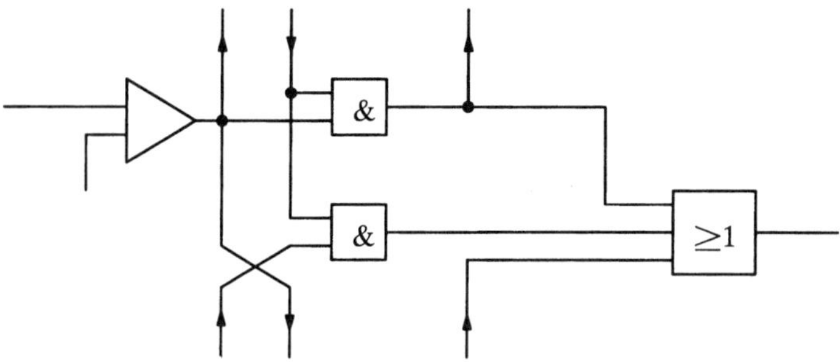

FIGURE 81. Median filterings.

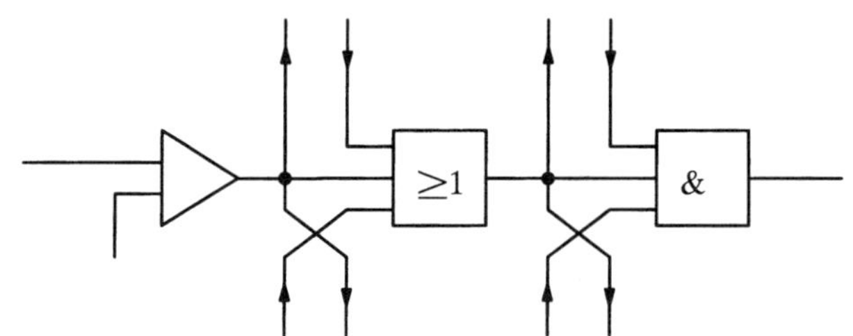

FIGURE 82. Grayscale morphological closing.

only one single pixel per row should be set when read out from the array. In practice, other cases may occur. The interesting thing about this approach is that the entire arsenal of NSIP operations can be applied during processing. This means that we can perform filtering during the discharge phase. If we want to perform a row-wise median filtering of the incoming signal, we recall from Section III.D that we need to apply the function:

$$\mathrm{OUT}_{\mathrm{mediam}} = \mathrm{SPE_L}\,\&\,\mathrm{SPE_C} + \mathrm{SPE_L}\,\&\,\mathrm{SPE_R} + \mathrm{SPE_C}\,\&\,\mathrm{SPE_R} \qquad (59)$$

Figure 81 shows how that can be implemented with only three extra logical gates per SPE. If we want to perform a closing operation instead, which is a dilation followed by an erosion, we add two extra logical gates as shown in Fig. 82.

XIII. Conclusion

The idea of near-sensor image processing (NSIP) originates from the development of LAPP1100. In those days, the VLSI constraints prevented the designers from incorporating a traditional analog to digital (A/D) converter. The solution was to keep only the comparator of the potential A/D converter and take a number of binary samples for different threshold voltages. Thereby, a limited form of grayscale processing capacity was obtained. It was then discovered that many image processing problems could be solved using a constant threshold, and the ideas underlying NSIP were conceptualized.

Our motivation for further developing the NSIP concept is that each SPE seems to require less space than a corresponding SPE with an ADC. And, as we have proved in this text, most of the typical low-level image processing operations can be implemented rather easily on an NSIP system. In many cases, for instance, grayscale morphology, the NSIP implementation is much simpler to implement than conventional methods. We have also described how other operations occur naturally with the NSIP concept. These include: locating the maximum intensity, gradient thresholding, rank-order filtering, histogramming, and adaptive thresholding.

The time from the precharge of the photodiode until the voltage passes the reference level is the available intensity measure in NSIP. It is inversely proportional to the light intensity. We have presented in this paper two different approaches to linearize the measurement. In the first approach we use variable sampling time to achieve different mappings between time and the NSIP loop index. The same idea could be used for histogram equalization. In the other approach we showed that by varying the threshold voltage and keeping the equidistant sampling, we can effectively perform both linear and nonlinear mappings.

For 2D NSIP the N1D architecture needs some modification. We have provided a programmer's 2D model as well as a gate-level CMOS design of the individual SPE. We have also shown in a few examples how such important 2D features as moments and shape factor can be obtained in this architecture.

The CMOS design needs approximately 100 transistors per SPE for its digital part. A 256×256 array would then require more than 7 million transistors given some control logic. This will be possible to implement in the near future. Also, as was shown by the SOL example, the basic NSIP idea may be viable in other more special purpose applications.

Acknowledgment

The authors would like to thank Professor Per-Erik Danielsson for valuable suggestions to this study. They also would like to thank CENIIT for financial support.

References

Åström, A., Forchheimer, R., and Eklund, J. E. (1996). Global feature extraction operations for Near-Sensor Image Processing, *IEEE Trans. Image Processing*, **5**, 1, 102–110, January.

Batcher, K. E. (1980). Design of a massively parallel processor, *IEEE Computer*, **C29**, 836–840.

Danielsson, P. E. (1978). A new shape factor, *Computer Graphics and Image Processing*, **7**, 2, 292–299.

Eklund, J.-E., Svensson, C., and Åström, A. (1996). Implementation of a focal plane image processor: A realization of the Near-Sensor Image Processing Concept, *IEEE Trans. VLSI Systems*, **4**, 3, September.

Forchheimer, R., and Åström, A. (1994). Near-Sensor Image Processing. A new paradigm, *IEEE Trans. Image Processing*, **3**, 6, 736–746, November.

Forchheimer, R., Chen, K., Svensson, C., Ödmark, A. (1993). Single chip image sensor with digital processor array, *J. VLSI Signal Processing*, **5**, 121–131.

Forchheimer, R., and Ödmark, A. (1983). Single chip linear array processor, in *Applications of Digital Image Processing*, SPIE, **397**.

Fountain, T. J. (1987). *Processor Arrays Architecture and Applications*, Academic Press, London, pp. 49–60.

Lyon, R. F. (1981). The Optical Mouse, and an architectural methodology for smart digital sensors, in *CMU Conference on VLSI Structures and Computations*. Computer Science Press, October.

Reddaway, S. F. (1973). DAP – a Distributed Processor Array, First Annual Symposium on Computer Architecture, Florida, pp. 61–65.

Sternberg, S. R. (1986). Grayscale morphology, *Computer Vision, Graphics, and Image Processing*, **35**.

Tucker, L. W., and Robertson, G. G. (1988). Architecture and application of the connection machine, *IEEE Computer*, August, 26–38.

Wendt, P., Coyle, E., and Gallagher, Jr., N. (1986). Stack filters, *IEEE Trans. ASSP-34*, 4, 898–911, August.

Wilson, S. S. (1988). One dimensional SIMD architectures – the AIS–5000, *Multicomputer Vision*, S. Levaldi, Ed., Academic Press, London, pp. 131–149.

Digital Image Processing Technology for Scanning Electron Microscopy

EISAKU OHO

Department of Electrical Engineering, Kogakuin University, 1-24-2, Nishishinjuku, Shinjuku-ku, Tokyo, Japan 163-8677

ISBN 0-12-014747-5
ISSN 1076-5670/99 $30.00

I. Introduction

For a number of years, scanning electron microscopy (SEM) has provided outstanding high-resolution images with very great depth of field in biophysics, material science, etc. In the early years, several digital image processing techniques as well as many analog techniques were introduced to the SEM field (White *et al.*, 1968; McMillan *et al.*, 1969). Because the SEM image is essentially an electric signal, it is very suited to use in image processing techniques. Analog image processing techniques were employed mostly for SEM signal enhancement in the early stages (Baggett and Glassman, 1974), because digital techniques were in the developmental stage and the cost of using them was extremely high. Analog techniques are still used as required. On the other hand, compared to transmission electron microscopy (TEM) images, SEM image information (digital data) could be easily acquired by digitizing through an analog-to-digital (AD) converter and storing in memory. Hence, performance of the on-line SEM image recording system improved rapidly (e.g., Oron and Gilbert, 1976).

At first, many digital techniques were simply introduced from the field of image processing (see Jones and Smith, 1978). These introductions were novel enough and significant in those days. For several purposes unique to SEM, the Cambridge group and other groups devoted their energy to their study from the 1970s through the 1980s (e.g., Unitt and Smith, 1976; Holburn and Smith, 1979; Erasmus and Smith, 1982). Unfortunately, for the last decade, the number of high-level studies in SEM image processing has decreased somewhat because those less difficult studies have already been done and the number of serious researchers has seen little increase. However, with the recent advances in computer technology, a high-performance and inexpensive personal computer applicable to digital image processing has fully taken root in the field of SEM and even a general-purpose SEM user can examine many digital image processing techniques (Oho *et al.*, 1996a; Postek and Vladar, 1996). In addition, the electron microscopy field as well as other fields is interested in the related technology of networking, for example, the Internet (Chumbley *et al.*, 1995; Chand *et al.*, 1997; Voelkl *et al.*, 1997). From the viewpoint of microscopists, this situation is one of the best opportunities for moving forward on new generation SEM technology that uses on-line digital image recording and processing as well as networking technology.

The results of several studies from among many others have survived and been utilized in commercialized and/or prototype SEMs (e.g., Erasmus, 1982; Edwards *et al.*, 1986; Oho *et al.*, 1995; Oho and Ogashiwa, 1996b). Contrary to general belief, there are few practical studies. Microscopists may not have realized yet how to make the best use of digital image processing technology as related to the SEM field. To use many techniques of image processing effectively, a successful combination of highly advanced SEM equipped with various functions for acquiring necessary data and concomitant techniques is the most important issue to be resolved. As one solution, a new concept has been proposed on "active image processing" (Oho *et al.*, 1997). This method gives priority to the development of various functions for acquiring SEM signals including sufficient information as well as image processing techniques.

Several important subjects closely relating to the SEM field are discussed here. Many techniques are suitably utilized and compared in the following sections. We have not tried to explain systematically all image processing techniques, because it may not be very helpful for microscopists. For further reference, for instance, see Rosenfeld and Kak (1982) and Gonzalez and Woods (1992).

II. Proper Acquisition and Handling of SEM Image

A. *Digital Recording and Processing System*

Following development of the SEM, various systems of image recording and processing have been used. In the 1970s, expensive minicomputer systems equipped with special hardware were used. In the early 1980s, microscopists began to use a combination of personal computers and off-the-shelf hardware to acquire and/or process SEM images. Performance for that sort of system was typically 256×256 pixels $\times$ 8 ($\sim$12) bit acquisition, 256×256 pixels $\times$ 4 bit on the display, RAM of $\sim$0.256 MB (Desai and Reimer, 1985; Joy, 1982), and high speed digital image processing was not possible.

A system similar in design soon began to deal with processing speed and more memory capability (Oho and Kanaya, 1990b). These older systems will not be referred to because they had outlived their usefulness. Today, we can easily find many commercial systems for image processing of SEM images based on a standard personal computer (pc) without any additional special hardware and equipped with an analog-to-digital (AD) converter, often with a performance of 2048×2048 ($\sim 4096 \times 4096$) pixels $\times$ 8 ($\sim$12) bit acquisition, 1280×1024 ($\sim 2048 \times 1536$) pixels $\times$ 8 bit on the display, RAM of 1 GB or more, and so forth. More improvements will follow in the near future.

B. Superiority and Problems in Quality of SEM Images Taken with the On-Line Digital Recording

1. *Superiority in image quality*

Many SEM users still utilize a conventional recording system consisting of a video monitor, with a resolution of approximately 2000 lines, and a high-performance camera. It is generally believed that the conventional system is satisfactory in image quality for the average SEM user. However, serious deterioration in information has already been confirmed by comparing it with an on-line digital recording system, a system that is closer to the ideal for SEM images (Oho *et al.*, 1986b; Oho and Kanaya, 1990b). It should be noticed that SEM images are essentially an electric signal.

We compare the difference in quality of on-line digital versus conventional recording of the SEM image. The micrographs (2048 scanning lines) in Figs. 1a and b are of digitized and conventional SEM images, respectively. The micrographs shown in Figs. 1a' and b' are extremely enlarged images of Figs. 1a and b obtained by the cubic convolution (interpolation) method based on the sampling theorem and a darkroom enlarger, respectively. Although surface structures are visible in Fig. 1a, those in Fig. 1b are disturbed by the film-grain noise and blur, which may originate from nonideal point spread function in the conventional recording system. It should be noted that the image degradation caused by conventional systems is more severe than expected. The validity of structures in digitized images can be confirmed by observing an image, identical view of Fig. 1a, but recorded at a much higher magnification (Oho *et al.*, 1986b). In addition, as some SEMs constructed by state of the art technologies have 4096 or more scanning lines, the differences may be remarkable. An ultrahigh quality SEM image of a rat kidney with 2745×3767 pixels (a part of 4096×4096 pixels) is shown in Fig. 2a (Oho *et al.*, 1995). A 26-fold enlargement from the original SEM magnification (identified by a square in Fig. 2a) is exhibited in Fig. 2b. This is equivalent to a 43-fold enlargement from 6×7 cm negative film. Although the original recording magnification indicated in the SEM instrument is only $500\times$, we can clearly observe a glomerular podocyte in the kidney in Fig. 2b (when conventional 2048 or 1024 scanning lines are used, the structure is severely deformed).

2. *Problems in image quality*

The scan coils of the SEM generally are used to perform a fast scan in the x direction and slow scan in the y direction. The former produces a continuous signal, while the latter gives what is called a sampled signal. In the present

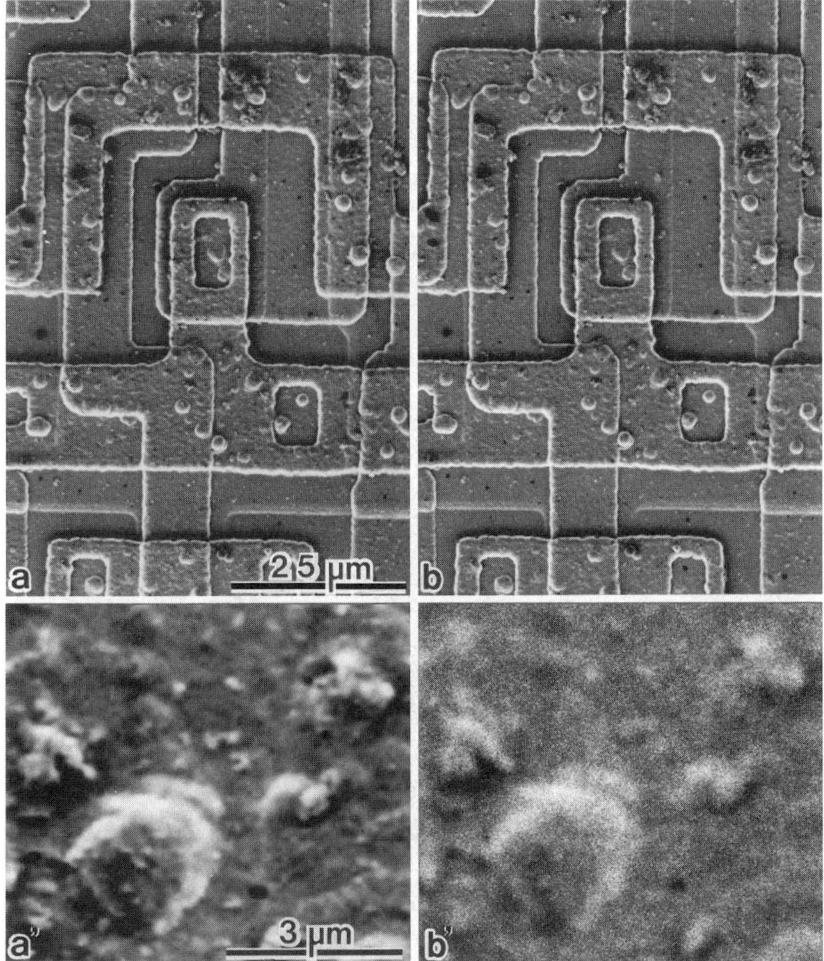

FIGURE 1. Superiority of a high-definition on-line digital recording system. (a) Digital SEM image of an LSI recorded with the on-line system; (b) conventional SEM image recorded on film together with the extremely enlarged versions (a') and (b'). It should be noted that image degradation caused by a conventional recording system is very severe.

section, we will first discuss the characteristics of the sampled signal (y direction). Next, the sampling of a continuous signal (x direction) is also considered.

The SEM is operated under various conditions that include electron beam size, incident current, number of scanning lines per frame, and magnification. In addition, the resolution, which is strongly related to the sharpness, signal-

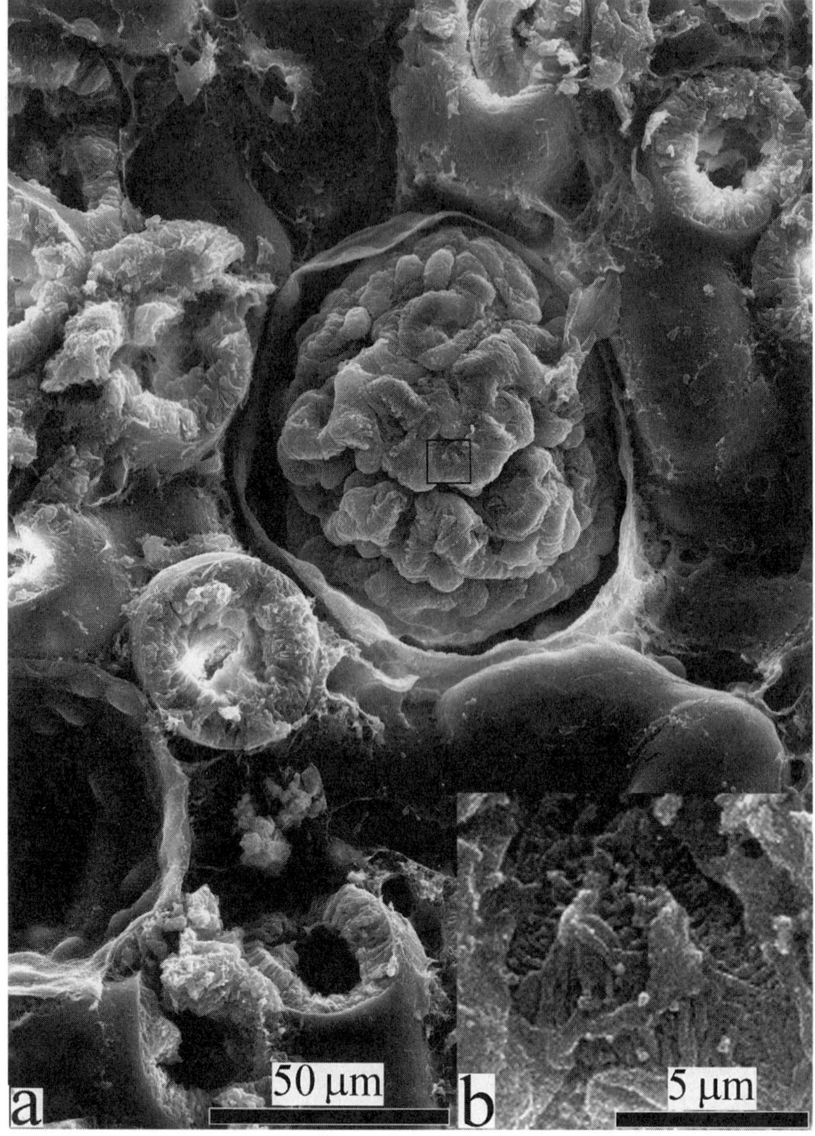

FIGURE 2. Ultra high-quality digital SEM image recorded with the on-line system. (a) SEM image of a biological sample with 2745 × 3767 pixels; (b) its expanded image.

to-noise ratio (S/N), contrast, and so on, of the SEM image is influenced greatly by the characteristics of the sample. Hence, in SEM image recording, it may be difficult to achieve the optimal scanning condition proposed by Crewe (1980) and by Crewe and Ohtsuki (1981). The concept of this optimal scanning is equivalent to the use of a fixed magnification (line spacing d) chosen to sample at the Nyquist sampling rate $2f_c(d = 1/2f_c$, f_c: cutoff frequency, which is mainly determined by the property of a specimen as well as by the resolving power of the microscope) in a direction perpendicular to the scanning line, that is, in the y direction. Generally, most SEM images are taken in over- or underscanning (sampling) condition (Oho *et al.*, 1996a).

Underscanning ($d > 1/2f_c$): SEM images taken with underscanning are contaminated by the aliasing error (artifact) to a greater or lesser extent. In other words, the fine structures of the specimen are not accurately converted into an analog SEM image. However, except for some particular specimens and conditions (Reimer 1985), it may not disturb observation of the specimen experimentally. If an expansion technique is used after digital acquisition of the SEM image, some problems occur in an expanded image; these will be discussed in this material.

Optimal scanning ($d = 1/2f_c$): As already mentioned, it is very difficult in routing work to find this condition for each SEM image although information included in an analog SEM image may have a validity and the largest areas can be recorded without aliasing error. However, from the viewpoint of S/N in the SEM image, this scanning might produce a noisier result than the overscanning condition (using a higher magnification).

Overscanning ($d < 1/2f_c$): In this scanning condition, an analog SEM image can generally be obtained without an aliasing error. However, excessive overscanning may aggravate the effects of radiation damage, contamination, vibration, stray magnetic fields, and/or charging problems from the specimen. It should be noted that these influences are likely to increase rapidly beyond our expectations as a SEM image is magnified.

On the other hand, in SEM signal (x direction) digitization through an AD converter, the sampling aperture (a sort of averaging filter) should generally have a width τ roughly equal to sample spacing (sampling interval) Δt. This has the effect of reducing noise and aliasing error (Castleman, 1979). This can be accomplished by using an analog integrating amplifier or a lowpass filter (antialiasing filter) at the input of the AD converter. However, it is very difficult to develop an ideal antialiasing filter (analog lowpass filter with a very steep cutoff frequency). Unfortunately, if that sort of filter is employed as an antialiasing filter, there will be a certain amount of distortion of the original waveform due to phase distortion in that filter.

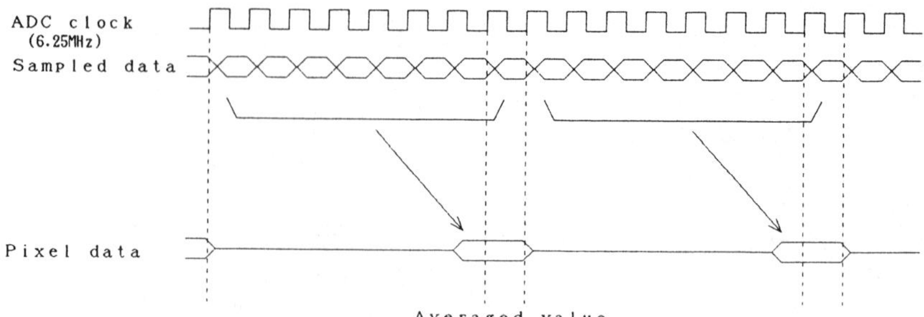

FIGURE 3. Explanation of a method for obtaining high-quality pixel data.

In our AD converter for SEM, a 6.25 MHz ultrahigh sampling rate is used with a slow-scan instrument in order to avoid producing aliasing error. Then a great many sampled data obtained from this sampling rate are reduced by proper averaging into new data (pixel data) consisting of, for example, 4096 pixels/line with 8-bit resolution, as depicted in Fig. 3 (Oho *et al.*, 1995). The effect of this operation is similar to that of $\tau \cong \Delta t$. Moreover, the present AD converter is easy to use (clearly selecting the optimal parameters for a variety of scanning speeds) and it is very effective for reducing both noise and aliasing error. An obtained digital image may be almost equal to the analog image in S/N (it might be better depending on digitization condition). This process is very important for SEM images, which usually do not have a high S/N. Of course, it is necessary to select the device that provides a digital SEM image of the highest possible quality (the performance of each commercial AD converter is not the same) because the image processing technology is utilized more effectively in a high-quality SEM image.

3. *Example of adverse effects of the undersampling and solutions in SEM images*

SEM images acquired with the underscanning (sampling) condition in x and/ or y directions are influenced by aliasing more or less as has been described. The effects of aliasing error are examined in this section (Oho *et al.*, 1996a). Figures 4a–d show SEM images of a mesh recorded at SEM instrument magnifications of 100, 400, 1600, 6400, respectively (a series of increasing magnifications). In this sample with a periodic structure, we may be able to pinpoint easily the effects of aliasing error. As the number of scanning lines was 512 and the measured beam diameter was approximately 2 nm (Oho *et al.*, 1986a), all SEM images in Fig. 4 are underscanned images. The condition of $\tau < \Delta t$ and low-density scanning of 512 lines was deliberately selected to show the severe effects of aliasing error. However, the effects of

FIGURE 4. Example of adverse effects of the undersampling in SEM images. See text for details.

aliasing cannot be seen at the original magnifications and the effects can be observed clearly only when these images are enlarged. Figures 4a_1, a_2, a_3 show 4-, 16- and 64-fold expanded images by the cubic convolution method for 5a, respectively. Also, Figs. 4b_1 and b_2 are 4- and 16-fold for 4b, and Fig. 4c_1 is 4-fold for Fig. 4c, respectively. In short, digital expansions (4 times) were performed as explained by the direction of arrows in Fig. 4. When we compare Fig. 4a_1 with Fig. 4b (of course, the identical view), periodic artifacts in Fig. 4a produced by the aliasing error can be specified. In a comparison between Figs. 4d, 4c_1, 4b_2 and 4a_3, the effects of aliasing in a common structure (not periodic) are recognized as a blur, but this complicated blur is essentially different from the effects of a lowpass filter. Thus, the effects of aliasing error are serious in digital expanded SEM images, which may be utilized increasingly in routine work.

To prevent a flood of artifacts by aliasing, we must utilize the high-performance AD converter. That is to say, when using both the scan generator attainable to an ultrahigh scanning density [in our case, 4096 (max. 8192) scanning lines/image] and the AD converter with an ultrahigh sampling rate (6.25 MHz sampling rate), SEM images without aliasing error may be obtained in all operating conditions. Subsequently, numerous data are reduced by the proper averaging into new data consisting of, for example, 2048×2048 pixels.

On the other hand, SEM noise with all frequency components is always undersampled. Therefore, a digital SEM image with considerable noise frequently has a serious problem (Oho *et al.*, 1995). To show a typical example of the effects, noisy SEM images of gold particles on carbon with 1024 scanning lines/frame (Fig. 5a) and with 4096 lines/frame (Fig. 5b) were obtained at the same recording time (80 sec) and area of scanning, and conditions of the AD converter were adjusted optimally in each case. In Fig. 5a, none of the gold particles retained their real structures due to the severe influence of undersampled noise. On the contrary, as the ultrahigh scanning density was applied in Fig. 5b for reducing the aliasing error of noise, gold particles can be seen even though the total electron dose was the same. However, decreasing the number of incident electrons per pixel (increase of the number of pixels per frame) affects our SEM images in S/N. If necessary, after image acquisition, the resolution of the 4096 lines can be reduced to 2048 or 1024 lines for improvement of S/N; moreover, we can use several methods to reduce noise in Fig. 5b. Processed images by a common averaging filter for Figs. 5a and b are shown in Figs. 5c and d, respectively. This conventional filter is effective for the removal of noise in Fig. 5b. However, all smoothing filters are not effective for reducing the severe aliasing error as shown in Fig. 5c. Here, we used enlarged images to show the difference clearly (eight times for Figs. 5a and c, twice for Figs. 5b and d).

4. *Proper Expansion of a SEM Image*

The cubic convolution method used in various sections is a highly precise expansion technique based on the sampling theorem. In other words, digital data that satisfy the sampling theorem can be interpolated (expanded) very accurately by this method. With the characteristics of the digital SEM image in mind, we should now discuss some useful expansion methods, because most SEM images will be treated as digital data in the very near future and these methods will be utilized increasingly in routine work.

The cubic convolution method used in the space domain is easily explained in the Fourier domain as illustrated in Fig. 6 (an example of three-times expansion). Bold arrows in Fig. 6 indicate the flow chart for getting expanded

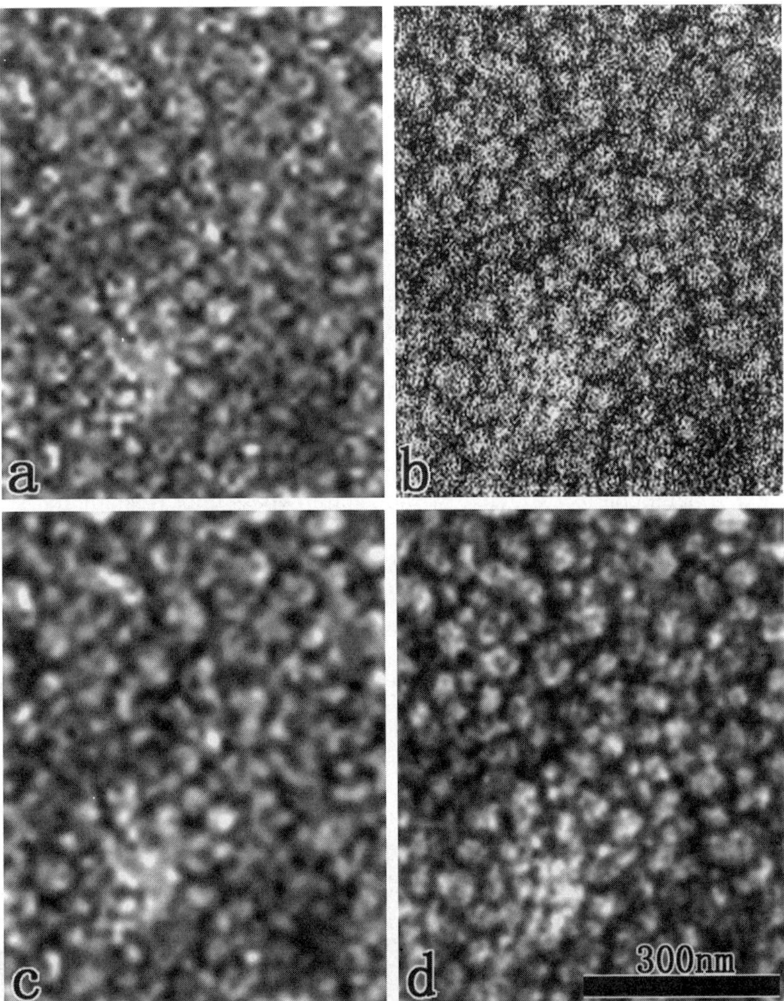

FIGURE 5. The difference between a digitized low electron dose SEM image (a) with 1024 scanning lines; (b) an image with 4096 lines; (c), (d) lowpass filtered images for (a), (b), respectively.

images. First, the digitized original waveform (a) is expanded three times by the insertion of zero samples in the computer as shown in (b). A power spectrum of (a) is calculated in (a') [sampling the analog signal makes its spectrum periodic by replicating the original spectrum at intervals $1/\Delta t$]. The (b') depicts a power spectrum of (b).

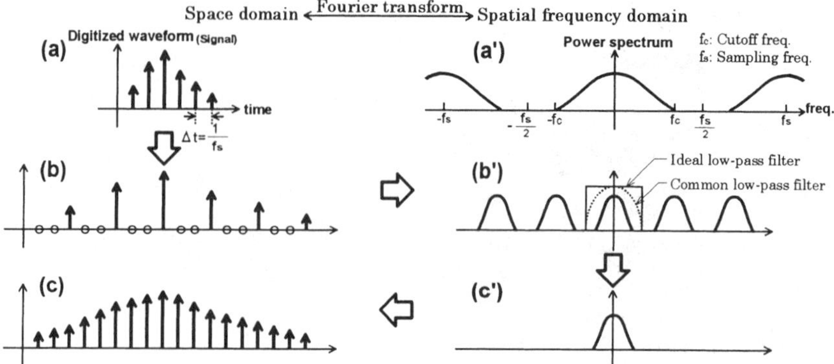

FIGURE 6. Explanation in Fourier domain of the cubic convolution (expansion) method.

Next, a net signal is extracted by the ideal lowpass filter [rectangular solid line in (b')] as indicated in (c'), and it is inverse-transformed to obtain the accurately expanded waveform shown in (c). This procedure is almost equivalent to the cubic convolution method using the sampling function as the interpolating function (The Fourier transform of the sampling function is a rectangular pulse, that is, the shape of the ideal lowpass filter). Figure 7 is an example of the use of the procedure in Fig. 6. An original SEM image of gold-coated magnetic tape and its power spectrum are shown in Figs. 7a and a', respectively. Figure 7a is a roughly optimal-scanned image with high S/N. A

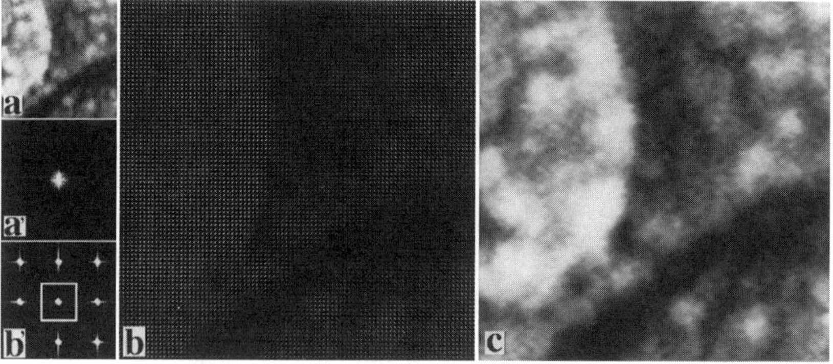

FIGURE 7. Example of an expanded SEM image obtained through the procedure shown in Fig. 6. (a) Original SEM image of gold-coated magnetic tape; (a') power spectrum of (a); (b) expanded image by the insertion of zero samples; (b') power spectrum of (b); (c) final result processed through the inverse Fourier transform from a net signal [inside of a small square in (b')]. Horizontal field width of (a) = 178 nm.

processed image corresponding to (b) in Fig. 6 is shown in Fig. 7b together with its power spectrum (Fig. 7b'). An inverse Fourier transformed image (Fig. 7c) is obtained from inside a small square (ideal lowpass filter) in Fig. 7b'. This is a nearly perfectly expanded image of Fig. 7a.

In the case where the data depicted in Fig. 6a are oversampled data and include appreciable noise (SEM images of this kind are seen frequently), we should utilize a common lowpass filter [dotted line in (b') of Fig. 6] for the ideal lowpass filter, because the noise has a spatial frequency component with infinite spread and is always undersampled. That is to say, the common lowpass filter is roughly equivalent to the well-known "bilinear interpolation" in the space domain, and has little effect in emphasizing the noise while maintaining high fidelity for expansion of the oversampled signal. By contrast, the ideal lowpass filter relatively enhances the undersampled noise.

As an example, a high-magnification (oversampling condition) image of latex balls coated with gold is shown in Fig. 8a. Images expanded by a factor of $5\times$ by the cubic convolution method and bilinear interpolation are indicated in Figs. 8b and c, respectively. Noise ("worms") is conspicuous in Fig. 8b as expected. By contrast with this result, the bilinear interpolation (Fig. 8c) can accurately enlarge Fig. 8a without noise enhancement. However, for some images with high S/N taken with roughly optimal scanning, for example, Fig. 7a, the bilinear interpolation may produce blurred results compared with cubic convolution. Nevertheless, because the bilinear interpolation has the advantage of insensitivity to noise, its ease of use is

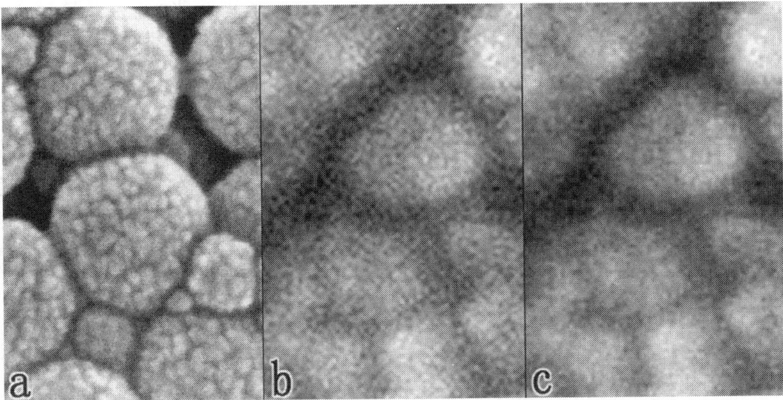

FIGURE 8. Optimal expansion method for an oversampled (scanned) SEM image with a small amount of noise. (a) Original SEM images of gold-coated latex balls; (b) five times expanded images by the cubic convolution method; (c) by the bilinear interpolation (optimal method). Horizontal field width of (a) = 400 nm.

helpful for fairly noisy images in all scanning conditions. Incidentally, nearest neighbor interpolation (simple expansion of each pixel) is not so useful for SEM images because of low precision. We can easily find these techniques in some well-known retouching software (e.g., "PhotoShop").

III. Quality Improvement of SEM Images

A. Generalization

SEM images are disturbed by noise, blur, excessively wide dynamic range, and other things. Because these are general problems in many of the fields that relate to digital image processing, we may be able to find many solutions (Rosenfeld and Kak, 1982; Gonzalez and Woods, 1992). In the early stage, many techniques for image enhancement were introduced in our field. For example, a lowpass (conventional averaging) filter was applied to an SEM image for noise removal (Yew and Pease, 1974). Several histogram-processing techniques were used for contrast improvement (Oron and Gilbert, 1976; Artz, 1983). The gradient, Laplacian or other derivative operators were utilized for image sharpening (Oron and Gilbert, 1976; Unitt and Smith, 1976). However, these operators may not be very practical for improvement of SEM image quality because SEM images are taken under various operating conditions and the operators have characteristics not found in other fields, that is, effects of charging, radiation (thermal) damage, contamination, stray magnetic field, vibration, and others. On the other hand, an SEM image has the ultrahigh scanning density (e.g., 4096 lines/frame) and this scientific instrument is frequently utilized for observing an object with unknown structures. Hence, the following basic processing requirements are needed for the enhancement of SEM image.

1. Smallest image details must be preserved. Many conventional processing methods allow enhancement of certain image details but often degrade the overall image, for example, producing a blurred or noisy image. As the SEM has many scanning lines, to use them effectively it is necessary to avoid these degradations.

2. Processing artifacts must be minimized. In general, most image enhancement methods produce artifacts to varying degrees. Because artifacts and intrinsic image details are generally not distinguishable in a processed image, useful processing methods must minimize spatial processing artifacts.

3. Processing parameters for enhancements must be eliminated as much as possible. Most conventional image enhancement methods require complex parameters that differ from image to image and depend on the varying visual perception of operators. Specifically, processing in space domain requires determination of optimal masks (size, mask shape, and number of iterations), and processing in Fourier domain requires definition of optimal filters (frequency characteristic). Unfortunately, it is very difficult to predict optimal processing parameters because SEM images, even when obtained from the same sample, vary considerably in image content (graylevel number and distribution, size of detail structures, extent of noise, etc.) if magnification, electron dose, accelerating voltage, or signal source are changed. Because routine microscopy requires a constant change of these imaging parameters, digital image enhancement is challenged in providing useful tools for image evaluation during a microscopy session. Although the rationale discussed here seems ordinary it becomes essential in SEM imaging.

In the following sections, several methods that satisfy the foregoing requirements are discussed, related to the SEM image characteristics. (However, some aspects of these methods do not fully satisfy the requirements.)

B. Noise Removal

Since development of the SEM, noise in SEM images has been one of the most difficult problems. By using a field emission gun, the S/N ratio of SEM images was dramatically improved. However, in both low magnification and high magnification conditions, we cannot yet settle this problem because image quality depends strongly on the characteristics of the specimen. In each field of SEM (Auger electron spectroscopy, Electron probe microanalysis, etc. as well as conventional observations of surface structures), users require, if possible, noise-free images for their work.

A few techniques for noise removal were introduced to the SEM field in the 1970s (Herzog *et al.*, 1974; Lewis and Sakrison, 1975; Oron and Gilbert, 1976; Yew and Pease, 1974). There are now several additional types of techniques for noise removal in the field of digital image processing (for a general review, see Wang *et al.*, 1983; Rosenfeld and Kak 1982; Gonzalez and Woods, 1992), but an ideal method for SEM noise removal does not yet exist because a nearly perfect separation of structures and noise is usually impossible. As a result, we have to submit to the side effects of processing, that is, the degradation of information. In addition, as most techniques have some processing parameters (e.g., mask size, shape, weight, and the number of iterations), the users have to determine them optimally for every original image according to their experience and knowledge. Otherwise, we may often see unfavorable results with many artifacts as well as a low degree of noise removal. This is a very difficult task for the SEM user. (Some operators may often utilize techniques for noise removal without adequate care.) Under these circumstances, conventional methods do not find wide application in conventional microscopy.

1. *Complex hysteresis smoothing (CHS)*

As a solution to forementioned problem, a very different idea for noise removal, complex hysteresis smoothing (CHS) has been proposed (Oho *et al.*, 1996c). This has essentially only one processing parameter, which can readily be determined. In addition, it does not intrinsically worsen the resolution of the original image. These characteristics are favorable for SEM images containing various-sized structures. This method is satisfied with the basic processing requirements already mentioned.

In order to explain the principle of CHS, it is necessary to illustrate the standard hysteresis smoothing (Ehrich, 1978), which is a 1D processing method, with an original waveform and its processed result (a thicker line) shown in Fig. 9. A hysteresis cursor (vertical line) is established whose width is at least equal to the size of the largest waveform peak or valley to be removed. The sole processing parameter is this cursor width (CW). The cursor is first placed over the left end of the waveform and is then pushed toward the right end. When the cursor moves to the right, it follows the waveform upward if the waveform reaches the top of the cursor (see ▲ in Fig. 9), and in the same manner it follows the waveform downward if the waveform has reached the bottom of the cursor (see ▽). The processed result is produced by recording the movement of a reference point at the center of the cursor as the cursor moves across the waveform.

Unfortunately standard hysteresis smoothing produces a severe artifact. Figure 10a (original image) is a noisy SEM image of latex balls 0.5 μm in diameter and coated with Pt–Pd. Its image smoothed by the standard hysteresis technique in question is shown in Fig. $10b_1$ (see remarkable artifact). It is not surprising that we can choose to process in any arbitrary

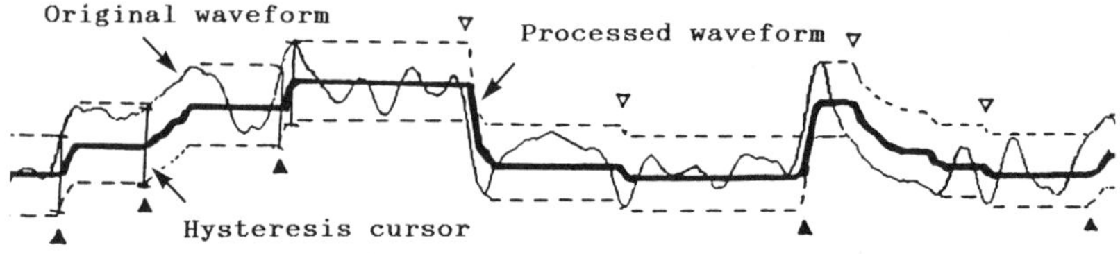

FIGURE 9. Explanation of the standard hysteresis smoothing.

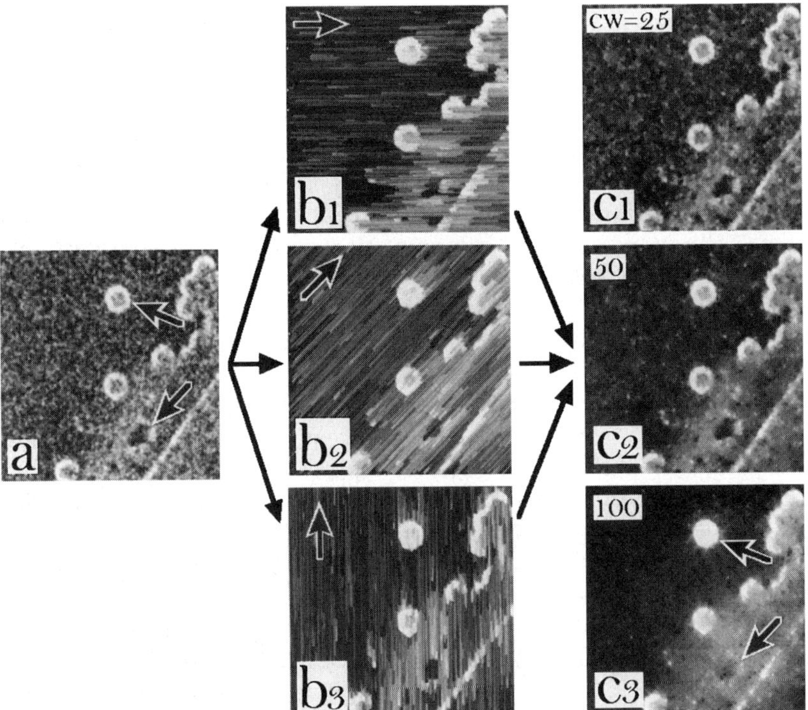

FIGURE 10. Principle of complex hysteresis smoothing (CHS). (a) Noisy SEM image of latex balls coated with Pt-Pd; (b_1)–(b_3) processed images by the standard hysteresis smoothing (arrows indicate the direction of processing); (c_2) image processed by CHS [cursor width (CW) = 50] through (b_1)–(b_3) and many processed images obtained from other directions; (c_1) and (c_3) are processed results by CHS with a different value of CW. See text for details.

direction, because this is a 1D technique. Figures $10b_2$ and $10b_3$ show processed results obtained from other directions and the arrows in Figs. $10b_1 \sim 10b_3$ indicate the processing direction. The CW used in processing Figs. $10b_1 \sim 10b_3$ was 50 (gray levels 256). In some experiments, we found that when the direction of processing was changed, that of the artifact was also changed according to the processing direction as shown in Figs. $10b_1 \sim 10b_3$. However, except for this, no obvious relation of the processing artifacts between the three images was seen. The artifacts are very similar to those of random noise in each SEM image obtained from multiple scanning (like a series of SEM images recorded with TV standards). We have applied this principle to 16 images (including Figs. $10b_1 \sim 10b_3$) with the particular artifacts obtained from 16 different directions of processing. As a result, the severe artifacts were nearly perfectly removed.

Figure 10c is a final processed result of Fig. 10a obtained by CHS (through 16 images). The noise as well as the artifacts are seen to be dramatically reduced. On the other hand, finding the optimal CW (processing parameter) and the controlling property of a processed image is usually not difficult. In the original image (Fig. 10a), it was assumed that the structures indicated by arrows in Fig. 10a were a part of the "important information". Therefore, based on the principle of hysteresis smoothing, we can easily find the optimal CW (CW = 50) by choosing values close to the information in the processed result. Here, we cannot see the difference in the processed results at all, even if the numerical value of the CW is changed to some degree. This is an advantage because it facilitates use. In the case where the magnitude of the CW is much changed, as shown in Figs. 10c_1 (CW = 25) and c_3 (CW = 100), we can easily recognize the effect of different CWs. A value of CW = 25 produced an insufficient effect of noise removal because the CW is smaller than the typical amplitude of the noise. A value of CW = 100 destroyed important signal details (increased secondary electron signal at edges caused by diffusion contrast, see arrows in Figs. 10a and c_3). Thus, it is easy to control the properties of a processed result.

A simple simulation can be performed to confirm the high ability of CHS in the preservation of structural details composed of a few pixels. The result was compared with a 3×3 weighted averaging filter (weight = 3, iteration = 1) and a 3×3 median filter (iteration = 1). We utilized these values for processing parameters because the two filters are not in a disadvantageous position (these combinations may make it possible to preserve structural details during the practical use of each filter). The former filter is a conventional smoothing one with unfavorable blurring effects; the latter one is also a common nonlinear-type filter and is highly rated from the viewpoint of its ability for edge preservation and noise smoothing in the field of digital image processing (Chin and Yeh, 1983).

Figure 11a is a simulated original image with an SEM noise and minimum size structures (written with single pixel width). To show the difference clearly, all images were 16 times expanded by nearest neighbor interpolation. Figure 11b shows the processed result of CHS (CW = 40). The noise was nearly perfectly removed while preserving structural details. On the other hand, the median filter (Fig. 11c) produced a terrible artifact, that is, two lines were perfectly removed and a new line appeared in a strange location. Also, by using the averaging filter, two lines became a single wider line from blurring effects. In addition, neither filter could remove the lower spatial frequency components of noise (a weak fluctuation of contrast in Figs. 11c and d), which are not conspicuous in the original image. When we use a larger mask size, the components may be somewhat reduced, but we must accept a

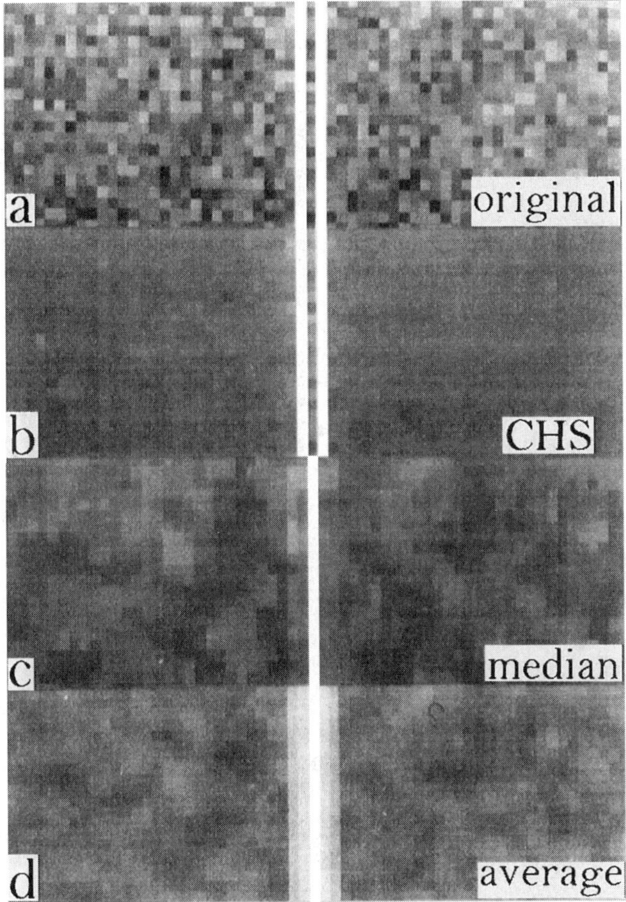

FIGURE 11. Comparison between conventional smoothing filters and complex hysteresis smoothing (CHS) in resolution of each processed result. (a) Simulated original image with a random noise and minimum size structures; (b) noise removed image of (a) by CHS (CW = 40); (c) 3×3 median filtered image; (d) 3×3 averaging filtered image.

processed image with a lower resolution as well as more severe artifacts compared with that of a 3×3 mask size.

Another advantage of CHS is shown here. In principle, an SEM image is characterized by resolution, contrast, edge sharpness, S/N, and structure sizes, which are dependent on both the operating magnification and the properties of the sample. These characteristics may differ in each SEM image, even though the same SEM instrument condition and specimen preparation techniques are utilized. In addition, the use of digital image processing allows us to obtain

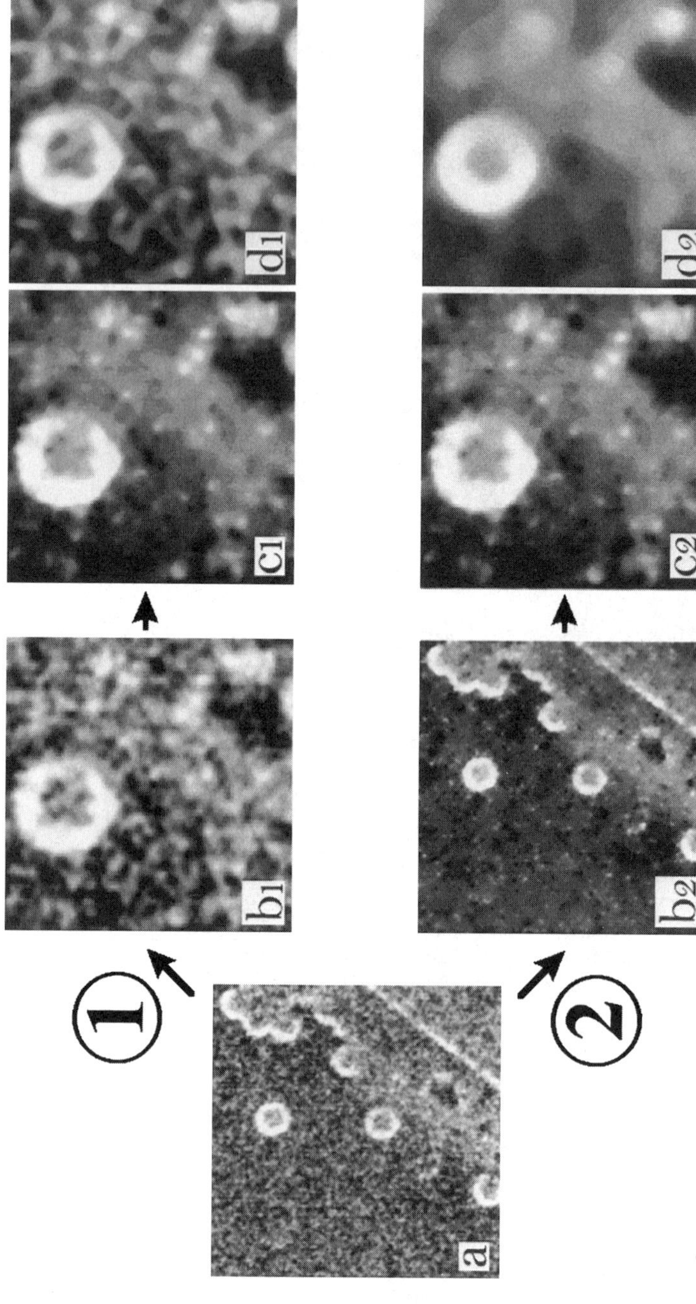

FIGURE 12. Special advantage of complex hysteresis smoothing (CHS) for scanning electron microscopy. (a) Noisy SEM image of latex balls; (b_1) expanded image of (a); (c_1) processed image of (b_1) by CHS; (d_1) median filtered image for comparison with (c_1); (b_2) processed image of (a) by CHS; (c_2) expanded image of (b_2); (d_2) another median filtered image for comparison with (c_2). See text for details.

various different magnifications after the acquisition of an SEM image. However, this produces additional changes in image characteristics. When using conventional techniques for noise removal in space domain or Fourier domain, the user has to choose optimally a few or several processing parameters for every SEM image with different characteristics. Microscopists may be unhappy about this situation. Fortunately, the CHS is mostly free from these difficulties. In fact, the processed results of CHS are rarely influenced by the change in characteristics (SEM magnification, size of various surface structures, magnifying power of digitized image) of the object image to be processed.

In order to show the practical advantages of CHS for SEM, the following experiment, including two procedures, was performed. Figure 12a is an original SEM image of latex balls including fine details and heavy noise (this is the same image that we used in Fig. 10a). Procedure ① in Fig. 12 consists of an expansion technique as the first step and CHS processing as the second step. Procedure ② is the same step performed but in reverse order. The expansion technique (similar to alteration of SEM magnification or observation of another structure with different size) is employed for changing image characteristics. When we use CHS as a noise removal technique, the same result is expected with each procedure in spite of change of characteristics of the object image to be processed. In procedure ①, the expanded noise and structures are shown in Fig. $12b_1$ (a part of Fig. 12a). Figure $12c_1$ is a successful noise-reduced image of Fig. $4b_1$ by CHS ($CW = 50$). In ②, a processed result of Fig. 12a by CHS and its expanded image are obtained in Figs. $12b_2$ and c_2, respectively.

Comparing Figs. $12c_1$ and c_2, as expected, no difference is observed. This demonstrates that everyone can easily utilize CHS for images with various characteristics. This advantage originates in the properties of CHS, which fairly satisfy the forementioned requirements (no degradation of resolution, only one easily chosen processing parameter and without processing artifacts). Conversely, the results of a 9×9 median filter (Figs. $12d_1$ and d_2) obtained with procedures ① or ② are very different. Procedure ① produced an insufficient amount of noise removal because of a mask size smaller than the noise structure size. Procedure ② destroyed important signal detail because of excessively large mask size for structural details. It was confirmed that common smoothing filters, represented by median filters, had difficulties when used for SEM images (e.g., the necessity of finding optimal parameters based on information from SEM magnification, size of structural details, amount of noise, contrast, and magnifying power of digitized image).

As an example, a very noisy SEM image of large-scale integration (LSI) chip and its noise reduced image by the CHS are shown in Figs. 13a and b, respectively. The surface structures, which have been buried in the noise up to

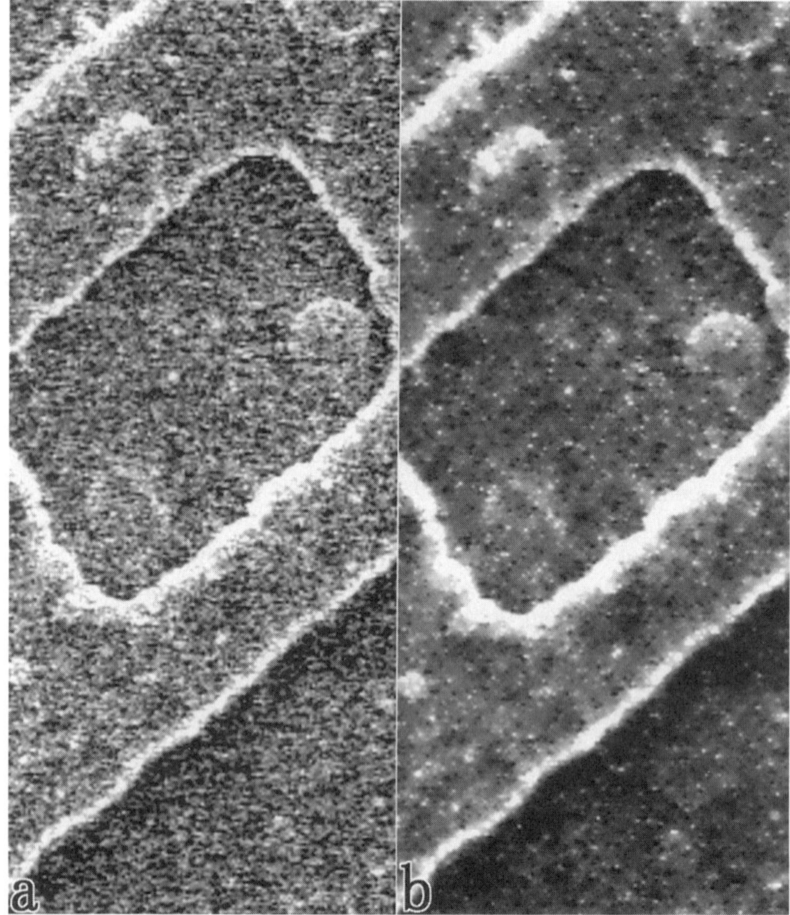

FIGURE 13. Very noisy SEM image of an LSI. (a) Unprocessed image; (b) processed image of (a) by CHS.

now, may be able to be observed clearly. Although noise with large amplitude remains in the form of many isolated points composed of just a few pixels, observation of the surface structures is not disturbed.

The CHS for noise removal is satisfied with the basic processing requirements already mentioned here. However, this cannot occasionally distinguish between signal and noise. It is necessary to develop a method with a better criterion for noise removal as well as having the basic processing requirements satisfied.

2. *Other nonlinear methods*

A few nonlinear methods for noise removal of SEM images have been proposed. Compared to a conventional lowpass filter in Fourier domain, a filtering technique performed in 2D autocorrelation function (Baba *et al.*, 1985) can successfully assort signal and noise. Smoothing by averaging along edges (Oho *et al.*, 1984, Oho *et al.*, 1987c) was introduced and improved to high accuracy for processing a noisy SEM image of an uncoated biological specimen (Fig. 14a). When an edge is present this method can take a directional average, involving only those neighbors that lie in a direction along the edge. As a result, the noise involved in an SEM image can be removed without blurring effects (Fig. 14b). Unfortunately, because these methods do not satisfy the forementioned requirements, it is not easy to use them in routine work (very small image details may be removed in Fig. 14b).

3. *Noise reduction in fast-scan SEM images*

Fast-scan, for example, TV scan images have some useful advantages in SEM. When adjusting the instrument and finding important objects, it is very convenient. In addition, this mode may be helpful to observe insulator and/or low melting point samples (Welter and McKee, 1972). Unfortunately, fast-scan images have a very low S/N due to the small number of electrons making up each pixel. In order to reduce noise in SEM images taken at TV scan rates, averaging over multiple digitized SEM images is effective (Erasmus, 1982).

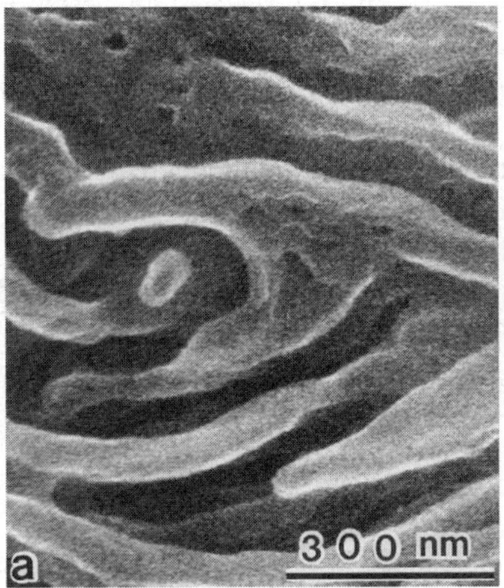

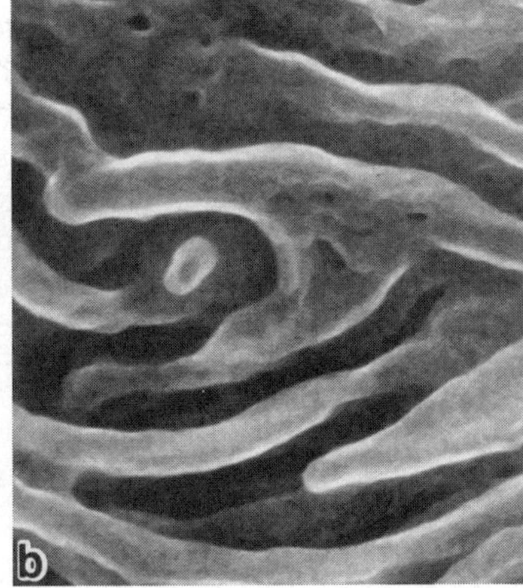

FIGURE 14. Effect of the smoothing by averaging along edges. (a) Noisy SEM image of an uncoated biological specimen (glomerular podocyte in rat kidney); (b) image following noise removal.

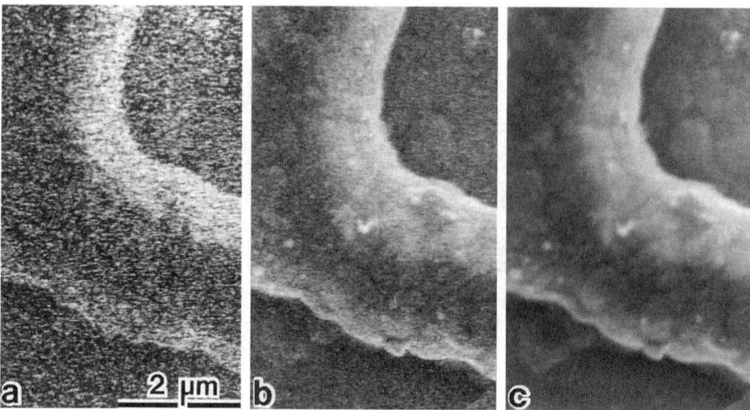

FIGURE 15. Example of a frame averaging technique in TV scan rate. (a) Result of 2 frames averaging; (b) 64 frames; (c) 1024 frames. Blurring effects are clearly seen in (c) although the S/N is improved remarkably.

The averaging is equivalent to acquiring a slow-scan image. When n images are averaged, the S/N improves $\sqrt{n}$ time.

One of the disadvantages of the averaging is that it produces an improved image with high S/N only once every n frames. The use of a recursive filter can solve this disadvantage, because it produces results continuously. The output of this filter is a weighted sum of all previous input frames with the most recent input frame having the largest weight and the weights decaying exponentially for earlier inputs. This technique has been employed by many SEM manufacturers. However, the ability of the S/N improvement of the common (first order) recursive filter is generally lower than that of an averaging filter. The recursive filter using Kalman filter theory can produce only the same noise reduction as with averaging. On the order hand, because specimen motion and/or deformation blurs the output, we may not be able to use long averaging times. In Figs. 15a, b and c, averaging 2, 64 and 1024 frames are utilized in TV scan rate, respectively. Blurring effects are clearly seen in Fig. 15c although the S/N is improved remarkably.

C. *Fine Details Enhancement*

It has generally been assumed that the limit of SEM resolution is determined by such factors as finite electron beam size and surface penetration effects. Only beam size may especially influence the secondary electron (SE) image of a sample coated with heavy metal. Therefore, we should first of all, use an electron beam size that is much smaller than the object, if possible. We should not rely only on the effects of digital image processing techniques. Even if these are utilized for deblurring some SE images, practical results will rarely be obtained due to the noise problem in the original image. However, in the case in which observation of an appropriately prepared SEM sample (i.e., prepared using a small enough electron beam) is disturbed by undesirable effects peculiar to SEM, digital techniques are necessary.

Many techniques were developed for detail enhancement (e.g., edge enhancement, sharpening). However, these techniques did not find wide acceptance in the field of SEM because they required first determinations of specific and image content-dependent processing parameters, determinations not easily accomplished. In addition, they may enhance the noise component, rather than only the structural information and have processing artifact problems. From these experiential and experimental results, it was understood that conventional methods did not satisfy the forementioned requirements in III.A. It seems clear that practical methods for digital image enhancement must provide specific new advantages for microscopy before being accepted and widely used in SEM.

1. *Highlight filter*

A method for fine detail enhancement of an SEM image is described (Oho, 1992). The method works best on detail-rich images as found in well-focused SEM images of various magnifications. In other words, this is useful for an SEM (SE) image including potentially sufficient high-frequency components (but it is obscured by some degradations). The method presents a "highlight" filter and satisfies most of the basic processing requirements already discussed; because it adjusts the overall contrast of the image under normal operating conditions, automatically and without need for processing parameters, it enhances the contrast of small details and produces only small image artifacts. Compared to a few widely used image enhancement techniques, the highlight filter for SEM images has some advantages that will be mentioned later.

Images contain two different image-related contrast types. The major and obvious contrast variations come from macrostructures and are summarized in the "brightness image." In the SEM, brightness information comes from large features and is enhanced by backscattered electron signals as well as by charging phenomena. The other contrast information comes from small-subdued signal variations generated at small surface features and at steep surface edges (microstructures). They are summarized in the "highlight image," which contains contrast that was contributed mainly by the secondary electron signal. In principle, the new filter separates from the

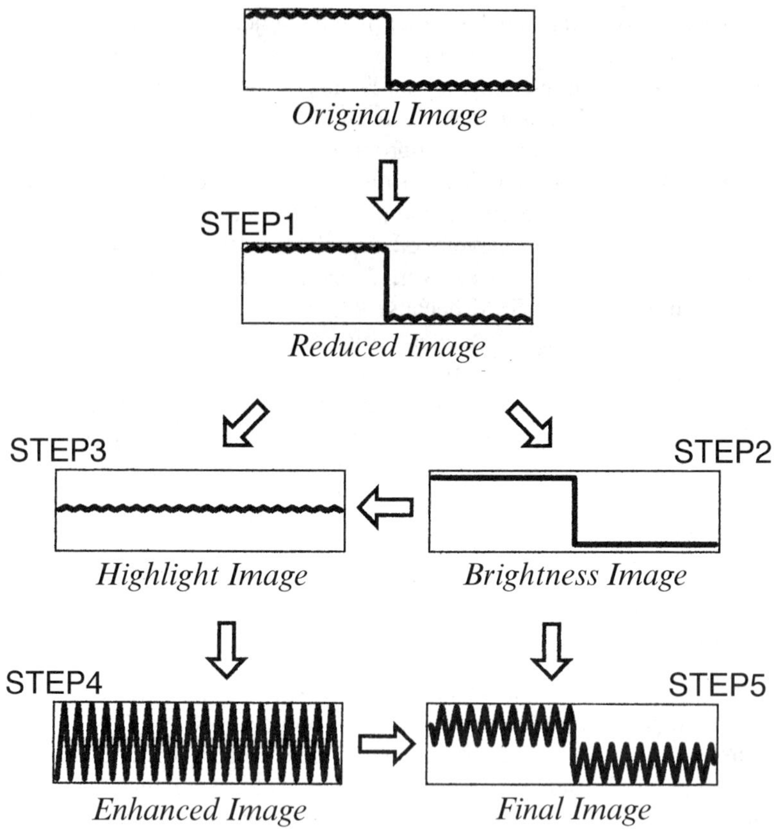

FIGURE 16. Schematic image processing diagram for the highlight filter.

image the brightness and the highlight information, enhances the contrast of the highlight image and mixes it with the unchanged brightness image at a preset ratio. The highlight filter uses five processing steps (Fig. 16).

Step 1

The image is digitally acquired through an AD converter with $4096 \times 4096 \times 8$ bit. An obtained image will not be influenced by the adverse effects of the aliasing error. Then, the original image is reduced to a

1024 × 1024 × 8 bit image for S/N improvement. The resulting image is the "reduced image."

Step 2

First, the "brightness image" is extracted from the "reduced image." This task can be established by using a median filter with an unusually large mask size (e.g., 19 × 19 pixels—depending on the employed system), because edge sharpness of the macrostructures is nearly perfectly preserved. Altering the mask size in the vicinity and applying the filter more than a few times did not significantly influence the final processing result. Therefore, a special step for determination of processing parameters (mask size and number of iterations) becomes obsolete.

Step 3

Then, subtracting the "brightness image" from the "reduced image" generates the "highlight image." The highlight image usually has very little contrast except for some cases.

Step 4

Next, the contrast of the "highlight image" is enhanced, using histogram equalization, to produce the "enhanced (highlight) image." This technique does not need parameters and the resulting enhancement is generally favorable. Enhancement limitation often occurs in contrast enhancement by histogram equalization of the reduced image including the brightness information (see Fig. 17d). Due to the possible wide range intensity distribution, the processing result would vary and be limited by the image content. The "enhanced image" provides valuable information on the highlight contrasts and the maximum detail enhancement obtainable by this filter. From the viewpoint of automatization, conventional contrast stretching is not currently applied due to the possible large variation in the maximal intensity range. However, it can also be utilized as an enhancer by modifying the problem.

Step 5

Finally, to regain the brightness information, the "enhanced (highlight) image" and the "brightness image" are mixed at a preset ratio (1:1), producing the "final image" (If necessary, the ratio can be easily changed). Thus, the whole procedure can be performed automatically without any input of processing parameters.

The highlight filter proved very valuable in routine microscopy on difficult samples. A typical problem specimen is found in noncoated semiconductor samples. Microscopy of large scale integration (LSI) chip was frequently hindered by lack of contrast at high accelerating voltages, or excessive charging and contamination deposition at low accelerating voltages. Conventional image processing could not produce a satisfactory image despite extensive search for and combination of different processing techniques. The automatic highlight filter produced immediately the pertinent image without any user interactions under routine microscopic imaging conditions and provided valuable information on the detail contrast content.

As an example, a noncoated LSI sample was investigated. Routine high-voltage (25 kV) imaging conditions for it allowed easy generation of micrographs because charging phenomena were mostly suppressed (Fig. 17a). However, only macrofeatures of the sample were revealed due to lack of detail features. According to expectation, the automatic highlight filter

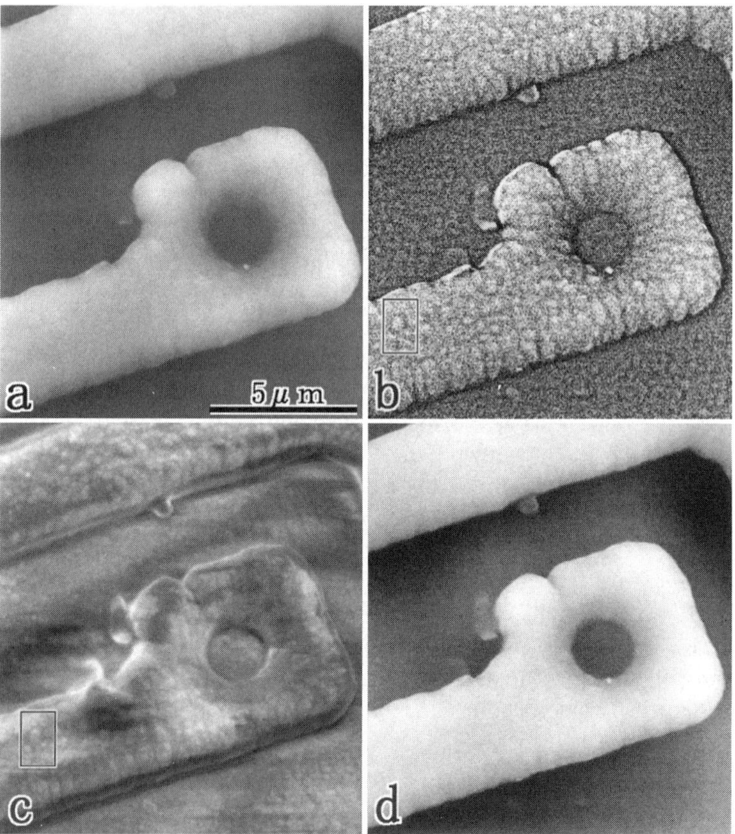

FIGURE 17. Comparison of SEM image enhancement using the highlight filter, histogram equalization, and SEM imaging techniques. (a) Partial image of a noncoated LSI sample acquired at 25 kV; (b) image enhancement through the automatic highlight filter for (a); (c) partial image acquired at 3 kV with no image processing applied; (d) image enhancement through histogram equalization for (a).

produced an image of balanced contrast that was rich in detail contrast (Fig. 17b). It should be noticed that effective contrast of structural details can dramatically improve maintaining accurate macrocontrast obtained by high-voltage condition. The enhancement revealed detail structures present in the data but visually inaccessible due to low contrast. To prove the existence of such structures on the sample, the same sample area was imaged with low-voltage microscopy (Fig. 17c). On this sample, microscopy was optimized at 3 kV but was cumbersome and severely limited by surface charging and high rates of contamination deposition. However, reasonable surface detail contrasts were obtained. The low voltage application provided a larger excitation volume for local detail contrast generation and thus increased the local signal and improved the local S/N (Fig. 17c: box). Although the macrostructures are obscured by the charging effects, low-voltage microscopy revealed a microstructure very similar to that revealed in high-voltage images after enhancement with the highlight filter (Fig. 17b: box). The image enhancement made possible a direct comparison of both low-voltage and high-voltage images and provided a new exciting tool for the analysis of detail contrast mechanisms.

On the other hand, the use of conventional contrast enhancement methods, for example histogram equalization (Fig. 17d) and contrast stretching, could not reveal the detail structures. They are only effective for images lacking a brightness component (SEM images often have a large brightness component).

The highlight filter is compared with unsharp masking, which is widely used for sharpening (deblurring) of blurred images. The unsharp masking does not fulfill the basic processing requirements. The filter is seemingly easy to use, but it needs some processing parameters and combination with other processing methods to achieve a "suitable" enhancement because it has a noise problem. The enhancement product is strongly dependent on the image content and is not predictable and thus requires a trial-and-error approach (there is no criterion for the optimal processing image). Figures 18a and b are an SEM image of gold-coated magnetic tape and its highlight filtered image, respectively. Many gold particles can be clearly seen in Fig. 18b. In addition, the processing result obtained by the highlight filter is of only one kind. On the contrary, unsharp masking (Fig. 18c) cannot demonstrate its maximum performance due to a failure in parameter setting (mask is too small and the enhancement ratio is excessive). Unfortunately, when improving these values, another problem occurs. Several limitations of conventional image enhancement have been demonstrated clearly (Oho and Peters, 1994), and several applications of the highlight filter were shown elsewhere (Oho, 1992).

As another example, an SEM image of an uncoated biological specimen influenced by weak charging phenomena (Fig. 19a) is processed by the high-

FIGURE 18. Comparison of SEM image enhancement using the highlight filter and unsharp masking. (a) SEM image of a gold-coated magnetic tape; (b) image processed by the highlight filter; gold particles can be observed clearly; (c) image processed by conventional unsharp masking.

light fiter. Enhancement of fine structures as well as reduction of the effects of charging phenomena are successfully indicated in Fig. 19b.

On the other hand, in order to settle a problem sensitive to noise in unsharp masking, Oho *et al.* (1990a) have proposed a nonlinear pseudo-Laplacian filter for enhancement of high-resolution SE images. However, it is not easy

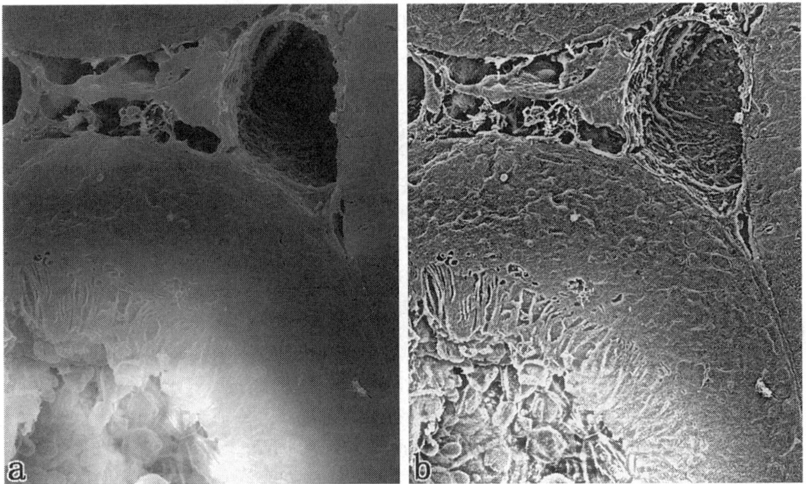

FIGURE 19. Suppression of the effects of weak charging phenomena by the highlight filter. (a) Unprocessed SEM image; (b) image processed by the highlight filter. Enhancement of fine structures as well as reduction of the effects of charging phenomena are successfully performed.

to use the method in routine work due to several processing parameters, although its performance improvement was attained.

2. *Enhancement of BSE images*

In principle, the resolution of backscattered electron (BSE) images can be little improved except for particular samples, even though an infinitely small beam size is achieved by various improvements in the intrinsic instrument. In other words, surface penetration effects of the incident beam greatly influence the resolution of a BSE image. If the resolution of BSE images could be improved beyond previously accepted classic limits, they would be a more attractive tool for many SEM users, as BSE images have superior advantages over SE images. We consider that the best way to improve the resolution of the BSE image is to utilize digital image processing techniques based on the characteristics of BSE images. However, the forementioned highlight filter will not be suitable for enhancement of a BSE image, because it may not be able to successfully separate from a BSE image disturbed by various blurs the brightness and the highlight information.

High-emphasis filters, which can improve the image resolution in principle, have not been used often in practical applications, due to the existence of noise in the SEM image. However, if an image without noise did exist, very useful processing results could be obtained by using refined image enhancement techniques. Fortunately, BSE images may not be degraded by contamination and charging phenomena (unlike in the situation for an SE image), and degradation caused by the radiation damage is not conspicuous at the level of resolution of normal BSE images for most specimens. Hence, a combination of ultrahigh performance BSE detector, lengthy recording time and appropriate image processing techniques may be able to produce a BSE image with extremely high S/N. As a result, a high-emphasis filter may be able to improve significantly the resolution of BSE images with extremely high S/N. The procedure to obtain a high-resolution BSE image follows the steps listed here (Oho *et al.*, 1991).

Step 1
The BSE image (e.g., 4096×4096 pixels) is stored through the AD converter in the image memory. Here, the BSE image must be recorded at a rather high magnification for attaining the oversampling condition and as long a time as possible for obtaining a very high S/N image without SEM shot noise and aliasing noise. Consequently, a noise reduced and blurred BSE image may be acquired.

Step 2
The BSE image consisting of many pixels is reduced into a new image consisting of, for example, 512×512 pixels. As each new pixel was obtained as a properly averaged value of many pixels, S/N increased. The information contained in the BSE image is little degraded by this reduction because the original blurred image is taken as the oversampling condition. In other words, the blurred image can be represented accurately by a comparatively coarse sampling.

Step 3
The reduced image now has an extremely high S/N, which allows enhancement of its high-frequency component. The image is subsequently processed by a high-emphasis filter in the Fourier or space domain, for example, unsharp masking. It should be noticed that the unsharp masking without appreciation in enhancement of SE images is useful for BSE and video microscope (see Section VI.) images which contain remarkable blur. Filters of this kind have been successfully applied to a blurred telescopic image (O'Handley and Green, 1972).

As an example, a BSE image obtained on paper from a word processor coated with 20 nm of A1 using steps 1–3, is shown in Fig. 20a. The image was recorded under the following conditions: accelerating voltage of 30 kV, incident current of 5.5×10^{-10} A and recording time of 320 s (use of a semiconductor BSE detector). The filtered image is shown in Fig. 20b. The images in Figs. 20a and 20b were enlarged to show the difference more

FIGURE 20. High-resolution enhancemnet for the BSE image. (a) BSE image with an extremely high S/N of a paper (use of a semiconductor BSE detector, incident current of 5.5×10^{-10} A and recording time of 320 s); (b) image processed with a high-emphasis filter in Fourier domain together with expanded images (a'), (b'); (c) SE image of the view identical with (a).

clearly (Figs. 20a' and 20b', respectively, as identical views). The processed image illustrates an impressive improvement in resolution due to ultrahigh S/N of original image. It is easy to produce a successful design of a high-emphasis filter, because the forms of the filter for the various BSE images under consideration closely resemble one another. An SE image of the same region is shown in Fig. 20c. Compared with Fig. 20c, the processed image (Fig. 20b) has a similar resolution but very different contrast information (Of course the sample coated with heavy metal may produce higher resolution in an SE image). Hence, both SE and BSE images can be effective tools for high-resolution studies. In the case of a BSE image with low S/N (Fig. 21a), we cannot obtain a successful result, shown very clearly in Fig. 21b.

In order to reduce an electron dose, a yttrium aluminum garnet (YAG) single crystal is used as a very highly efficient scintillator. A BSE image of

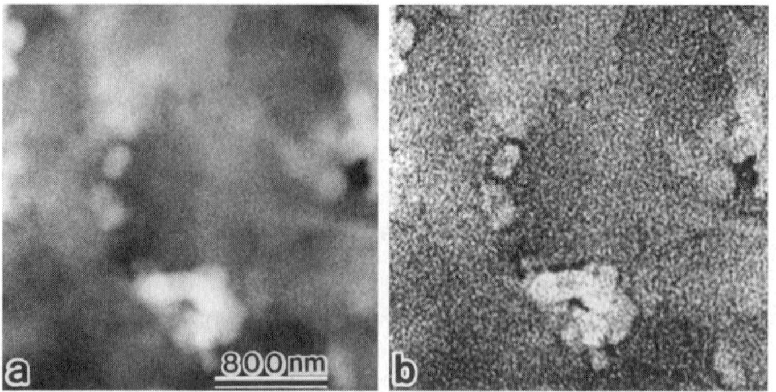

FIGURE 21. Example of an unsuccessfully processed result. (a) BSE image with a low S/N; (b) image processed with a high-emphasis filter excessively disturbed by the effects of noise.

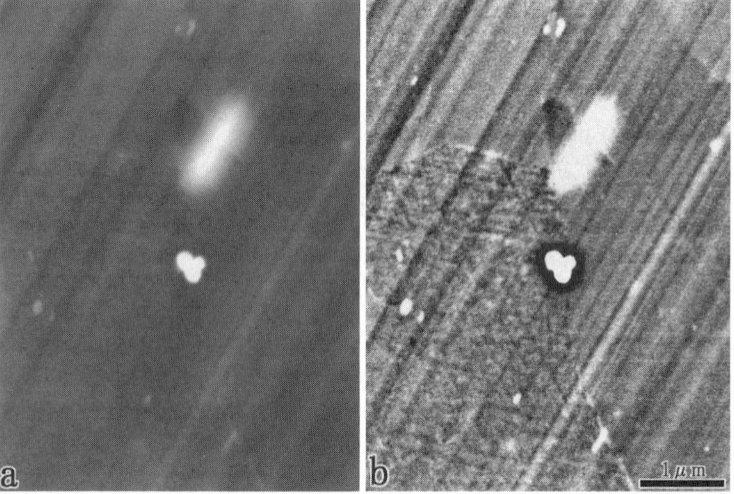

FIGURE 22. Improvement of recording conditions (use of a YAG scintillator detector, incident current of 1×10^{-10} A and recording time of 160 s). (a) BSE image of aluminum foil; (b) enhanced image of (a).

aluminum foil (Fig. 22a) was acquired with the YAG detector (accelerating voltage of 30 kV, incident current of 1×10^{-10} A and recording time of 160 s). The electron dose was reduced to 1/9. Its enhanced image is shown in Fig. 22b. Fine structures can be seen dramatically, because Fig. 22a has a sufficient S/N. Incidentally, an SE image of the same sample did not produce important additional information.

On the other hand, the Wiener filter has been used for restoration of blurred SE images (Lewis and Sakrison, 1975). However, as the properties of the signal and noise in an image must be known for the development of an optimal image filter, it is not easy to utilize the Wiener filter. Moreover, it may not be possible to obtain more information than in Figs. 20b and 22b, because the current BSE images originally had a sufficient S/N.

3. *Reduction of unfavorable effects*

Noise, contrast and blur problems in SEM images have been fairly improved as already mentioned here. However, these have various characteristics not found in other fields, for example, effects of vibration, contamination, and field-emission noise. After these adverse effects have been received, how much they can be improved will be described in this section.

As the recording magnification is increased, the visual effect of vibration on an observed image may increase. Hence, in the case where vibration is a serious problem, the image should be recorded at the lowest magnification that satisfies the sampling theorem and, if necessary, the image subsequently should be expanded by the interpolation technique. However, in this process, because the expanded image differs from an image recorded originally at higher magnification, in terms of the number of incident electrons per unit area in the specimen, the S/N of the expanded image may be degraded. Therefore, it may be necessary to average a few SEM images with the

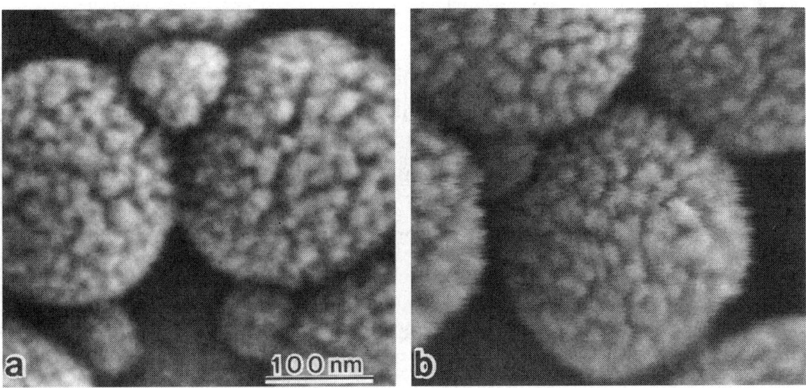

FIGURE 23. Reduction of the effect of vibration in SEM image. (a) Vibration reduced image obtained by using the optimal scanning and digital expansion technique; (b) high-magnification image with the effect of vibration taken under the same conditions. See text for details.

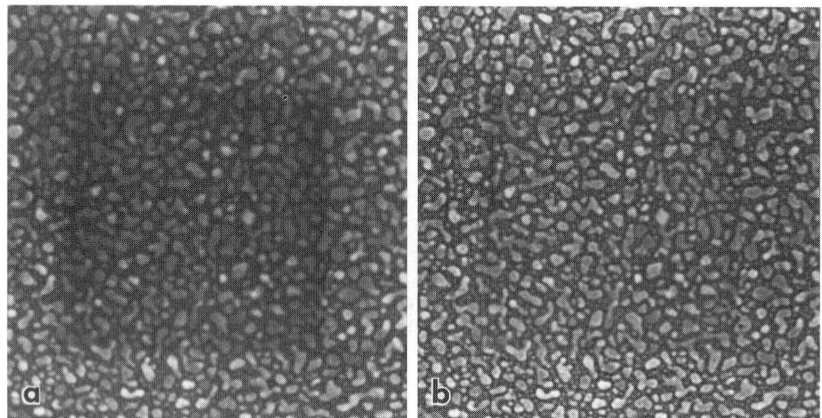

FIGURE 24. Removal of the effects of contamination. (a) Contaminated SEM image of gold particles deposited on a carbon; (b) processed image by homomorphic filtering.

identical view for increasing the S/N (decreasing random noise). This averaging as well as recording at low magnification can also reduce the effects of vibration, because no obvious relation of the effects of vibration between a few images may be seen like the SEM noise (Oho *et al.*, 1986b). As an example, using the forementioned techniques, a processed ($4\times$ expanded) image without the adverse effects of vibration is shown in Fig. 23a (gold-coated latex balls). For comparing the quality of this image, an originally high-magnification image (Fig. 23b) is taken under the same conditions except for recording magnification. It can be seen that the image is severely disturbed by the effects of vibration.

Contamination is another serious problem in SEM. A method using homomorphic filtering, which is a combination of logarithmic transformation of the gray scale and high-emphasis filter in Fourier domain, was applied for reducing the effects of contamination, in the case where an SEM image is once more observed at a lower magnification after a high-magnification observation (Oho *et al.*, 1985; Oho *et al.*, 1987a). Figure 24a is an SEM image of gold particles. We can see the contaminated region easily. The homomorphic filtered image is shown in Fig. 24b. The effects of contamination are successfully reduced even as high-resolution enhancement is achieved.

Noise in the field emission (FE) source is sometimes remarkable in SEM images taken by the FE-SEM (Fig. 25a). This noise may have a frequency characteristic as shown by the arrows in a power spectrum (Fig. 25b) [the scanning is a horizontal direction]. When successfully removing this component in Fourier domain, we can obtain a noise removed image (Fig. 25c) without the degradation of fine structures.

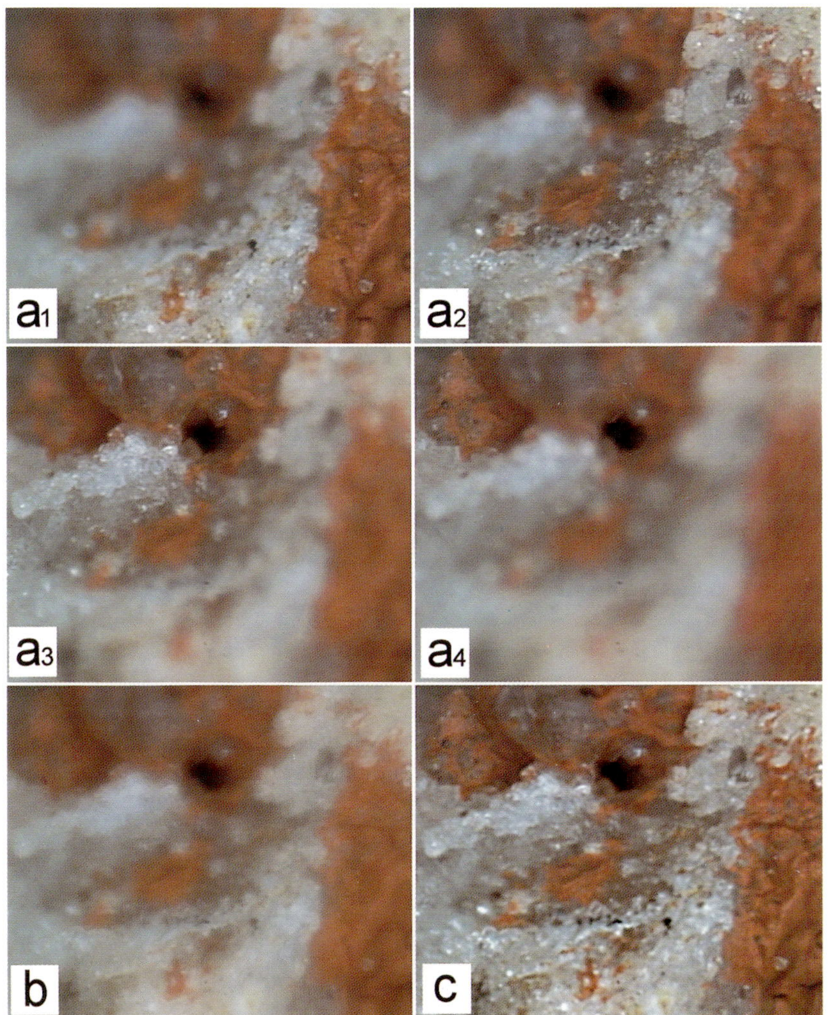

COLOR PLATE 1. VM image with improved depth of focus. (a1) ~ (a4) A series of VM images obtained through moving the specimen; (b) superimposed VM image; (c) enhanced image of (b). Horizontal field width = 800 μm.

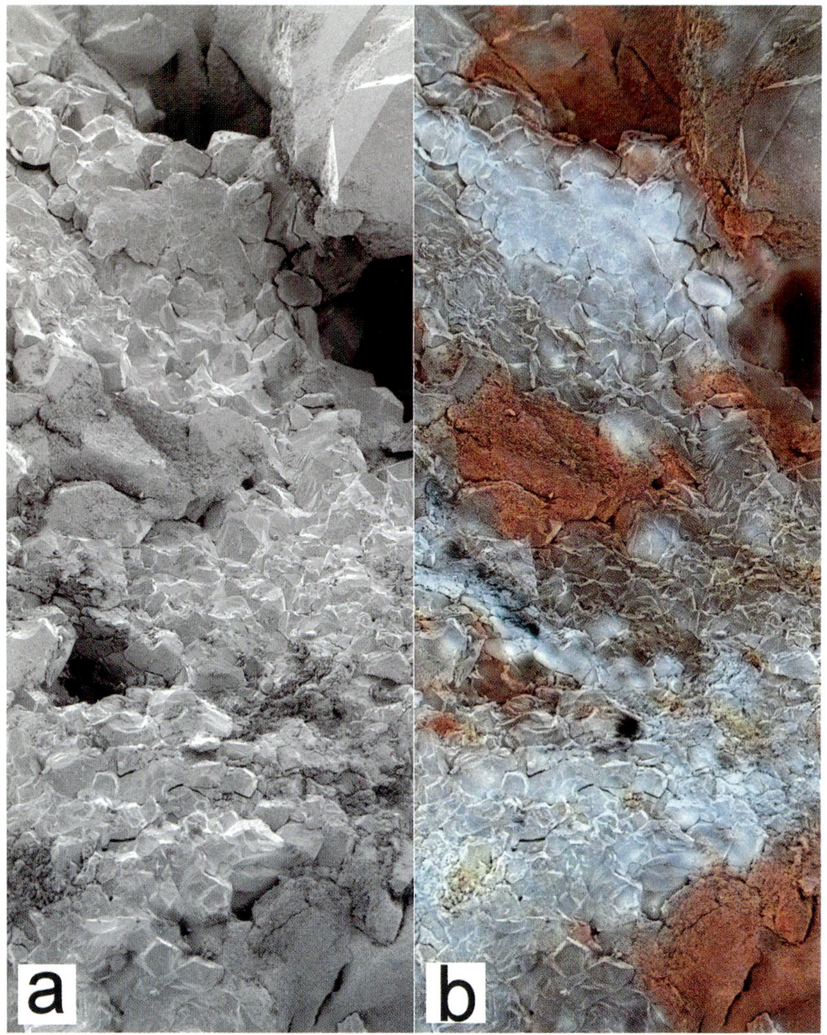

COLOR PLATE 2. Comparison between a common SEM image and a natural color (NC)SEM image. (a) SEM image of gold ore recorded under the low voltage of 2 kV; (b) NCSEM image obtained through the method shown in Figs. 31 and 32. Horizontal field width = 245 μm.

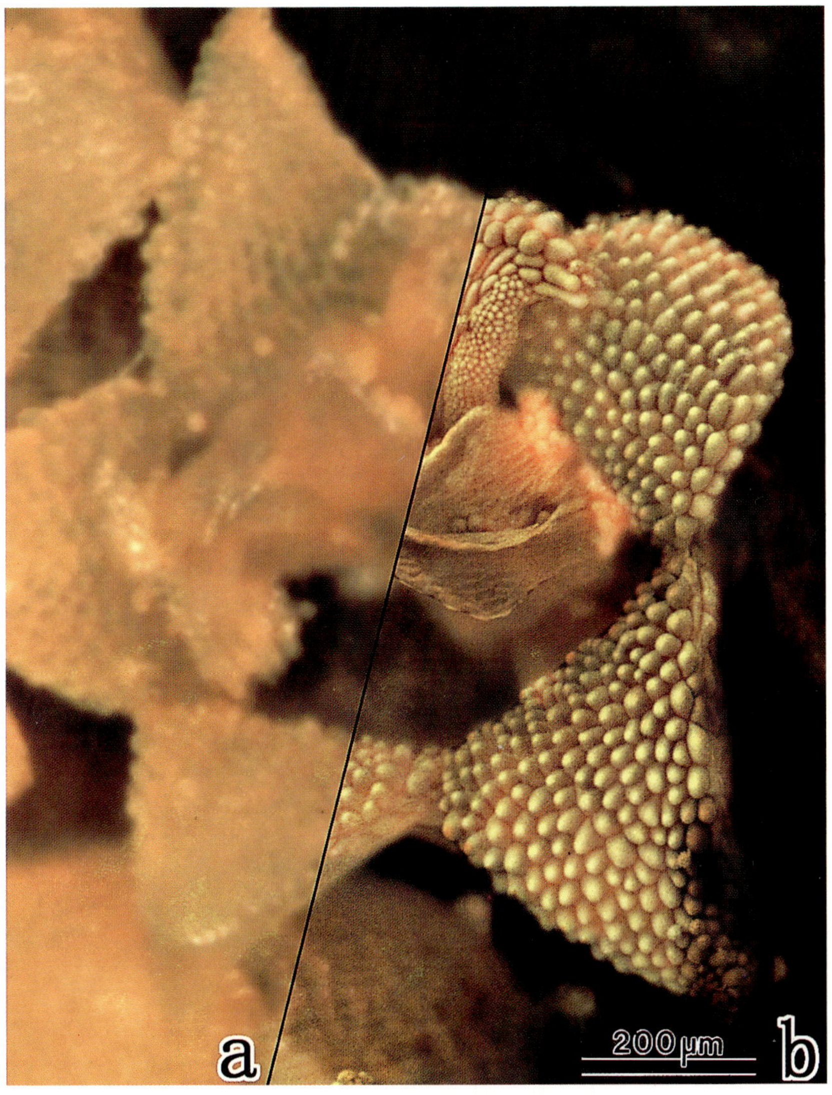

COLOR PLATE 3. Comparison between a VM image and a natural color (NC)SEM image. (a) VM image of a tubular flower of composite; (b) NCSEM image.

Color Plate 4. High-magnification NCSEM observation. (a) SEM (BSE) image of color toner powder. (b) NCSEM image. Horizontal field width = 115 μm.

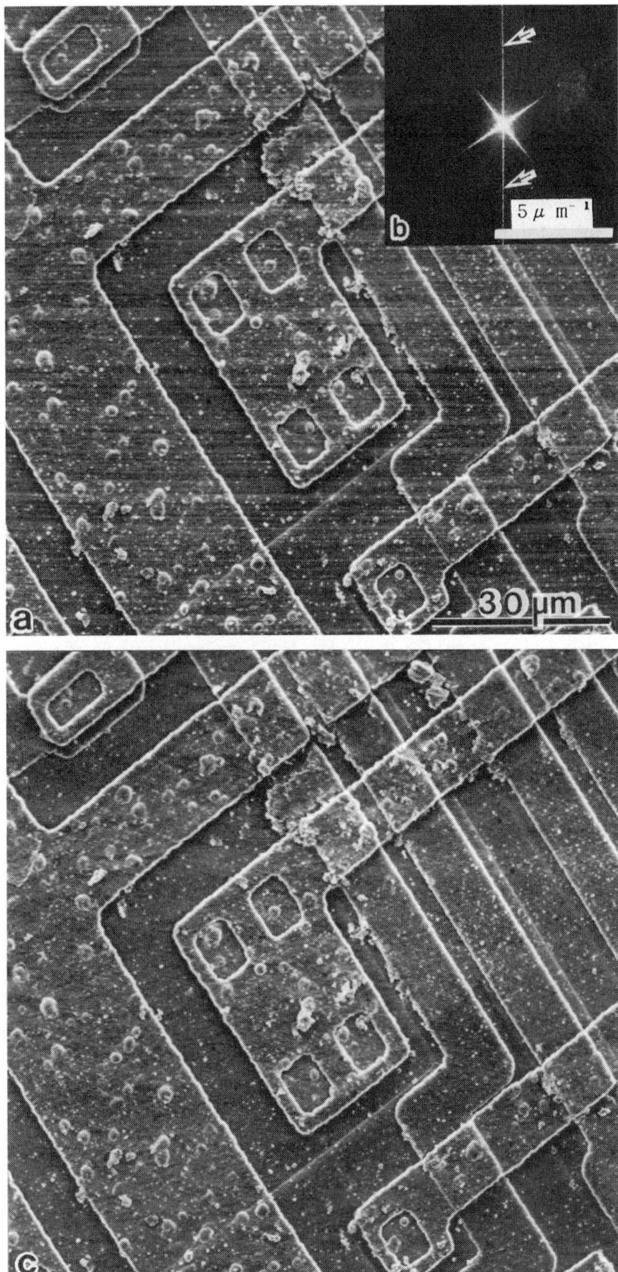

FIGURE 25. Removal of the effects of noise in the field emission (FE) source. (a) Noisy FE-SEM image together with its power spectrum (b); (c) noise removed image.

Unfortunately, those methods are not the best strategies. In other words, these are methods to patch over each problem temporarily when it arises. A fundamental improvement is necessary for each case. However, the mastery of these techniques and knowledge, that is, sampling theorem, averaging, filtering in Fourier domain, may be useful for many microscopists.

IV. Image Measurement and Analysis

A. Precautions for Effective Use of Conventional Statistical Measurement

After acquiring SEM images using an on-line computer system, they are frequently analyzed with a statistical measurement method. There are many commercial systems for image statistical analysis of SEM images based on a standard PC. However, this kind of measurement might not be very useful, because it was introduced from the field of pure digital image processing without considering the characteristics of SEM images. To utilize this method effectively, it is necessary to perform some solutions.

When SEM images to be analyzed by statistical measurement are disturbed by noise, blur, excessively wide dynamic range, and so on, some preprocessing techniques (symptomatic therapy) have to be employed to reduce these unfavourable effects. For example, it may be helpful for determination of threshold in binary processing to remove "brightness image" (described in Section III) from a SEM image with wide dynamic range. This technique is similar to shading correction.

On the other hand, it is more important to choose suitable SEM operating conditions for statistical measurement. Naturally, it is necessary to decrease the noise and blur in a SEM image as much as possible. For example, it is effective for noise reduction to utilize a comparatively long recording time. And it may be helpful for many samples with large unevenness to employ a condition that produces great depth of focus. Otherwise, a fine structure size in out-of-focus areas will be different from a true size.

B. Critical Dimension Measurement and Foreign Material Observation on the Wafer for Semiconductor Process

The semiconductor industry uses quarter-micron design rules in ultralarge-scale integration (ULSI). Many specialized high-performance SEMs are employed for semiconductor process evaluation. These are designed considering fully the characteristics of the sample [wafer] (Otaka *et al.*, 1995). In these instruments, statistical measurement techniques are utilized for precisely measuring the size of line and hole patterns.

Foreign material (particle), and/or wafer defects, which are found through other systems based on a certain light microscope with a high sample throughput but low resolving power, are inspected and analyzed by the SEM used for detailed observation (statistical measurement) of them as required. In this procedure, the measurement coordinates in light microscopy are sent, using digital processing techniques, to the SEM. The automatic foreign material and defect classification system using a variety of image processing techniques will become a very important field (Chou *et al.*, 1997).

The SEMs will be built in the total system to facilitate better semiconductor process evaluation. The feature size undoubtedly will be reduced for higher device integration and density in the near future. With the improvment of the SEM instrument that will follow, it may gradually become more difficult to deal with resolving power; as a result, digital image processing technology will be more and more important in this field. However, as there are few references concerning this field, it is necessary to confirm the current state of the system by observing and experiencing a commercial state-of-the-art SEM.

C. Surface Topography Measurement

In the field of SEM, we can find several techniques for obtaining height information. Surface topography measurements using digital image processing fall roughly into one of three categories, that is, multiple detectors, focusing, and stereometric methods. As these methods have some advantages, respectively, they are used in each method.

By using multiple detectors, detected intensities (either BSE or SE signal) may be related to the surface slope of the area being scanned by the incident beam. The profile is obtained by numerical integration of the slopes in the direction of the scan line (Lebiedzik, 1979). In the reconstruction process of surface topographies, the noise in the digitized detector signals accumulates in an unpredictable way during the course of the integration, thereby leading to artifacts that heavily distort the resulting surface. In order to settle this problem, Carlsen (1985) used least-square-techniques, which are types of smoothing techniques of an image. This type of system is suitable for specimens with protrusions, or depressions with less steep slopes [such as compact disks] (Suganuma, 1985).

The stereometric technique is used to acquire stereo-pair images and determine the height by measuring deviations of corresponding points in the two images (Boyde, 1975). Hence, it is necessary to find the points in two digitized images. For example, Koenig *et al.* (1987) employed a combination of the normalized cross-correlation and the least-squares solution to very accurately determine homologue pairs of points. Unfortunately, enormous processing time and detection failure are involved in this process.

The focusing method uses the fact that the focus of the objective lens is a monotonic function of lens excitation. This characteristic can be applied as an absolute measure of the height at any point by focusing the electron beam there. Holburn and Smith (1982) used a digital method for focusing based on the 2D gradient of acquired data. However, as the SEM has great depth of focus, the measurement accuracy of this method may be inadequate.

As already mentioned, there are some disadvantages peculiar to each method. In order to improve the accuracy of the surface topography, a combination of multiple detectors and stereoscopy was performed with some digital image processing techniques (Beil and Carlsen, 1990). As a result, stereo mismatch and wrong detector calibration were reduced. As well, the focusing method was successfully combined with the stereometric method to improve precision (Thong and Breton, 1992). This is a method using measurement of the parallax between a stereo pair for more accurate focusing on a specimen surface. Cross correlation was used to determine the parallax in a small area including the point of interest.

With the development of a recent computer graphics technology, the data obtained through each of the forementioned methods can be easily expressed in three dimensions, for example, a computed tomography (CT) scanner.

V. SEM Parameters Measurement

The performance of SEM and a scanning transmission electron microscope (STEM) is roughly determined by the incident electron beam size involving a sufficient electron current (to obtain an image with sufficient S/N). The resolution of a STEM image is limited by the scanning beam diameter, and that of a SEM image is influenced by both the beam diameter and the interactions of electrons with a specimen. Although theoretically these are easily calculated, the result of calculation is frequently not suitable for an actual situation as a result of many indefinite factors. Hence, both STEM and SEM images should be measured from a recorded image. When the measured values can be effectively used, it is useful for various work and for study of SEM.

A. *Electron Beam Diameter*

The electron beam diameter is conventionally measured from the risetime in a transmitted electron signal, or secondary electron signal, when the beam is scanned across a suitable target with a sharp edge (e.g., Joy, 1974). However, as the beam diameter decreases, it is difficult to measure it, due to insufficient edge sharpness, a low S/N, the build-up of contamination layers, and several other problems. Unfortunately, at the resolution limit, this method would tend to produce inaccurately measured values.

In order to settle a part of these problems (insufficient edge sharpness, the build-up of contamination layers), a crystalline specimen is examined with a SEM or a STEM to obtain a sharp edge and 2D information, instead of measuring the risetime (Reimer *et al.*, 1979). If this is done, we must make certain that no "clipping" of the width of an edge takes place by the signal saturation in the AD converter.

As a next step, the data on the blurring of an edge obtained from the SEM image is transformed into reliable data by suitable digital processing techniques for reducing the effect of a low S/N (Oho *et al.*, 1985). A series of procedures described here gives an accurate value of the beam diameter. To confirm the accuracy of this method, a simple simulation can be performed. This simulation is based on the assumption that scanning images are approximately formed by convoluting an image function s(x,y) of the specimen scanned by an infinitely small beam with a practical electron beam distribution defined as a Gaussian function.

Figure 26a is an image of the simulated specimen with a sharp edge scanned by an infinitely small beam together with its 3D representation (Fig. 26a'). Figure 26b is a simulated image scanned by the Gaussian spot of $2r_s$ diameter. Figure 26c is a simulated image, which represents Fig. 26b plus SEM noise. This image is considered to be roughly equal to an image taken under a high-resolution condition in SEM or STEM. We must then attempt to measure the scanning beam diameter ($=2r_s$) from information contained in Fig. 26c. For the measurement of beam diameter, the noise involved in Fig. 26c must be appropriately removed as preprocessing. In this processing, note that a change of ramp steepness (edge sharpness) after reducing the noise is closely related to a measured value of the beam diameter. From the viewpoint of the extent of unchanged ramp steepness after processing, median filtering is used in the present method. It is a nonlinear-type smoothing technique.

A mask size, which is roughly equal to the statistical size of the noise, was utilized in the present study (Oho *et al.*, 1984). Because median filtering does not blur edges, it can be repeated. Further, as a result of several median filterings, there would be "stationary states"; the processing no longer improves or degrades the image. On the other hand, other smoothing techniques generally change the state of blurring of the edge whenever they are repeated. Hence, judging from the ease of use, median filtering may be the best technique for the present study.

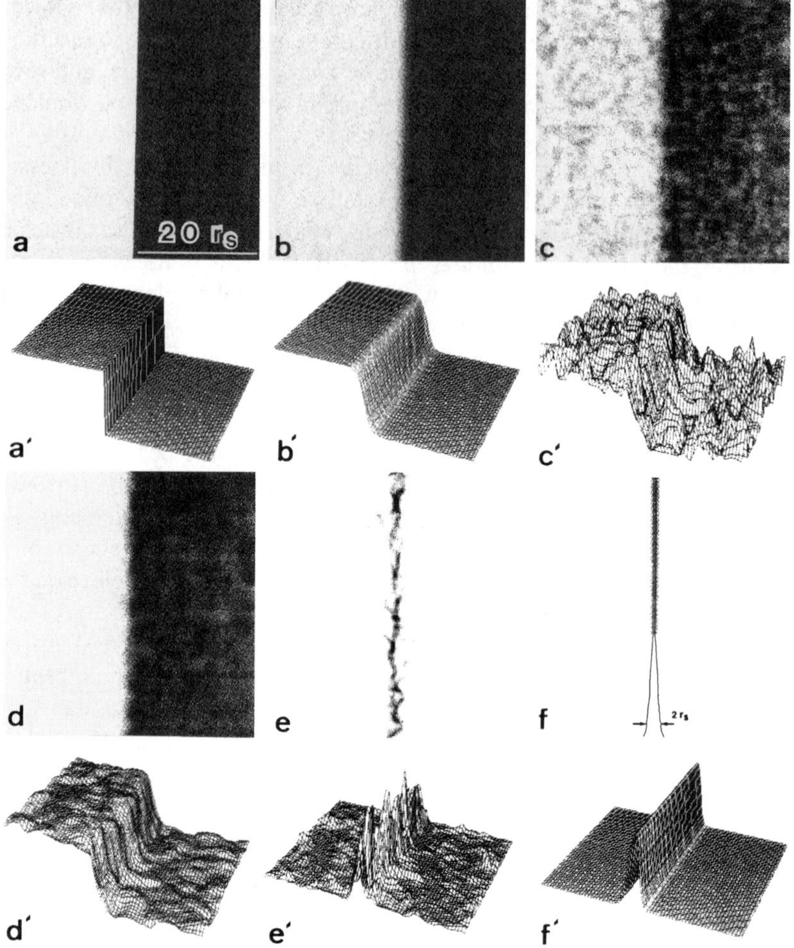

FIGURE 26. Simulation for confirming the validity of measured electron probe diameter. See text for details.

The processed image shown in Fig. 26d is obtained by removing the noise contained in Fig. 26c using median filtering. This result of noise removal is sufficient to observe the edge. Nevertheless, as may be obvious from the 3D image shown in Fig. 26d', the ramp steepness at each position is different from the true one shown in Fig. 26b. Hence, inaccurate measured values will be obtained from each line profile in Fig. 26e (it shows the square of the differential image for Fig. 26d). However, as these measured values are distributed around the true value ($=2r_s$), a synthetic image (Fig. 26f), which

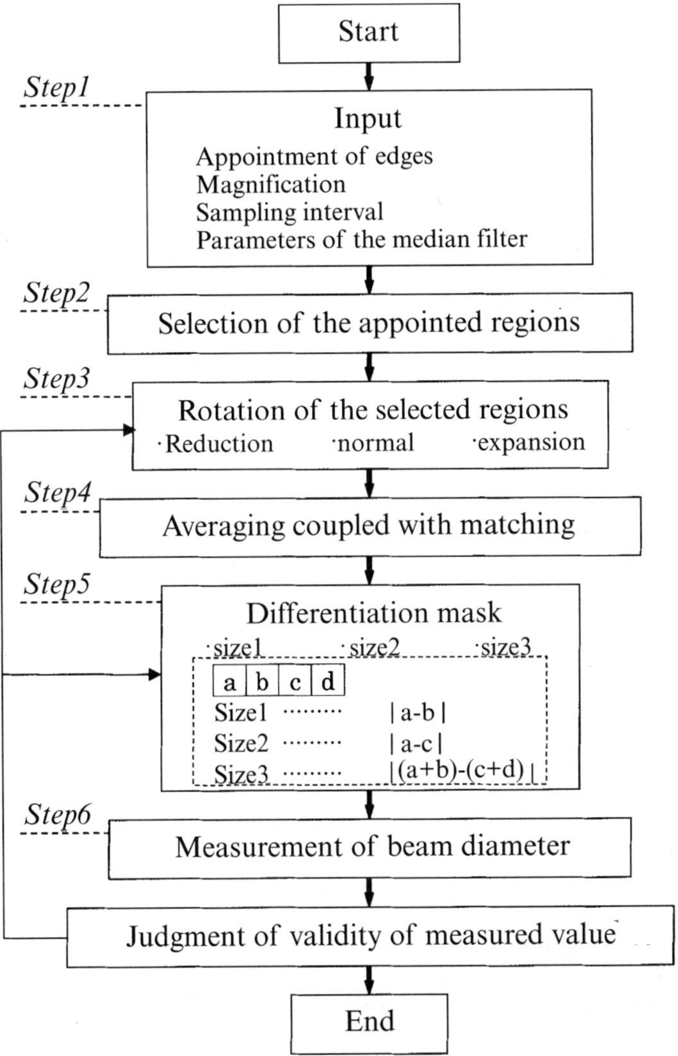

FIGURE 27. Procedure for the automatic measurement of scanning beam diameter.

is averaged in all lines with matching peak positions of all line profiles in Fig. 26e, can indicate the real beam diameter ($=2r_s$). As a result of simulation, we are able to measure the diameter of a fine electron probe beam under a high-resolution condition using present method.

An automatic measurement has been developed based on the forementioned study (Oho *et al.*, 1986a). The procedure for automatic measurement is shown in Fig. 27. The main steps can be summarized as follows. The STEM image of a crystalline hole in a gold thin film (use of field emission gun, focal length 7.5 mm (working distance 0 mm), accelerating voltage 30 kV, objective aperture semiangle 5×10^{-3} rad, detector aperture 7×10^{-3} rad (Oho *et al.*, 1987c) has already been stored in the image processing system.

Step 1
Some necessary parameters for this measurement, for example, SEM instrument magnification, are input. Next, the users must appoint several edges in the image; these should be measured. An ideal sharp edge may be included in these edges. The STEM image of crystalline holes and its photographically enlarged image of a part of an edge are shown in Figs. 28a and a', respectively.

Step 2
Several square regions including each appointed edge are selected from the stored image as processing regions.

Step 3
A rotating program, which does not change the edge sharpness, is performed to align the edges in the same direction. Under a certain condition of relatively low magnification and high resolution, the aliasing error would increase. The computer can detect this by searching the number of pixels that compose the edge. If this condition exists, the number of pixels is increased by an interpolation technique that decreases the adverse effects of error.

Step 4
As a preprocessing, the noise involved in the STEM image is removed by median filtering. The noise removal image consists of many line profiles. The statistical edge profile is obtained by averaging all lines coupled with matching.

Step 5
Its differentiated value is calculated, representing the beam size. Here, a proper mask size for differentiation is chosen automatically. This statistical processing result includes the information on the scanning beam diameter.

Step 6
The scanning beam diameter is measured from the statistical data. Compared to the real profile of a scanning beam, the profile of the obtained

FIGURE 28. STEM image of crystalline holes in a gold thin film (a); its enlarged image (a') for detecting blur of edge.

data may be deformed by the interaction of incident electrons with the specimen. For the ideal sharp edge, and ignoring the noise, interactions, and other items, the distance $d = 2r_{(1/e)}$ between 37% intensity points may correspond to the scanning beam diameter (Fig. 29). Under high-resolution conditions, only a particular part of the profile may be equal to the scanning beam profile having the Gaussian distribution. According to our study, the part between 50 and 80% of the maximum value (shown in Fig. 29 as oblique lines) is equivalent to that part. From an equation in Fig. 29, the relationship between beam diameter d and the distance X_c between certain intensity points C (50 ~ 80%) is obtained. Consequently, the measured values can be calculated from X_c at each intensity point, and the average of the measured values, except obviously erroneous values involved in those measured values, is adopted as the candidate of the scanning beam diameter. Ultimately, the minimum value of candidates obtained from several processing regions, which are appointed in Step 1, is adopted as the

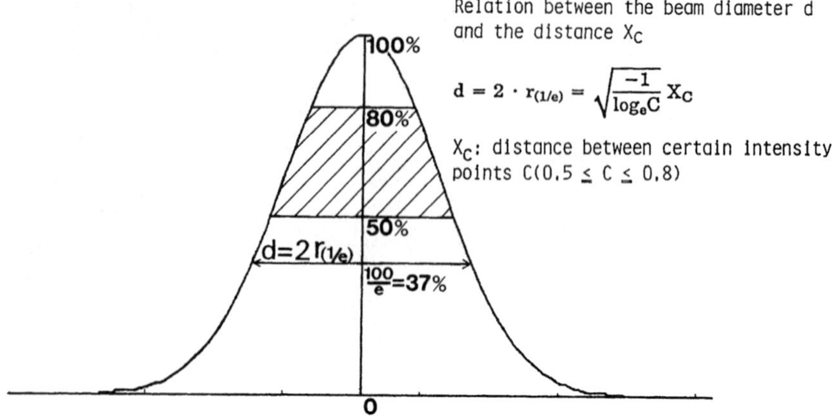

FIGURE 29. The profile of statistical data obtained by Steps 1 ~ 6. The part between 50 ~ 80% of the maximum value shown as oblique lines contains the information on the scanning beam diameter.

final measured value of scanning beam diameter. The current method accurately measures the beam diameter of 1.5 nm from the STEM image shown in Fig. 28 (Oho, 1987b).

B. Resolution (Maximum Spatial Frequency)

As a well-known method in electron microscopy, diffractogram (equivalent to the 2D Fourier transform) can be used for measuring the "maximum spatial frequency" included in an image, which is closely related to image resolution. Concerning objectivity and repeatability of the measured value, this statistical method must be superior to conventional edge-to-edge or point-to-point resolution in space domain. However, as it is very difficult for the diffractogram to detect accurately the maximum spatial frequency in a noisy image, it is inadequate for a common SEM image including noise. Decreasing this disadvantage was experimented, explored using the superposition diffractogram method. This method was proposed by Frank (1972) and applied to SEM images (Reimer *et al.*, 1978; Erasmus *et al.*, 1980). However, as far as it is possible to tell, it could not show its real ability fully (why this is so will be explained later). In this section, the latent ability of the superposition diffractogram method is successfully demonstrated by the improvement obtained.

In order to successfully obtain a superposition diffractogram used in resolution measurement, first an odd-looking SEM image with incorrect aspect ratio is recorded by twice scanning each line, and two images with correct aspect ratio are reconstructed by thinning out lines correctly from the SEM image. These images obtained through this unusual scanning possess nearly a perfectly identical signal. However, SEM shot noise in two images is not correlated between them. Second, one of two images is shifted by some pixels in the x-(horizontal) or y-(vertical) direction and next the shifted image and another image are superposed. Finally, the result (superposed image) is Fourier-transformed to obtain a superposition diffractogram. On the other hand, we can make good use of two images in question, because an SEM image with higher S/N is successfully obtained without a blur from them by averaging. If we employ the usual scan (repeated scans) for image recording, an averaged SEM image obtained from these scans and integration technique is usually blurred to a greater or less extent.

For this work, a digital scan generator equipped with some useful functions has been constructed. This is useful for "active image processing," which will be mentioned later. Functions of the present digital scan generator are summarized in Table 1.

The superposition diffractogram can distinguish better between structural detail and noise in an SEM image because structural detail is present in both SEM images, but noise, although it is present in two images, is not correlated between them. Consequently, a part of the diffractogram corresponding to structures included in the SEM image is modulated by a periodic interference fringe pattern (like Young's interference fringes) as shown in Fig. 30c, while that of noise is not changed. The number of shifted pixels is equal to the number of periodic patterns, if these fill the whole image.

TABLE I

Some of the Available Commands in Digital Scan, with their Principal Functions

Command	Parameter Range	Function
START		Start scan
STOP		Stop scan
HOFFSET	64n ($n=1\sim128$)	Set horizontal offset (pixels)
VOFFSET	64n ($n=1\sim128$)	Set vertical offset
HSCANNO	64n ($n=1\sim128$)	Set HOFFSET + the number of horizontal pixels
HSCALE	2^n ($n=6\sim14$)	Set horizontal scale
VSCANNO	64n ($n=1\sim128$)	Set VOFFSET + the number of vertical pixels
VSCALE	2^n ($n=6\sim14$)	Set vertical scale
SCANTIM	4, 6, 12, 30, 60, 120	Set scan time per pixel (μS/pixel)
LINEREP	$n=1\sim15$	Set the number of scan repetition at each line
SCANCND		Synchronize the scan speed to the 50 Hz line frequency, etc.

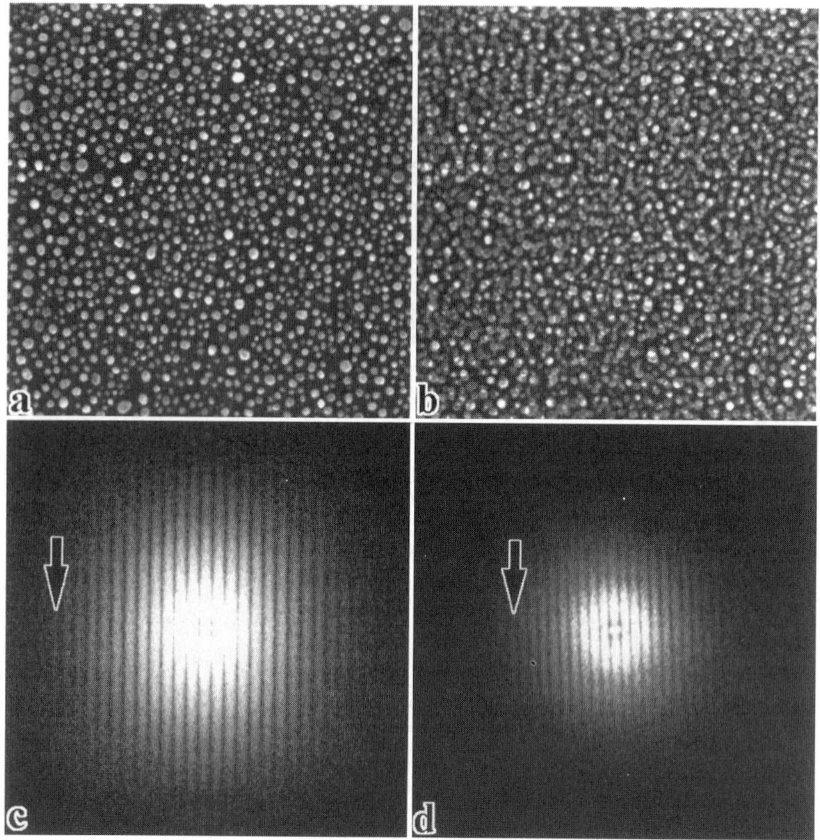

FIGURE 30. Highly precise resolution measurement of SEM images. (a) One of two SEM images of gold particles on carbon by twice scanning at each line (unusual scanning); (b) superposed image obtained from (a) shifted by 32 pixels in x-direction and another image; (c) superposition diffractogram of (b); (d) superposition diffractogram obtained through the iteration of usual scanning. Horizontal field width of (a) = 1.2 μm, (c) = 0.41 nm^{-1}.

Figures 30a, b, and c show one of two high-resolution SEM images of gold particles on carbon with the identical view (accelerating voltage of 30 kV, working distance of 4 mm, 512 × 512 pixels × 8 bit), the superposed image and its superposition diffractogram, respectively. Here, Fig. 30a shifted by 32 pixels in x-direction and another image were used for Fig. 30b. Visually we can measure the maximum spatial frequency q_{max} of 0.16 nm^{-1} in Fig. 30c (see arrows). This is equivalent to the minimum specimen periodicity $L_{min}(= 1/q_{max})$ of 6.3 nm. The significant figures of the measured value depend only on the accuracy of SEM instrument magnification. In the current case, we obtained the measured value on the assumption that an accuracy of about 1% is possible (of course the magnification has to be checked with a calibration specimen any time precise measurements are required).

This method fundamentally allows us to measure resolution accurately. However, a highly precise technique for finding the maximum spatial frequency in superposition diffractogram has to be developed to use it more practically. A certain method using autocorrelation function is now under construction. On the other hand, this measured value is closely related to the electron beam size when we utilize the ideal sample for high resolution observation. However, the electron beam size is not numerically in agreement with the current measured value, as they each have different definitions (Erasmus *et al.*, 1980). The minimum specimen periodicity $L_{\min}$ may be of the order of twice the electron beam diameter or the conventional edge-to-edge resolution (Reimer, 1985).

In contrast with the forementioned recording method, it is noticed that two images obtained by the iteration of usual scanning are not identical because of the effects of specimen deformation (damage), drift, stray magnetic field, and other things, even though their images are recorded under the same SEM operating condition. Hence, these images are not suitable for the present purpose due to inaccurate information. This is the reason why the superposition diffractogram method could not show its real ability fully. A superposition diffractogram obtained through the same operating condition used in Fig. 30c except for the usual scanning is shown in Fig. 30d. The measured value of the minimum periodicity $L_{\min}$ included in Fig. 30d is 9.5 nm, which is incorrectly estimated. Moreover, this measured value is so unreliable, because it is unstably changed by the influence of differences of specimen property and/or the extent of disturbances.

C. Signal-to-Noise Ratio (S/N)

SEM images are usually recorded in a slow scan mode for obtaining a suitable S/N (TV rate SEM images have a poor S/N due to the small number of electrons in each frame). These images are complicated by noise problems to a greater or less extent due to the unfavorable effects peculiar to SEM. Therefore it is important to know the S/N of an SEM image in microscopy. In order to measure the S/N of an SEM image, Erasmus (1982) used two images (i_1 and i_2) of the same specimen area and the following equation:

$$\mathrm{S/N} = \sqrt{\frac{\text{covariance }(i_1, i_2)}{\sqrt{\text{variance}(i_1)\,\text{variance}(i_2)} - \text{covariance}(i_1, i_2)}}$$

The definition of this S/N is equivalent to (the signal standard deviation)/(the noise standard deviation) in an SEM image. It is assumed that the SEM noise is additive, has zero mean, and is independent of the signal. Of course, two images must share the identical signal (view) with each other except for noise components. Otherwise, we may obtain an inaccurate measured value. The forementioned digital scan generator can acquire the ideal images for this work.

When we can accurately measure both the maximum spatial frequency and the S/N in a SEM image, "quantitative evaluation of SEM image quality," which was impossible until now, may be attempted. Studies of this kind are very important in the field of SEM. Fortunately, as the forementioned special SEM image has the information needed for producing both of them, it is possible not only to take time but also accurately to measure them. This study is in the developmental stages.

VI. Color SEM Image

The SEM is much superior to the conventional light microscopy (LM) in resolution, depth of focus, ease of altering magnification, and so on. However, unfortunately we have obtained these excellent advantages in exchange for a complete lack of color information, which may be needed for observation, interpretation, and analysis of various samples.

In order to support the SEM image (monochrome), the pseudocolor (color coding) has been discussed and often utilized in many papers and works. This may be effective for enhancement of a weak contrast, for example, voltage contrast, electron-beam-induced current (EBIC), and superimposing of SE and BSE images. Many new commercial SEMs are equipped with this function. On the other hand, the cathodoluminescence mode produces a "real" color image in SEM. However, this color differs from a natural color perceived by the human eye. The principles and applications of cathodoluminescence and color-coding have been reviewed by Saparin and Obyden (1988). In addition, a cathodoluminescence mode, which can show 2D and 3D information, has been developed by Saparin *et al.* (1997).

A. Background of the Generation of the Natural Color Scanning Electron Microscopy Images

SEM images with the natural colors determined by the nature of the light reflected from the object [natural color (NC) SEM images] have recently been obtained by a successful combination of the low vaccum or low voltage

SEMs, a very small-sized video microscope (VM), and a high-performance personal computer for digital image processing (Oho and Ogashiwa, 1996b). As the human eye is essentially the organ for observing color and can discern thousands of color shades and intensities compared with roughly only two-dozen shades of gray and, moreover, the human brain judges various things according to color information from the eyes, the NCSEM images in question are helpful for many samples.

The backgound to the generation of the NCSEM images is described as follows:

1. The low vacuum or low voltage SEM has a fairly long history (Robinson, 1975; Welter and Coates, 1974), and, recently, every SEM manufacturer has produced them according to the demand for them. Because most samples may be observed in these instruments without metal coating, the specimen retains the natural colors of the surface. However, such instruments are generally worse in resolution than conventional SEMs.
2. Very small-sized VM systems combine recent advances in color CCD camera technology with the long light microscopy history. The depth of focus of the VM may be intermediate between that of SEM and that of conventional LM and this allows acceptable 3D viewing. In addition, the VM may be able to be easily incorporated in a SEM specimen chamber due to its very small size.
3. Only several years ago, we could not find a high-performance personal computer that could easily process full-color images (256^3 colors) with a large number of pixels. However, today's readily available personal computers can speedily perform various tasks for huge-sized images (Oho *et al.*, 1995; Oho *et al.*, 1996a). Thus the necessary conditions to conceive and develop the NCSEM have been satisfied very recently.

B. *Method for Obtaining a NCSEM Image*

For obtaining a NCSEM image, we employ a superimposition of the "highlight image" (Oho, 1992; Oho and Peters, 1994) from the SEM and the "color image" produced from the VM image to be described later. However, some problems in depth of focus and resolution between the SEM and the VM must be settled first.

Certainly, the SEM used in low vacuum or low voltage operating condition has a resolution fairly higher than that of the VM, although it is much inferior to that of a conventional SEM. In addition, from the viewpoint of the characteristics of the human visual system, we do not have to use an

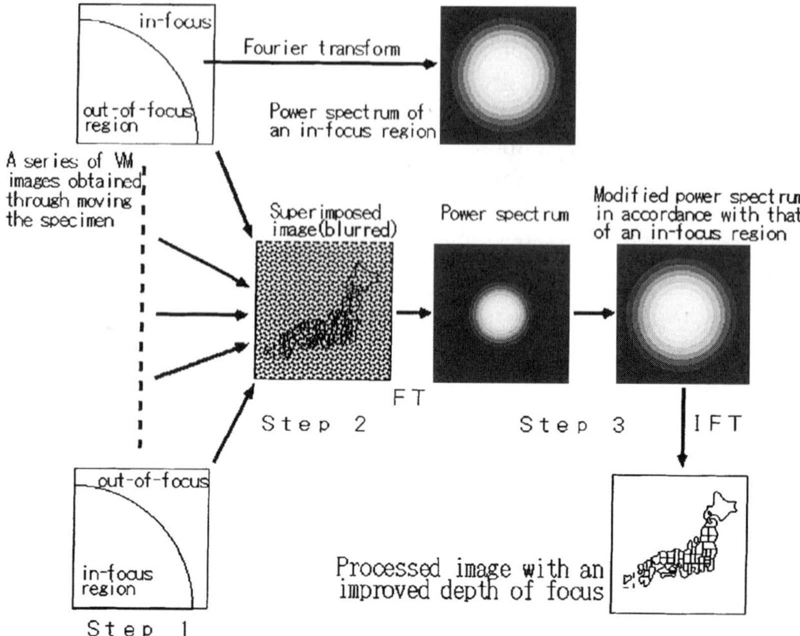

FIGURE 31. Flow diagram for the improvement of the depth of focus in VM image.

excessively high resolution VM image for obtaining color information. In other words, because the human visual system has a lower sensitivity to changes in hue or saturation than to changes in luminance (Gonzalez and Woods, 1992), a blurred VM image can be accepted in this study. Therefore, we do not have to consider the difference in resolution between those microscopes up to magnifications of about several thousand times, which is the upper limit for including effective color information.

It is only necessary for obtaining the NCSEM image to improve the depth of focus in VM image. The processing steps for improvement of the depth of focus are summarized in a flow diagram in Fig. 31 and are explained and discussed in the following.

Step 1

A series of VM images—red, green, blue (RGB)—are acquired with 512×480 pixels (or larger size) $\times$ 24 bit by moving the specimen (axial scanning). These have a different portion of the object brought into focus. Here, in order to obtain a VM image with high S/N, several or more frames obtained from the TV rate are automatically averaged (acquisition time for the VM image is within 1 s). From averaging theory, averaging N frames increases the S/N by $\sqrt{N}$ times.

Step 2
The series of VM images obtained by Step 1 are then superimposed to improve depth of focus. Techniques of this kind are successfully utilized in scanning confocal microscopy (Wilson, 1985). However, as the video microscope is not a confocal scanning type, our superimposed image is blurred throughout.
Step 3
The (blurred) superimposed image obtained from Step 2 must include sufficient information for the present study. By reducing signals from out-of-focus regions in each image, we may be able to gain most of the information obscured within the superimposed images. For this purpose, working in Fourier space, the power spectrum of the superimposed image is modified in accordance with that of an in-focus region and a processed result with an improved depth of focus is then obtained through the inverse Fourier transform (IFT). A very similar result can be obtained by using the combination of a simple unsharp masking and an averaging filter. Designing an optimal filter is very easy work because the resolution, S/N, magnification of VM images is little changed with respect to those of SEM images. A more rigorous analysis of the improvement of image resolution can be done based on the somewhat complicated Wiener filter theory (Lewis and Sakrison, 1975). However, it may be difficult to obtain additional information, because our superimposed image already had adequate S/N.

As an example, a series of VM images (Color Plate 1a1 $\sim$ a4) of a gold ore were processed by the method depicted in Fig. 31. A superimposition of four images and its enhanced image are shown in Color Plate 1b and 1c, respectively. Color Plate 1c is utilized for obtaining an NCSEM image (Color Plate 2b).

We have already settled the problems of depth of focus and resolution. Next, a procedure for properly superimposing a SEM image and a VM image with the improved depth of focus is summarized in a diagram in Fig. 32 and is explained in Steps 1–3.

Step 1
First, a SEM image (monochrome) and a VM image (RGB) are recorded, respectively. As the SEM usually produces a different size of image compared to the VM, we have to adjust them beforehand by clipping and interpolation (expansion) techniques in order to be able to superimpose both images.
Step 2
The RGB (VM image) is converted to hue, saturation, intensity (HSI), a mode that is frequently utilized for color image manipulation. The

"intensity" is decoupled from the color information in the image, and the "hue" and "saturation" are closely related to the color information (Gonzalez and Woods, 1992). In order to produce an NCSEM image with both the fine-scale structures included in the SEM image and the color information included in the VM image, we alter the HSI model. That is, "VM-brightness" and "VM-highlight" are separated from the "intensity" as shown in Fig. 32. Also, "SEM-brightness" and "SEM-highlight" are done from the SEM image (intensity). The former comes mainly from macrostructures while the latter comes from microstructures. These details are described in the previous section.

Step 3

The image is reconstructed by only H (hue), and S (saturation) and VM-brightness components in the VM image indicate the color information of macrostructures. We call it "color image." As a final result, an NCSEM image is successfully obtained through superimposition of the "SEM-highlight," including fine structures and the "color image," and it may include much more effective information on various specimens. Fine structures are conspicuous by emphasizing the SEM-highlight component.

As an example, a part of the previously processed VM image (Color Plate 1c) and a SEM image with the identical view (Color Plate 2a) are superimposed using the method depicted in Fig. 32. Unfortunately, as our somewhat old field-emission SEM can perform only a coarse change of voltage, it does not have sufficient ability to observe our uncoated sample. Hence, in this experiment, first the VM images were recorded and next, after metal coating for the avoidance of charging effects, the SEM sample was recorded at a low voltage of 2 kV. An NCSEM image with both very fine structures and color information is shown in Color Plate 2b. It should be noticed that each important information is completely indicated in the NCSEM image without the obstruction, that is, adverse effects of blur in color image and change in the color.

As another example, VM and NCSEM images of a tubular flower of composite are indicated in Color Plate 3a and b, respectively. (An SEM image utilized for Color Plate 3b was recorded by using the low vacuum SEM.) It is understood that the NCSEM is much more useful even in a low magnification observation, which must be suitable for VM.

A high-magnification NCSEM observation was attempted to confirm the performance of the present system. BSE (low vacuum) and NCSEM images of color toner powder spread on a paper for copying machine are shown in Color Plate 4a and b, respectively. To obtain the image of b, a VM image recorded under the instrument magnification of 2000× was utilized. In the

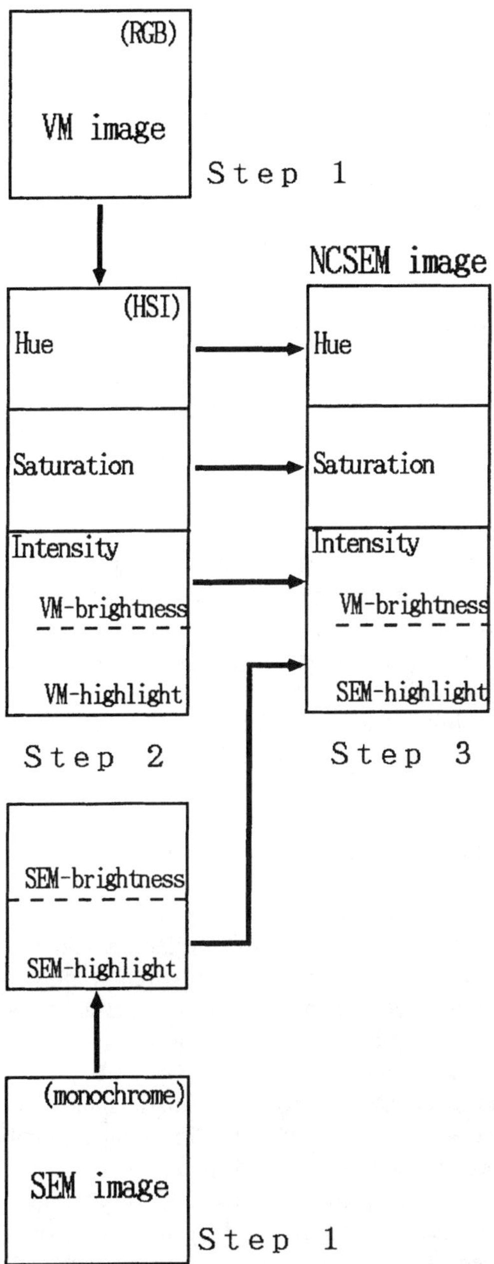

FIGURE 32. Explanation of the procedure for obtaining an NCSEM image from SEM and VM image with the improved depth of focus.

present NCSEM system, which is in developmental stages, this magnification might be the highest magnification at a practical level.

The NCSEM images may be able to simplify the observation, interpretation, and analysis of various samples, and it suggests several advantages for various samples. When we have successfully constructed an easy-to-use acquisition system for SEM and VM images used in this work, the NCSEM will be widely accepted in the field of SEM and many data from the NCSEM will be discussed more quantitatively.

VII. Automatic Focusing and Astigmatism Correction

Accurate focusing and correcting astigmatism of an SEM is time consuming and difficult. Of course we cannot obtain the maximum resolving power of the SEM without accomplishing these works. In addition, because the demand for automation of various measurements has increased, for example, in SEM for semiconductor process evaluation, this kind of study will become more and more important. Hence, automatic focusing and astigmatism correction using digital image processing technology is one of the most interesting areas in SEM. So far as we know, SEM manufacturers have supplied SEMs equipped with those functions because they have adequate room for the various improvements expected over the next decade or so.

Automatic focusing and astigmatism correction systems generally fall into one of two categories. One is a method that applies the derivative or gradient of the SEM signal or image, and another is a method that analyzes the power spectrum of the SEM image. When a certain solution is not considered, both methods may be influenced by the adverse effects of the noise component in SEM signal (SEM images are frequently very noisy at high magnification and at TV scan). SEM manufacturers have generally employed a method based on differentiating the video signal or similar technique due to reducing processing time and production costs, although it may be a dissatisfied performance in high-resolution microscopy. In a certain SEM instrument, the optimal values of exciting current of each lens are automatically searched out from among one million combinations.

In order to more precisely perform automatic focusing and astigmatism correction, Erasmus and Smith (1982) developed an automatic method based on computing the covariance of two SEM images with the identical view. This method can operate under the condition of very noisy images because covariance is not essentially influenced by noise. According to them, the accuracy and speed of the method were at least as good as those results that could be achieved manually by an experienced operator; these findings came mostly from a previous study.

VIII. Remote Control of SEM

In the microscopy domain, several applications of the networking technology have been attempted to find its potential. It might be unknown up to now whether it is actually useful. However, attractive data and the possible use to microscopists have already been shown in some studies. According to some recent reserarch, remote-controlled SEM using the networking technology may be helpful in remote control, remote diagnosis, virtual microscopy for education, real-time collaboration, and so forth. On the other hand, in the field of medicine, the prototype system, which has been used to diagnose illness from a distant location has already aproached practical use.

In order to construct a system for remote-controlling SEM, some combinations from among the modern SEM (PC-SEM), which is fully controlled by sophisticated software packages and works on the PC, local area network (LAN), Internet, Intranet, integrated services digital network (ISDN), among others, and a physical layer of networking, for example, modem and Ethernet, are considered according to the purposes of the system. Moreover, we may need the *de facto* standard software packages, for example, DigitalMicrograph to include some commands for microscope control, TimbuktuPro to provide remote control capabilities, CUSeeMe for videoconference and necessary hardware.

Examples of actual applications include the following. Chumbley *et al.* (1995) developed the multiuser SEM environment for supporting SEM teaching within a university. Chand *et al.* (1997) constructed a World Wide Web (WWW)-controlled SEM system. It can fully remote-control SEM operation through the Internet, and request SEM diagnostic parameters (e.g., beam current, magnification, etc.) and receive responses. Of course, the remote user can see and acquire small-sized live SEM images. When transferring a large-sized image, for example, $1024 \times 768 \times 8$ bit, a few minutes or more are required. Voelkl *et al.* (1997) has also tried remote operation of SEM and TEM moreover, to gain a perspective on the current state of SEM and TEM instruments, networking technology, and other areas.

For the forementioned merits of remote-controlled SEM to be truly advantageous, it is necessary to improve both the microscope and the network as follows: a) a much higher speed is required for smooth communications; and sufficient performance requires a huge investment. b) it is necessary to redesign the electronics and the interface of present microscopes to control the main parameters of electron optics. These improvements are likely to follow very quickly as society demands them. Then, as many general SEM users become interested in the utility value of technology related to the Internet and it can actually be used, the true value of this kind of study and work will very quickly be understood. In the near future, an excellent SEM

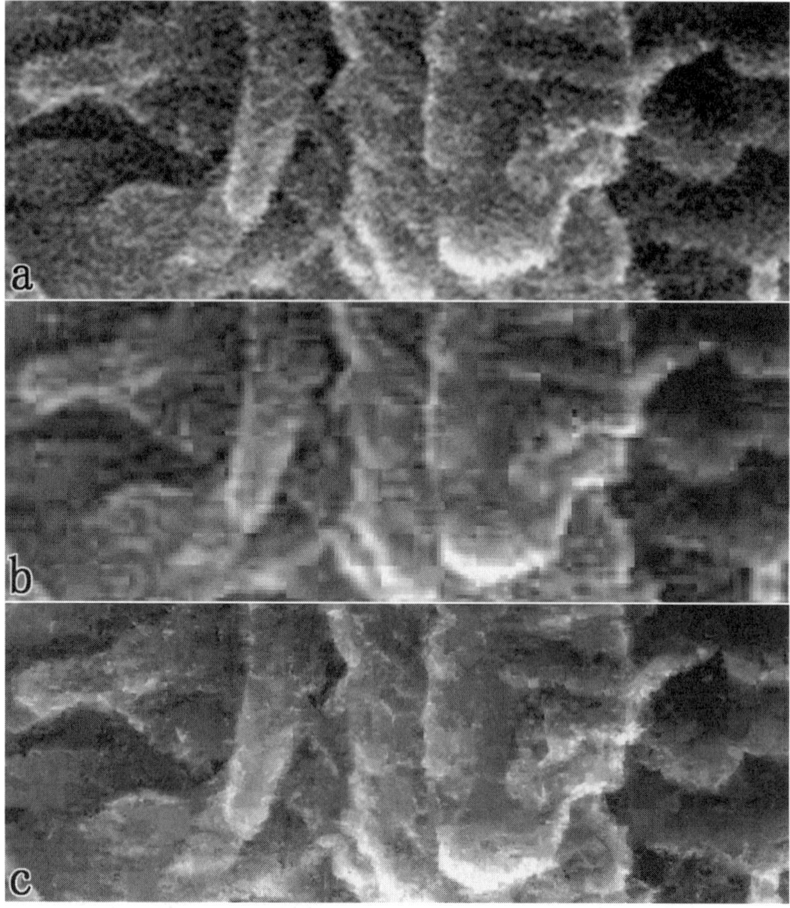

FIGURE 33. Image compression of SEM image. (a) Original image; (b) image compressed into 1/14 by JPEG; (c) image compressed into 1/14 by fractal technology.

operator with an ultrahigh-performance SEM, optimal sample making and scholars far away, will be able to have virtual meetings for high level real-time collaboration. Of course, this system may be useful for electron microscope-related education, for example, an expensive SEM would be bought by an academic institution and then successfully connected to many inexpensive computers located in various places. The number of teachers required would be reduced. On the other hand, if remote diagnosis of SEM instrument problems becomes routine, service may improve, for example, quick repair, reduction in repair cost, and warning of problems with SEM instruments beforehand.

As the limited bandwidth of networks, especially the Internet, is improved, these possibilities move into a practical realm. Also, the performance improvement of image compression technology, for example, Joint Photographic Experts Group (JPEG) fractal, wavelets, and so on, will promote advancements in this field. For instance, fractal compression might produce better results. Figure 33a is an original SEM image. This data file is compressed into 1/14 by conventional JPEG (Fig. 33b) and fractal technology (Fig. 33c), respectively. Although many unfavourable artifacts can be observed in Fig. 33b, an image compressed by fractal technology with seemingly only few artifacts is produced in Fig. 33c (however, SEM noise in Fig. 33a cannot be shown in Fig. 33c). These are expanded images to show differences clearly. Although it is in the developmental stages, advancement in this important field is very rapid.

IX. Active Image Processing

As discussed in the previous sections, after acquiring a digital SEM image through the AD converter, we use digital image processing techniques for a designated purpose. In other words, a digital SEM image is first conventionally recorded, and then we determine a purpose for the image processing as the need arises; finally, a suitable image processing technique is chosen for the recorded image. Image processing techniques of this kind are called "passive image processing." To be concrete, contrast stretching, histogram equalization, lowpass filter, median filter, unsharp masking, among others, were utilized in the present paper as passive image processing techniques. We can easily find these techniques in some image analysis or retouching software.

Most passive image processing techniques often require fairly complex parameters, which differ from image to image and depend on the varying visual perception of operators. According to circumstances, these may adversely affect the measured value (recorded image) obtained through a scientific instrument. Some low-quality SEM images are especially disturbed by those techniques, that is, change of the measured value, dissatisfaction with the degree of image quality improvement and/or processing artifacts. Therefore, the adaptation of passive techniques to SEM image processing was not successful. However, if we can use SEM images that always have a constant characteristic, for example, concerning resolution, sharpness, S/N, contrast, the foregoing problems may be reduced because some processing parameters can be determined easily. In addition, as increasing the information included in the recorded image is impossible through image processing, it is necessary beforehand to secure information content sufficient for utilizing a certain image processing technique. These may be the roots of the concept of "active image processing."

In contrast with passive image processing, when using active image processing, we must first decide on the "purpose of the image processing in electron microscopy" before acquiring a digital SEM image. Then in order to obtain "sufficient information for accomplishing the purpose," various operating conditions and the performance of every device related to the SEM instrument are examined and, if necessary, altered and improved. Although we have recently introduced the digital scan generator, an ultrahigh efficiency YAG detector, and a video microscope in order to obtain additional information [as part of multimodal microscopy (Glasbey and Martin, 1996)], it is still insufficient for the successful development of various methods based on active image processing. In parallel with this consideration, some "digital image processing techniques utilized in active processing" are also determined. Of course we purposely design these techniques so that the properties of obtained data will fit those properties required by the techniques. Although this technology might seem ordinary, it becomes essential in SEM digital imaging (Oho *et al.*, 1997).

In the previous sections, several methods based on the concept of active image processing have already been discussed. To put it concretely, suppression of the aliasing error, enhancement of BSE image, measurements of resolution and S/N, proposal of natural color (NC)SEM image, are categories of active image processing. For example, in measurements of resolution and S/N (see Section V), "purposes of image processing" are measurement of these parameters. In order to obtain "sufficient data for attaining the purpose," we have newly developed the forementioned digital scan system equipped with special functions. When utilizing this system, "image processing methods" for the measurements are adequate. As a result, we can obtain very high-performance methods for measuring the resolution and S/N of a SEM image. On the contrary, if we employ the concept of passive image processing, for example, use of the SEM without remodeling and conventional power spectrum for the measurements of resolution and S/N, we obtain an unfavorable result.

The successful combination of highly advanced SEM equipped with various functions for obtaining necessary data and digital image processing techniques useful for the data is an essential part of the present study. With appropriate employment of this technology, we can probably benefit from various designs because specially acquired data may always have sufficient information for attaining whatever purpose is determined. The author believes that, as advanced new methods based on the concept of active image processing are developed, a new generation of SEM technology will follow.

References

Artz, B. E. (1983). Examples of image processing using a computer controlled SEM. *Scanning*, **5**, 129–136.

Baba, N., Oho, E., and Kanaya, K. (1985). A method of noise removal using the inverse transform of an image autocorrelation function. *J. Electron Microsc. Tech.*, **2**, 431–438.

Baggett, M. C., and Glassman, L. H. (1974). SEM image processing by analog homomorphic filtering techniques. *Proc. 7th Ann. Conf. Scanning Electron Microscopy Symp.*, (Ed. by O. Johari), IIT Res. Inst Chicago Part I, 199–206.

Beil, W., and Carlsen, I. C. (1990). A combination of topographical contrast and stereoscopy for the reconstruction of surface topographies in SEM. *J. Microscopy*, **157**, 127–133.

Boyde, A. (1975). Measurement of specimen height difference and beam tilt angle in anaglyph real time stereo TV SEM systems. *Proc. 8th Ann. Conf. Scanning Electron Microscopy Symp.*, (Ed. by O. Johari), IIT Res. Inst Chicago, 189–198.

Carlsen, I. C. (1985). Reconstruction of true surface-topographies in scanning electron microscopes using backscattered electrons. *Scanning*, **7**, 169–177.

Castleman, K. R. (1979). *Digital Image Processing*, Prentice-Hall, Inc., pp. 226–249.

Chand, G., Berton, B. C., Caldwell, N. H. M., and Holburn, D. M. (1997). World Wide Web-controlled scanning electron microscope. *Scanning* **19**, 292–296.

Chin, R. T., and Yeh, C. L. (1983). Quantitative evaluation of some edge-preserving noise-smoothing techniques. *Computer Vision Graphics and Image Processing*, **23**, 67–91.

Chou, P. B., Rao, A. R., Sturzenbecker, M. C., Wu, F. Y., and Brecher, V. H. (1997). Automatic defect classification for semiconductor manufacturing. *Machine Vision and Applications*, **9**, 201–214.

Chumbley, L. S., Meyer, M., Fredrickson, K., and Laabs, F. (1995). Computer networked scanning electron microscope for teaching, research and industry applications. *Microsc. Res. Tech.*, **32**, 330–336.

Crewe, A. V. (1980). Imaging in scanning electron microscopy. *Ultramicroscopy*, **5**, 131–138.

Crewe, A. V., and Ohtsuki, M. (1981). Optimal scanning and image processing with the STEM. *Ultramicroscopy*, **7**, 13–18.

Desai, V., and Reimer, L. (1985). Digital image recording and processing using an Apple II microcomputer. *Scanning*, **7**, 185–197.

Edwards, R. M., Lebiedzik, J., and Stone, G. (1986). Fully automated SEM image analysis. *Scanning*, **8**, 221–231.

Ehrich, R. W. (1978). A symmetric hysteresis smoothing algorithm that preserves principal features. *Computer Graphics and Image Processing*, **8**, 121–126.

Erasmus, S. J., Holburn, D. M., Smith, K. C. A. (1980). On-line computation of diffractograms for the analysis of SEM images. *Scanning*, **3**, 273–279.

Erasmus, S. J. (1982). Reduction of noise in TV rate electron microscope images by digital filtering. *J. Microsc.*, **127**, 29–37.

Erasmus, S. J., and Smith, K. C. A. (1982). An automatic focusing and astigmatism correction system the SEM and CTEM. *J. Microsc.*, **127**, 185–199.

Frank, J. (1972). Observation of the relative phase of electron microscopic phase contrast zones with the aid of the optical diffractometer. *Optik*, **35**, 608–612.

Glasbey, C. A., and Martin, N. J. (1996). Multimodal microscopy by digital image processing. *J. Microsc.*, **181**, 225–237.

Gonzalez, R. C., and Woods, R. E. (1992). *Digital Image Processing*, Addison-Wesley Publishing Company, Reading, Mass., pp. 1–304.

Herzog, R. F., Lewis, B. L., and Everhart, T. E. (1974). Computer control and the scanning electron microscope. *Proc. 7th Ann. Conf. Scanning Electron Microscopy Symp.*, Ed. by O. Johari, Part I, pp. 175–182, IIT Res. Inst., Chicago.

Holburn, D. M., Smith, K. C. A. (1979). On-line topographic analysis in the SEM. *Scanning Electron Microscopy 1979/II*, IITRI, Chicago, 46–52.

Holburn, D. M., and Smith, K. C. A. (1982). Topographical analysis in the SEM using an automatic focusing technique. *J. Microsc.*, **127**, 93–103.

Joy, D. C. (1974). SEM parameters and their measurement. *Proc. 7th Ann. Conf. Scanning Electron Microscopy Symp.*, Ed. by O. Johari, Part I, pp. 327–334, IIT Res. Inst., Chicago.

Joy, D. C. (1982). Microcomputer control of a STEM. *Ultramicroscopy*, **8**, 301–308.

Jones, A. V., and Smith, K. C. A. (1978). Image processing for scanning microscopists. *Scanning Electron Microscopy 1978/I*, IITRI, Chicago, 13–26.

Koenig, G., Nickel, W., Storl, J., Meyer, D., and Stange, J. (1987). Digital stereophotogrammetry for processing SEM data. *Scanning*, **9**, 185–193.

Lebiedzik, J. (1979). An automatic topographical surface reconstruction in the SEM. *Scanning*, **2**, 230–237.

Lewis, B. L., and Sakrison, D. J. (1975). Computer enhancement of scanning electron micrographs. *IEEE Trans. Circuits and Systems*, **CAS-22**, 267–278.

McMillan, R. E., Johnson, Jr., G. G., and White, E. W. (1969). Computer processing of binary maps of SEM images. *Proc. 2nd Ann. Conf. Scanning Electron Microscopy Symp.*, (Ed. by O. Johari), pp. 439–444, IIT Res. Inst., Chicago.

O'Handley, D. A., and Green, W. B. (1972). Recent developments in digital image processing at the Image Processing Laboratory at the Jet Propulsion Laboratory. *Proc. IEEE*, **60**, 821–828.

Oho, E., Baba, N., Katoh, M., Nagatani, T., Osumi, M., Amako, K., and Kanaya, K. (1984). Application of the Laplacian filter to high-resolution enhancement of SEM images. *J. Electron Microsc. Tech.*, **1**, 331–340.

Oho, E., Sasaki, T., and Kanaya, K. (1985). Removal of the effects of contamination in SEM image by homomorphic filtering. *J. Electron Microsc.*, **34**, 427–429.

Oho, E., Kobayasi, M., Sasaki, T., Adachi, K., and Kanaya, K. (1986a). Automatic measurement of scanning beam diameter using an on-line digital computer. *J. Electron Microsc. Tech.*, **3**, 159–167.

Oho, E., Sasaki, T., and Kanaya, K. (1986b). A comparison of on-line digital recording with conventional photographic recording for scanning electron microscopy. *J. Electron Microsc. Tech.*, **4**, 157–162.

Oho, E., Sasaki, T., Ogihara, A., and Kanaya, K. (1987a). An improvement in digital homomorphic filtering and its practical applications to SEM images. *Scanning*, **9**, 173–176.

Oho, E. (1987b). *The image quality improvements based on digital image processing in SEM and the conversion of a field emission SEM to a high performance STEM.* D.Eng. Dissertation, Kogakuin University.

Oho, E., Baba, M., Baba, N., Muranaka, Y., Sasaki, T., Adachi, K. Osumi, M., and Kanaya, K. (1987c). The conversion of a field-emission scanning electron microscope to a high-resolution, high-performance scanning transmission electron microscope, while maintaining original functions. *J. Electron Microsc. Tech.*, **6**, 15–30.

Oho, E., Ogihara, A., Kanaya, K. (1990a). A new non-linear pseudo-Laplacian filter for enhancement of secondary electron images. *J. Microscopy*, **159**, 33–41.

Oho, E., and Kanaya, K. (1990b). The utility of an on-line digital image recording system for SEM. *Scanning*, **12**, 141–146.

Oho, E., Ogihara, A., and Kanaya, K. (1991). A method using on-line digital computer for improvement of resolution of backscattered electron images. *J. Microsc.*, **164**, 143–152.

Oho, E. (1992). Automatic contrast adjustment for detail recognition in SEM images using on-line digital image processing. *Scanning*, **14**, 335–344.

Oho, E., Peters, K-R. (1994). Practical methods for digital image enhancement in SEM. *J. Electron Microsc.*, **43**, 299–306.

Oho, E., Ichise, N., Ogashiwa, T. (1995). The necessity and utility of a scanning electron microscope with 4096 lines. *J. Electron Microsc.*, **44**, 390–398.

Oho, E., Ichise, N., Ogashiwa, T. (1996a). Proper acquisition and handling of SEM images using a high-performance personal computer. *Scanning*, **18**, 72–80.

Oho, E., and Ogashiwa, T. (1996b). A natural color scanning electron microscopy image. *Scanning*, **18**, 331–336.

Oho, E., Ichise, N., Martin, W. H., and Peters, K-R. (1996c). Practical method for noise removal in scanning electron microscopy. *Scanning*, **18**, 50–54.

Oho, E., Hoshino, Y., and Ogashiwa, T. (1997). New generation scanning electron microscopy technology based on the concept of active image processing. *Scanning*, **19**, 483–488.

Oron, M., and Gilbert, D. (1976). Combined SEM-minicomputer system for digital image processing. *Proc. 9th Ann. Conf. Scanning Electron Microscopy Symp.*, (Ed. by O. Johari), Part I, pp. 120–127, IIT Res. Inst., Chicago.

Otaka, T., Mori, H., Yamada, O., and Todokoro, H. (1995). Critical dimension measurement scanning electron microscope for ULSI. *Hitachi Review*, **44**, 113–118.

Postek, M. T., Vladar, A. E. (1996). Digital imaging for Scanning Electron Microscopy. *Scanning*, **18**, 1–7.

Reimer, L., Volbert, B., Bracker, P. (1978). Quality control of SEM micrographs by laser diffractometry. *Scanning*, **1**, 233–242.

Reimer, L., Volbert, B., Bracker, P. (1979). STEM semiconductor detector for testing SEM quality parameters. *Scanning*, **2**, 96–103.

Reimer, L. (1985). *Scanning Electron Microscopy*, (Springer Series in Optical Sciences, **45**), Springer-Verlag, Berlin, Heidelberg, pp. 13–56.

Robinson, V. N. E. (1975). Backscattered electron imaging. *Proc. 8th Ann. Conf. Scanning Electron Microscope Symp.*, (Ed. by O. Johari), IIT Res. Inst, Part I, pp. 51–60.

Rosenfeld, A., Kak, A. C. (1982). *Digital Picture Processing* (2nd edition) **1**: Academic Press, New York, pp. 290–352.

Saparin, G. V., Obyden, S. K. (1988). Colour display of video information in scanning electron microscopy: Principle and applications to physics, geology, soil science, biology, and medicine. *Scanning*, **10**, 87–106.

Saparin, G. V., Obyden, S. K., Ivannikov, P. V., Shishkin, E. B., Mokhov, E. N., Roenkov, A. D., and Hofmann, D. H. (1997). Three-dimensional studies of SiC polytype transformations. *Scanning*, **19**, 269–274.

Suganuma, T. (1995). Measurement of surface topography using SEM with two secondary electron detectors. *J. Electron Microsc.*, **34**, 328–337.

Thong, J. T. L., and Breton, B. C. (1992). *In situ* topography measurement in the SEM. *Scanning*, **14**, 65–72.

Unitt, B. M., and Smith, K. C. A. (1976). The application of the minicomputer in scanning electron microscopy. *Proc. 6th European Cong. on Electron Microsc.*, Tal Int'l Pub. Co., Jerusalem, **1**, 162–167.

Voelkl, E., Allard, L. F., Nolan, T. A., Hill, D., and Lehmann, M. (1997). Remote operation of electron microscopes. *Scanning*, **19**, 286–291.

Wang D. C. C., Vangnucci, A. H., and Li, C. C. (1983). Digital image enhancement: A survey. *Computer Vision, Graphics and Image Processing*, **24**, 363–381.

Welter, L. M., and McKee, A. N. (1972). Observations on uncoated, nonconducting or thermally sensitive specimens using a fast scanning field emission source SEM. *Proc. 5th Ann. Conf.*

Scanning Electron Microscopy Symp., (Ed. by O. Johari), Part I, pp. 161–168, IIT Res. Inst., Chicago.

Welter, L. M., and Coates, V. J. (1974). High resolution scanning electron microscopy at low accelerating voltages. *Proc. 7th Ann. Conf. Scanning Electron Microscope Symp.*, (Ed. by O. Johari), IIT Res. Inst., Part I, pp. 59–66.

White, E. W., McKinstry, H. A., and Johnson, Jr., G. G. (1968). Computer processing of SEM images. *Proc. Ann. Conf. Scanning Electron Microscopy Symp.*, (Ed. by O. Johari), pp. 95–103, IIT Res. Inst., Chicago.

Wilson, T. (1985). Scanning optical microscopy. *Scanning*, **7**, 79–87.

Yew, N. C., and Pease, D. E. (1974). Signal storage and enhancement techniques for the SEM. *Proc. 7th Ann. Conf. Scanning Electron Microscopy Symp.*, (Ed. by O. Johari), Part I, pp. 191–198, IIT Res. Inst., Chicago.

ADVANCES IN IMAGING AND ELECTRON PHYSICS, VOL. 105

Design and Performance of Shadow-Mask Color Cathode Ray Tubes

E. YAMAZAKI

LG Electronics Inc. Display Device Research Lab, 184 Kongdan-dong, Kumi, Kyoung-buk, 730-030 Korea

ISBN 0-12-014747-5
ISSN 1076-5670/99 $30.00

I. Introduction

The shadow-mask color cathode ray tube (CRT), introduced in about 1950, is now almost 50 years old. Since then a large body of literature has appeared covering the numerous technical advances made in this area. The first comprehensive discussion of the subject was provided with the publication in 1974 of the book *Color Television Picture Tubes* by Morrell *et al.* However, in view of the important technical advances made since then it is felt that an up-to-date survey paper on this subject would be of value.

It should be noted that the impetus for the development of the shadow-mask tube initially came from the needs of the television industry. Although black-and-white television became commercially viable largely as a result of the introduction of the modern high-vaccum CRT in 1929 (particularly in the form developed by Zworykin), subsequent attempts to open the way for color television as a medium for general use were largely impeded by the unavailability of a satisfactory direct-view color display medium.

In recognization of this situation, a massive research program was initiated at the RCA Research Laboratories in 1949 in which a variety of novel types of CRTs were simultaneously investigated as possibilities for producing color images. Unfortunately, compared to black-and-white CRTs, all of these schemes involved relatively complex structures with stringent mechanical tolerances. Of the various schemes, however, the shadow-mask approach was selected as the most likely to lead to a successful color tube. (Looked at otherwise, it appeared to be the scheme with the objectionable features.)

As a result of intensive effort on the shadow-mask approach a successful tube was publicly demonstrated by RCA within a period of about six months. Although this tube had a brightness of only 24 cd/m^2 (7 fl) its performance was widely praised. In particular, since it was capable of producing color images in a so-called dot-sequential manner it opened the way for a new broadcast standard to be adopted in the United States, in which color pictures could be received on the same 6 MHz channel previously used for black-and-white TV while, at the same time, black-and-white TV receivers could continue to receive a monochrome version of the color broadcasts. Since then, the fact that no other form of direct-view color tube has succeeded in supplanting the shadow-mask tube for television would seem to confirm that its initial choice was justified.

Since 1974 an important new application for such tubes emerged, namely the display of computer-generated information. To satisfy this requirement the developers of color tubes were faced with a somewhat formidable set of new requirements considerably more severe than those imposed by television. In particular, these involved increasing the scanning uniformity, reducing the flicker, and increasing the resolution (especially at the corners of the screen). A measure of the success achieved in satisfying the new requirements can be appreciated if one considers that the number of color tubes manufactured today for computer applications is almost as great as the number manufactured for television.

One purpose of this article is to provide background on the problems posed in the design of shadow mask tubes and the methods employed in solving them. Another purpose is to bring the reader up to date on the present performance capabilities of such tubes. Aside from the important progress made in electron guns (covered in the accompanying article "Electron-Gun System for Color Cathode Ray Tubes" by Suzuki) the material presented here covers a wide variety of topics including: the deposition of multicolor phosphor screens; methods for enhancing image contrast; geometric requirements imposed on the mask and screen for obtaining proper electron-beam color selection; and the special requirements imposed on the magnetic deflection field to maintain the electron beams converged to a single spot. In addition, the problems related to the design of large bulbs with flat screens are discussed as well the methods used for minimizing the deleterious effects of X-ray generation.

II. Basic Structure of Shadow-Mask Tube and Principle of Operation

Figure 1 shows a horizontal cross section of a typical shadow-mask tube (the arrangement shown here assumes the use of an in-line gun structure, which is described in more detail in Section III.A and B). As shown, the electron gun

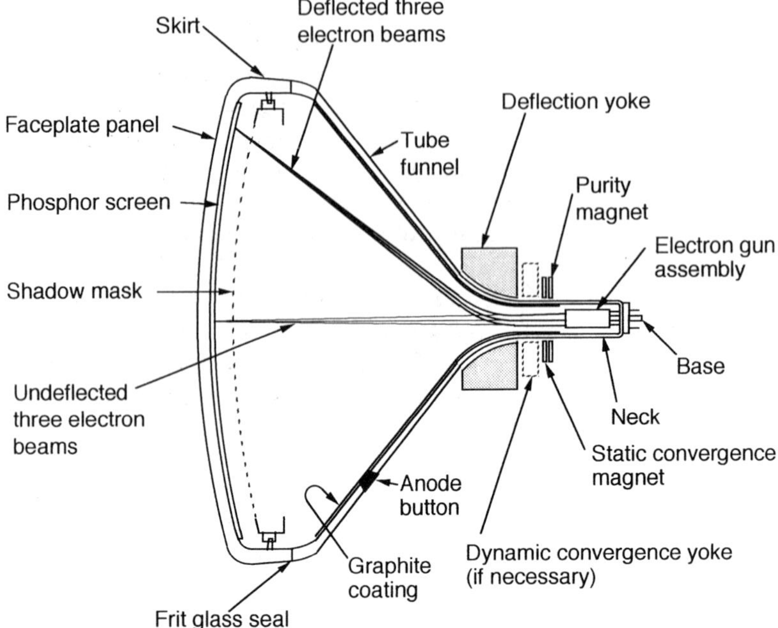

FIGURE 1. Section of typical shadow-mask tube.

assembly in the tube neck emits three focused electron beams, which are maintained converged to a single spot at all points of the phosphor screen. The screen consists of an array of red, green, and blue phosphor stripes on the inner surface of the face-plate panel. The outer edge of the panel is provided with a surrounding skirt that is sealed to the tube funnel. Adjacent to the phosphor screen a shadow mask is attached to the skirt. This has an array of suitably located slits that allow electrons from each beam to land only on phosphor elements of a given color. (Figure 2 shows in greater detail a horizontal cross section of a portion of the mask and screen.)

Since the phosphor stripes are narrow and normally viewed from a distance where they cannot be resolved by the eye, a spot on the screen (that includes a number of phosphor stripes) has a resultant brightness and color determined by the additive light emission produced by the three beams (Morrell *et al.*, 1974). As shown in Fig. 1, wires from the gun electrodes are brought out to external base pins. A high potential, such as 25–30 kV or more, is supplied to the anode button that is connected to the graphite conducting coating on the inner surface of the tube funnel, which extends to the gun structure. Because

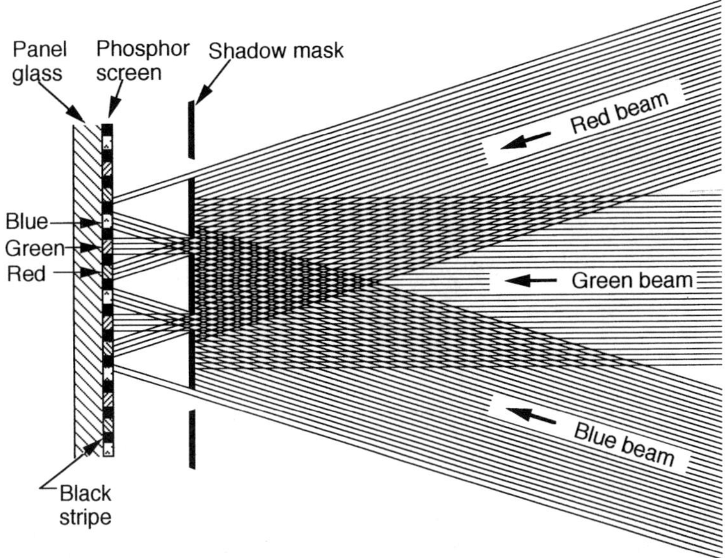

FIGURE 2. Detail of Fig. 1 showing convergence of beams and color selection.

the shadow mask is also internally connected to this coating, the entire space between the end of the guns and the phosphor screen is field-free.

The gun structure is designed so that the three beams are converged to form a single spot at the screen center in the absence of any deflecting field. This condition is known as static convergence. (It should be noted here that the three beams, although overlapping, are not converged at the plane of the shadow mask as often assumed.) To compensate for manufacturing errors a small adjustable permanent-magnet assembly is mounted on the tube neck just beyond the electron guns to ensure static convergence. Mounted around the tube neck in the region where it joins the funnel is the deflection yoke. This simultaneously deflects the three beams both horizontally and vertically to scan a raster pattern on the phosphor screen. To maintain the beams converged when they are deflected, a specially designed deflection yoke having an appropriate field distribution is generally used (see Section VII). (In some tube designs where a self-converging yoke cannot be used, a special dynamic convergence yoke is inserted as indicated by the dashed lines of Fig. 1.) Aside from converging to a common spot, it is essential that the beams arrive at the shadow mask at the correct angle in order to land on the correct color phosphor elements. For this purpose a small purity-magnet assembly is mounted on the tube neck between the yoke and gun as shown, which allows the three beams to be moved transversely so that they are properly positioned.

III. Alternative Gun and Screen Arrangements

In practice tubes can be made with several different combinations of gun arrangement, types of shadow mask structure, and phosphor screen pattern. This makes possible three basic systems as follows:

a. In-line gun structure with striped screen
b. In-line gun structure with dot screen
c. Delta gun structure with dot screen.

A. In-Line Gun with Striped Screen

In this arrangement, indicated in Fig. 3 (a portion of which was shown in cross section, Fig. 2), the three electron guns are placed side by side in a horizontal plane. The phosphor screen consists of an array of red, green, and blue phosphor stripes running vertically, with each group of three stripes placed in proper registry with a particular slit in the shadow mask as indicated. For correct color selection proper registration of the mask only in the horizontal direction is necessary. As also shown in Fig. 3, the converged spot extends over several groups of red, green, and blue phosphor stripes (each group being referred to as a color triad).

A primary advantage of the in-line gun arrangement is that it allows a self-converging type of deflection yoke to be used (Haantjes and Lubben, 1959; Barbin and Hughes, 1972; Morrell *et al.*, 1974), which can maintain the convergence of the three deflected beams without the need for the separate dynamic convergence yoke mentioned in Section C here. For static convergence a magnet assembly is provided as shown in Fig. 4. A combination of a pair of 6-pole and a pair of 4-pole ring magnets can move the outer beams in the same or opposite directions, respectively, without moving the center beam. By rotating the magnets of each pair with respect to each other the strength of each pair can be adjusted.

The in-line arrangement of electron guns, however, limits the diameter of the guns that can be accommodated by a tube of given neck diameter (compared to the delta gun arrangement discussed in Section C). Because of the smaller diameter of the electron lenses, the minimum attainable spot size of the electron beams is increased (Suzuki, 1998). However, in recent years various gun designs have been developed that, in spite of their reduced diameter, have an increased effective lens diameter, thereby allowing small spots to be obtained without any sacrifice in beam current.

Because of the special nonuniform magnetic field required by a self-convergent yoke an added distortion and enlargement of the electron beam

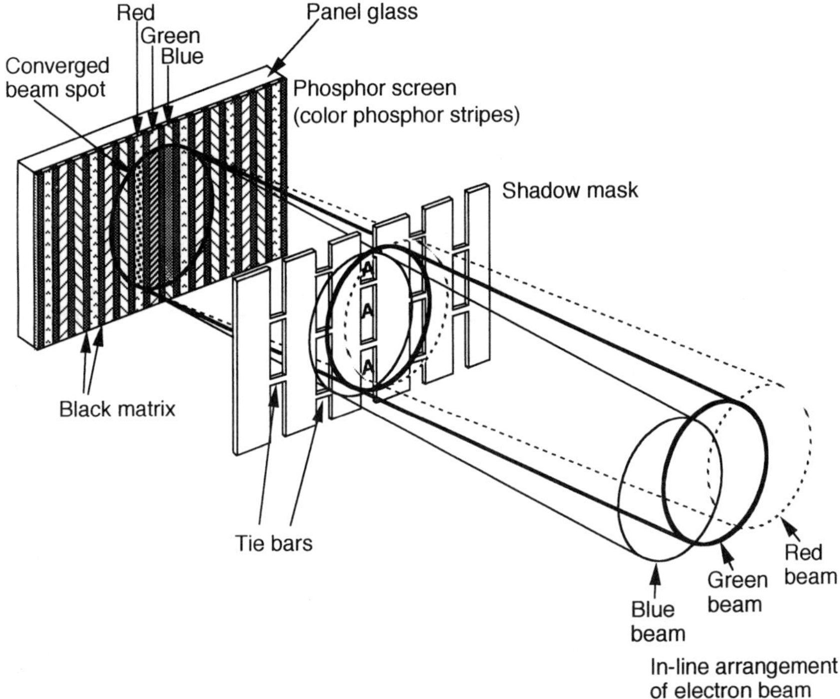

FIGURE 3. Three-dimensional view of in-line gun and stripe screen arrangement showing color selection principle and convergence of three beams of shadow-mask tube.

spot results when it is deflected. To overcome this, special gun designs were developed (Shirai *et al.*, 1987). As a result, the in-line electron gun system (initially developed for TV applications), which required only moderate resolution, is now widely used in all types of shadow-mask tubes, even those intended for ultrahigh-resolution.

In most tubes the shadow mask has a spherical shape as shown in Fig. 1. To enable the mask to retain its shape tiny horizontal tie bars are incorporated between adjacent mask slots as shown in Fig. 3. Such bars are not required, however, in tension mask tubes (see Section IX.C) where the shadow mask has a flat or cylindrical surface and is maintained under tension. Although the mask openings or slits are interrupted by the tie bars, the phosphor screen itself is deposited in the form of continuous stripes since the shadowing action of the bars has no effect on the color selection process. However, the shadowing effect of the tie bars may create a moiré problem (see Secton XI).

An important advantage of the striped phosphor screen is the ease of manufacturing. As discussed in Section VI.B, a line type of light source can

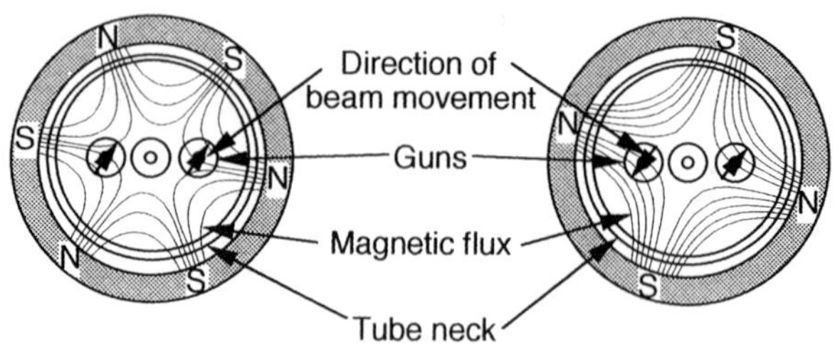

(a) Movement of outer beams in same direction in 6 pole field

(b) Movement of outer beams in opposite directions in 4 pole field

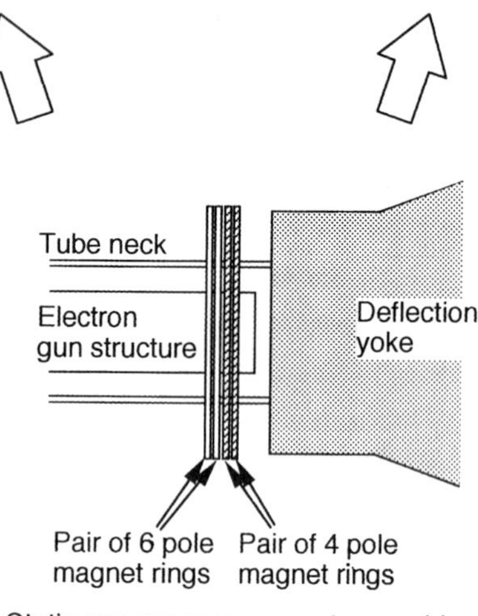

Static convergence magnet assembly

FIGURE 4. Static convergence magnet assembly. A combination of a pair of 6-pole and a pair of 4-pole ring magnets can move the outer beams in the same or opposite directions, respectively, wihtout moving the center beam. By rotating the magnets of each pair with respect to each other the strength of each pair can be adjusted.

be employed for the optical exposure process used in the photolithographic printing of the phosphor stripes. In the case of dot screens, such as will be discussed, a point light source is required that requires a much longer exposure time. Also, because the phosphor elements of striped screens are continuous, they are less likely to detach from the glass substrate and produce screen defects.

For ultrahigh-resolution displays the striped phosphor screen may not always be desirable since it results in a different resolution in the vertical and horizontal directions. (In the vertical direction the resolution is limited only by the spot size of the electron beams while in the horizontal direction it is further limited by the screen structure.) However, if the pitch of the mask slits is made smaller to improve the horizontal resolution, the required slit width becomes correspondingly smaller, being limited by the etching process used in fabricating the mask (see Section V).

B. In-Line Gun with Dot Screen

In this arrangement, as shown in Fig. 5, an in-line gun system is used but the phosphor screen is printed as an array of dots instead of stripes. Although originally proposed for TV applications (Buchsbaum, 1966), this arrangement is now mainly used only for special high-resolution applications where the striped phosphor screen is not always suitable.

As explained in Section IV, the diameter of a shadow-mask aperture in the case of the dot structure can be 1.7 times larger than the width of a mask slit, assuming the same pitch for both masks. [Compare Eqs. (10) and (20) of

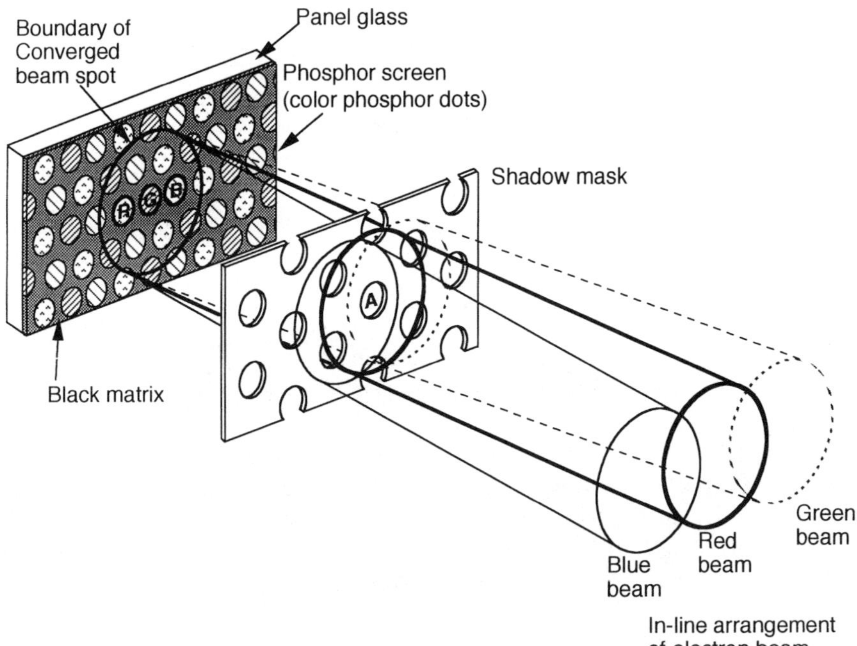

FIGURE 5. In-line gun structure with dot screen. The RGB triad is excited by electrons from red, blue and green beams passing through aperture A of mask.

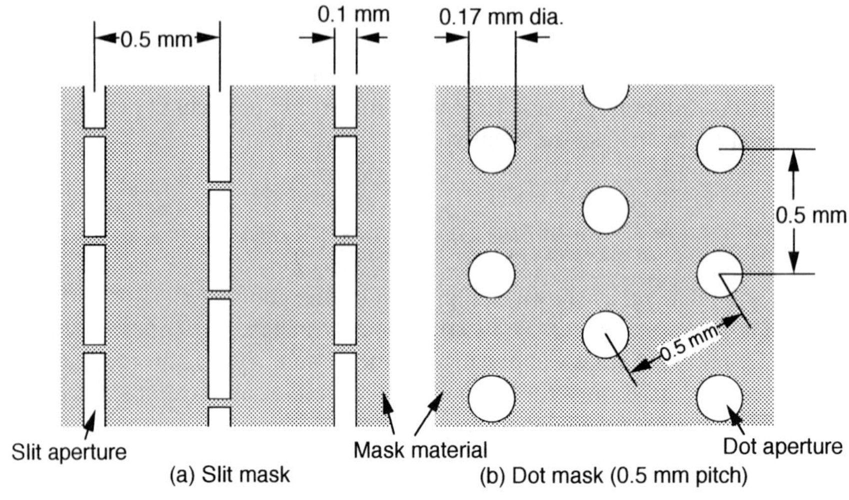

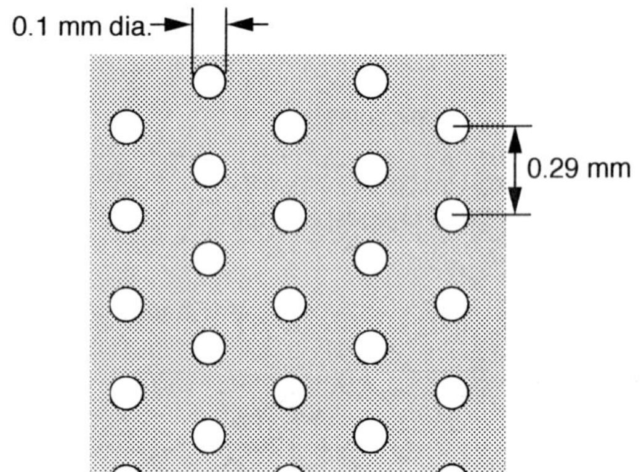

(c) Dot mask (0.29 mm pitch)

FIGURE 6. Dimensional comparison of slit mask and dot mask.

Section IV.] This in turn allows a much smaller pitch to be used for the dot structure if the dot aperture is made as small as the slit width. An example of this is shown in Fig. 6. Figure 6a shows the case of a 0.1 mm thick mask whose slits are also 0.1 mm wide and whose pitch is 0.5 mm. Figure 6b shows the case of a dot mask with the same pitch of 0.5 mm and an aperture size widened to 0.17 mm. This dot diameter, along with its pitch, however, can be made considerably smaller before being limited by etching problems. An example of this is shown in Fig. 6c where the pitch has been reduced to 0.29 mm.

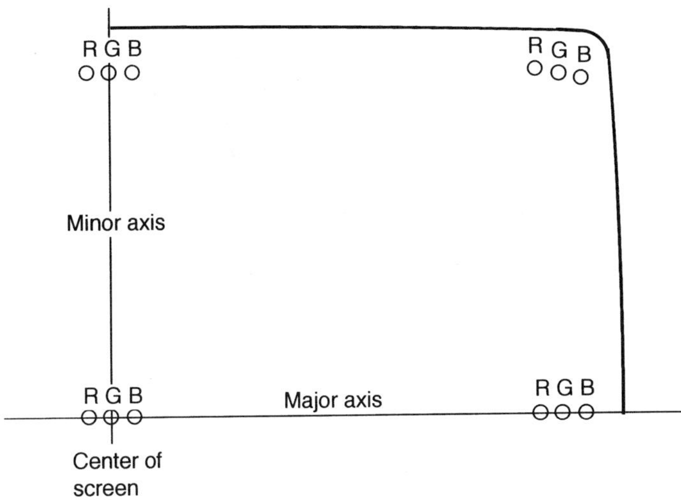

FIGURE 7. Beam landing triads produced by single mask aperture.

The use of a dot phosphor screen, however, results in other technical problems. If one considers the electron-landing triads, consisting of the electron spots on the screen produced by the three beams that pass through a single mask hole, the triads will be aligned horizontally at the screen center and on the screen axes. However, at the screen corners they become inclined, as shown in Fig. 7, because of geometric reasons, thereby disturbing the ideal packing or nesting of adjacent triads and preventing full utilization of the phosphor screen. (Further details of this phenomenon are discussed in Section C.)

An example of the effect resulting from the inclination of the triads is shown in Fig. 8. To avoid exciting undesired color dots when the triads are inclined at the screen corners, the mask apertures and phosphor dots must be reduced in size. Also, to offset the effect of the triad inclination, the shadow-mask dot array can be inclined along the triad inclination as shown in Fig. 9. (For comparison, Fig. 10 shows a beam-landing triad at both the center and corner of the screen for an in-line gun with a striped screen. Although the inclination of the triads, or electron beamlets, causes a vertical misregister at the corner, this does not result in the beamlets landing on the wrong color stripes.)

In actuality, the beam-landing spots and phosphor dots do not match precisely because of various reasons such as the deflection yoke characteristics, extraneous magnetic fields, and deformation of the phosphor panel and mask during processing. As discussed in Section VI.B, to minimize this

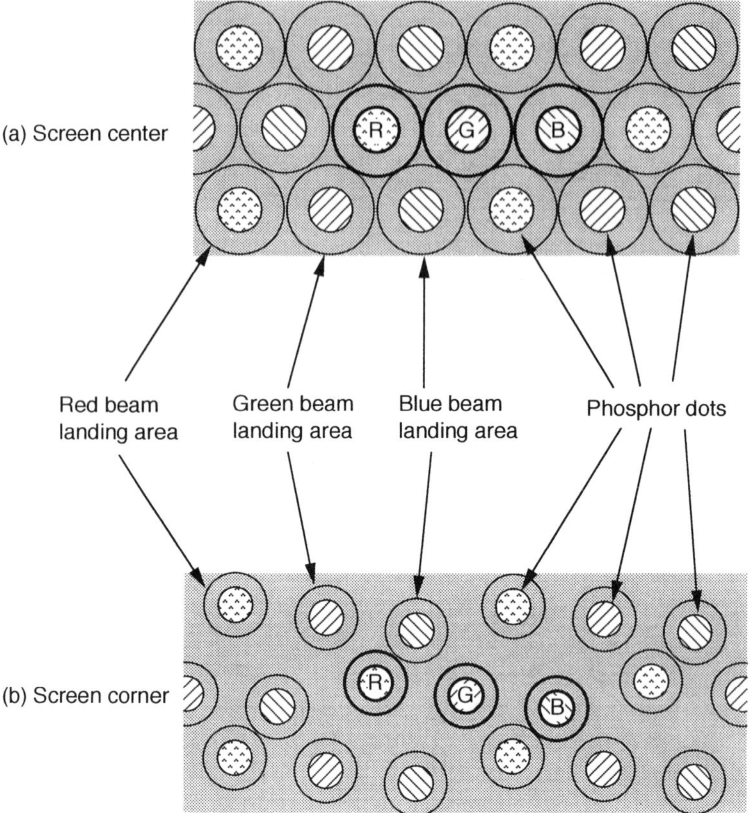

FIGURE 8. Beam landing of rays from three beams passing through the same mask aperture (shown by darker boundaries).

mismatch, a correction lens is used during the photolithographic printing of the phosphor dots on the screen. However, in the case of a dot screen, because this mismatch occurs in two dimensions it may not always be possible to entirely correct for it using a normal type of lens system, as will be explained in Section VI.B.

C. Delta Gun with Dot Screen

The delta arrangement of the gun structure combined with a dot screen and dot mask was used in the first shadow-mask tubes. However, it is rarely used now. Figure 11 shows the basic geometry of this system (Morrell *et al.*, 1974).

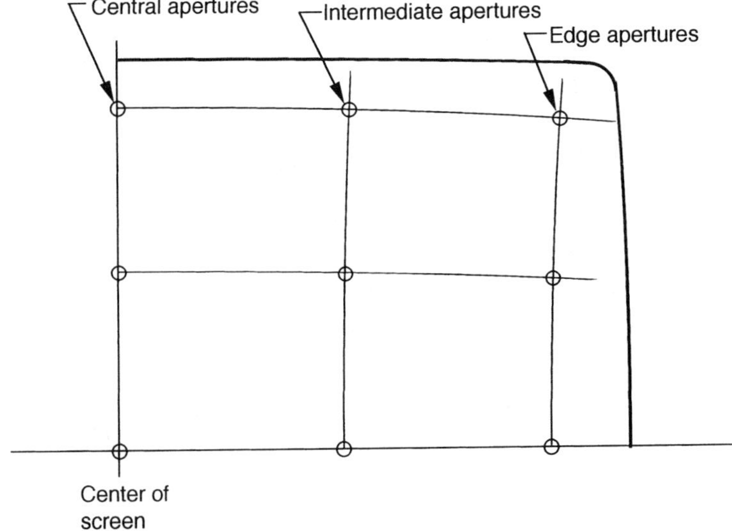

FIGURE 9. Location of mask aperture to produce better nesting.

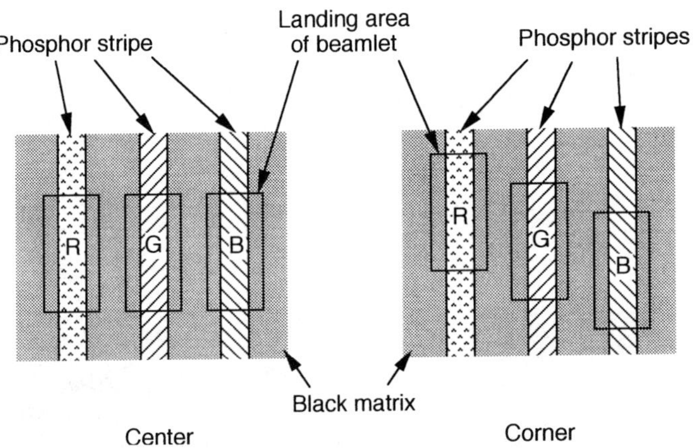

FIGURE 10. Inclination of beam triads at corner of striped screen.

For a given neck diameter, by arranging the guns in a triangular or delta configuration as shown in Fig. 12a each gun can have a 40% larger diameter compared with the in-line arrangement shown in Fig. 12b. Because of the greater lens diameter and reduced lens aberrations in the delta system, making possible a smaller electron beam spot size, this gun arrangement was employed in early color picture tubes.

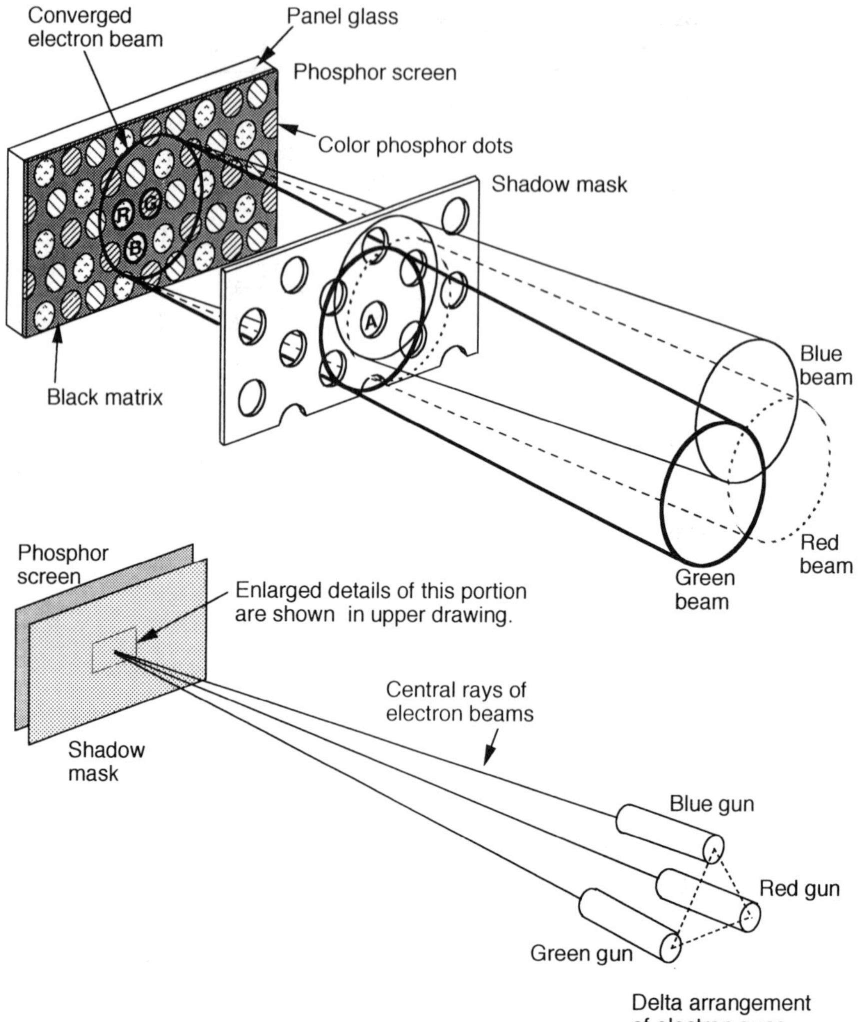

FIGURE 11. Delta gun structure with dot screen.

An important limitation of this system, however, is the fact that a self-converging magnetic deflection system cannot be designed that can maintain the beams converged in the horizontal and vertical directions simultaneously as required. To accomplish this it was necessary to insert a special convergence yoke between the deflection yoke and color purity magnet as shown in Fig. 1 to correct dynamically the deflection of each beam during

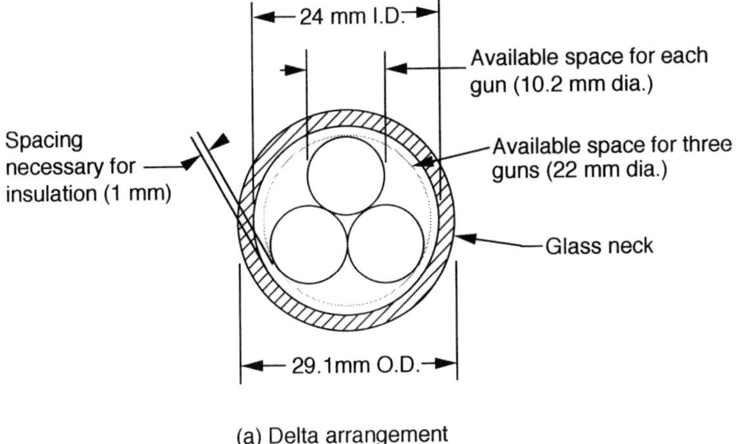

(a) Delta arrangement

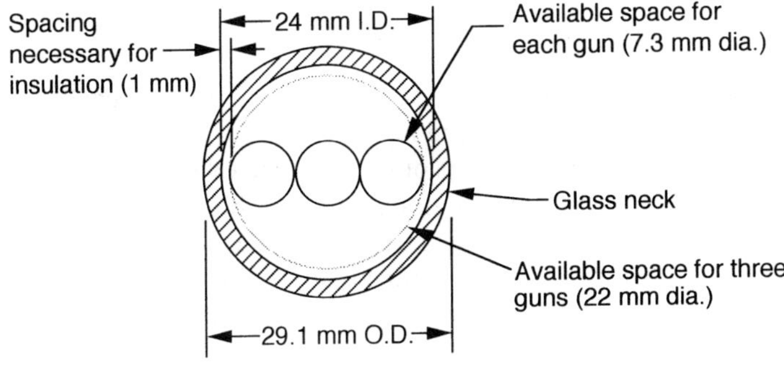

(b) In-line arrangement

FIGURE 12. Delta and in-line arrangements of electron gun.

scanning. The cross section of such a convergence yoke is shown in Fig. 13. As indicated, three external electromagnets are coupled to corresponding pairs of magnetic pole pieces inside the tube neck. The resulting field across these pairs produces a radial deflection of the individual beams as they emerge from the guns. In operation, suitable time-varying currents are passed through the coils in synchronism with the currents through the horizontal and vertical coils of the main deflection yoke, thus maintaining the beams converged over the entire screen. The circuitry to produce the correct wave-

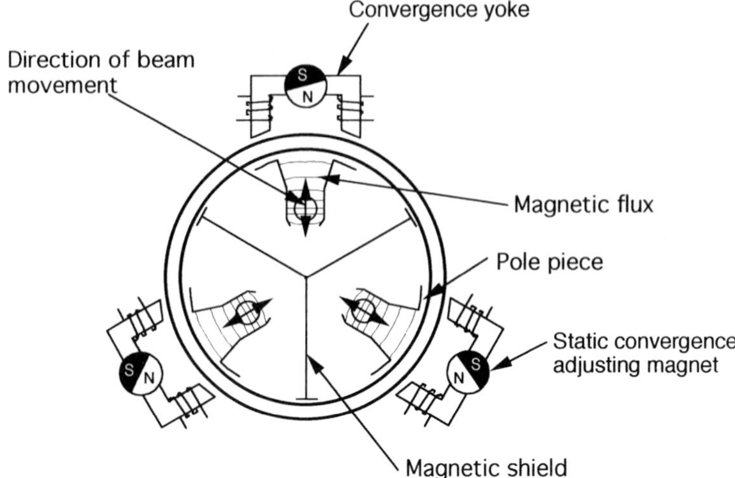

FIGURE 13. Cross section of dynamic convergence yoke for radial correction of beam positions in delta gun system.

forms for the convergence yoke, however, is complex and costly and requires a complicated adjustment procedure.

To produce static convergence of the undeflected beams at the center of the screen a small, permanent, rotationally adjustable magnet is also placed in series with the external magnetic structure of the convergence yoke as shown in Fig. 13, allowing radial adjustment of the individual beams. To obtain complete static convergence, however, it may also be necessary to laterally move one of the beams, for example, the blue beam, with respect to the other two beams. For this purpose two adjacent multipolar ring magnets as shown in Fig. 14 are mounted around the tube neck a short distance beyond the electron gun. By varying their angular position relative to each other the combined strength and direction of the magnetic field in the neighborhood of the electron beams can be varied, allowing the blue beam to be moved horizontally in either direction with respect to the red and green beams. This two-ring combination is referred to as the blue lateral magnet.

Even such complicated structures and their associated adjustment procedures do not necessarily guarantee good convergence. Because of this the delta gun arrangement is now rarely used, particularly since the initial drawbacks of the in-line gun arrangement have largely been overcome, as mentioned in Sections A and B above.

Although the delta gun arrangement is usually combined with the dot phosphor screen it can also be combined with the striped phosphor screen.

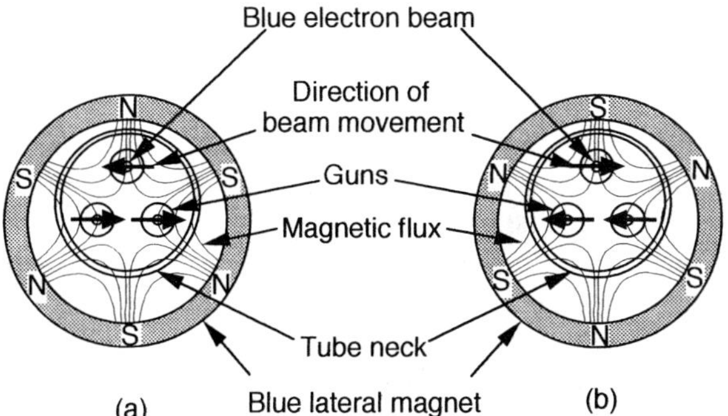

FIGURE 14. Ring magnet for correction of lateral position of blue beam with respect to other beams.

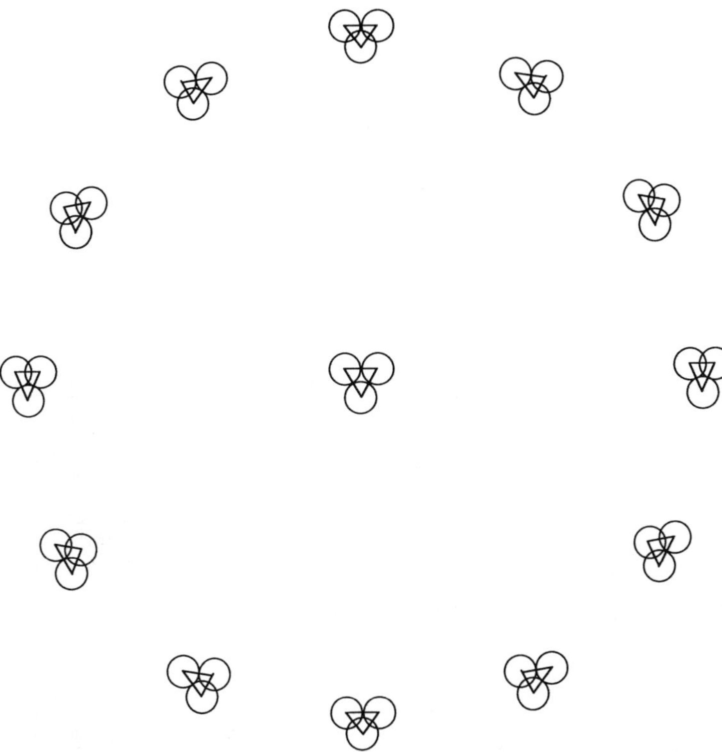

FIGURE 15. Triad crowding in tube with curved mask and screen.

However, such a combination is not used in practice since it offers no technical benefit.

Another problem with the delta gun arrangement is the serious distortion of the beam landing triads as shown in Fig. 15, particularly near the screen edges. This compression of the triads in the radial direction is referred to as crowding. Unlike the in-line gun arrangement, such distortion is very difficult to correct by distorting the shadow mask dot array. This is a further reason why the delta gun arrangement is rarely used.

As a matter of fact, as shown in Fig. 16, this beam triad distortion or crowding and the beam triad inclination already mentioned in Section B are basically caused by the same geometric reason. In Fig. 16 both the white and black spots denote an in-line and delta arrangement of the guns, respectively. At the center of the screen (Fig. 16a) the beam triads projected onto the screen through a mask aperture retain the same position with respect to each other as the guns themselves. However, at the top right corner of the screen (Fig. 16b) the beam triads for both gun arrangements are compressed in the radial direction as shown by the arrows. The distortions of the triads are the inclination and crowding, respectively, of the beam triplets in the case of the in-line and delta arrangements.

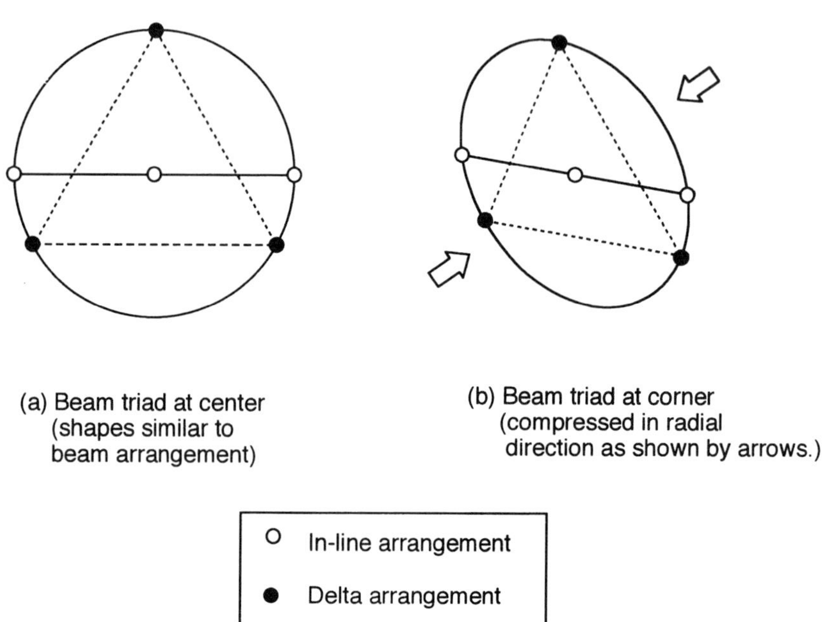

FIGURE 16. Beam triad distortion for in-line and delta beam arrangements.

IV. Geometric Considerations

As discussed in Section II each beam is broken into a set of tiny beamlets as it passes through the shadow mask openings. In this section the triad patterns of the beamlets landing on the phosphor screen as a function of the mask-screen geometry are discussed in detail showing the condition necessary for exciting maximum phosphor area without losing color purity. In all cases a black nonluminescent material is deposited between adjacent color phosphor elements to enhance the contrast as discussed in Section X.D.

A. *In-Line Gun with Striped Screen*

Figure 17 shows a horizontal cross section containing the tube axis of a picture tube between the phosphor screen and deflection points of the beams. The shadow mask and the phosphor screen are assumed to be flat planes. The

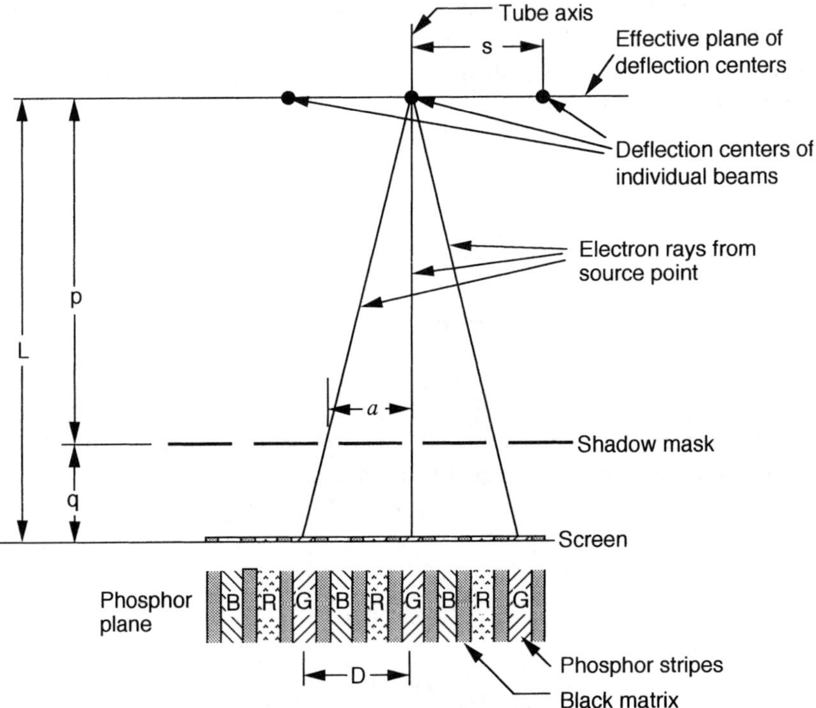

FIGURE 17. Basic geometry of in-line gun with striped screen.

virtual source point from which each of the three electron beams is assumed to originate is defined as the deflection center of each beam. A plane perpendicular to the tube axis, which includes all of these deflection centers, is defined as the plane of deflection centers. As shown, the deflection centers are located either on-axis or at a distance s from the axis. The shadow mask is located at a distance p, and the phosphor screen at a distance L from the plane of the source points, with the distance between the phosphor screen and the shadow mask designated as q. By definition,

$$L = p + q \tag{1}$$

When rays from an electron beam (assumed to follow straight lines) are projected onto the phosphor screen from the source point, the magnification λ of the shadow mask at the screen will be

$$\lambda = \frac{L}{p} = \frac{p+q}{p} \tag{2}$$

and

$$D = \lambda a \tag{3}$$

where a is the slit pitch of the shadow mask, and D is the projected pitch on the phosphor screen. In order for the three beams to fall at the optimum positions on the phosphor screen (i.e., for optimum nesting), equal spacing of the beams is required so that

$$\frac{D/3}{s} = \frac{q}{p} \tag{4}$$

$$\frac{D/3 + s}{s} = \frac{p+q}{p} = \lambda \tag{5}$$

Combining Eq. (3) with Eq. (5) we get

$$D = \frac{3as}{3s - a} \tag{6}$$

As shown in Fig. 18 the beams on the phosphor screen will just touch each other when their widths are equal to E, that is,

$$E = \frac{D}{3}, \quad \text{or}$$
$$E = \frac{as}{3s - a} \tag{7}$$

From Fig. 17:

$$\frac{D}{a} = \frac{L}{p} \tag{8}$$

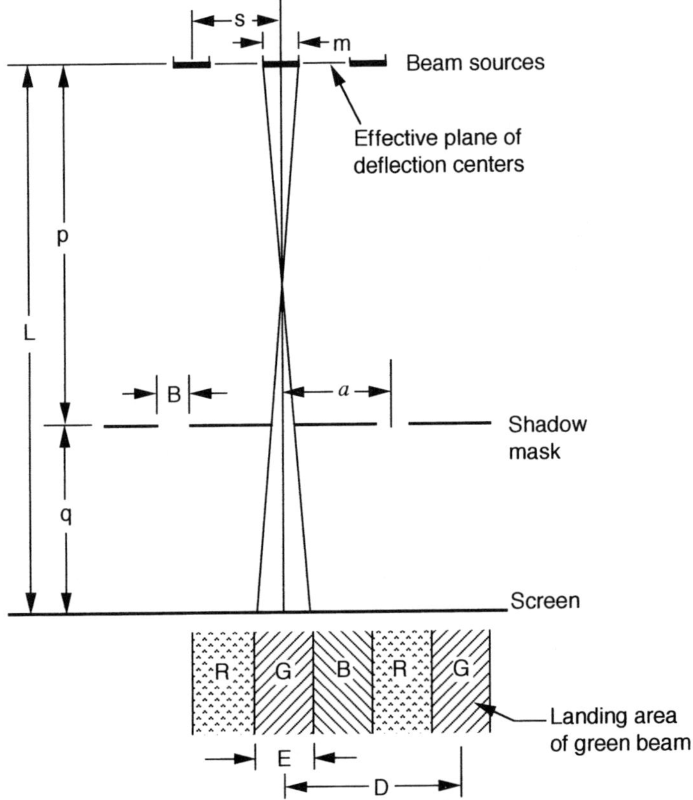

FIGURE 18. Beam landing on phosphor screen (in-line gun with striped screen).

Substituting Eq. (4) with this, we get

$$q = \frac{aL}{3s} \tag{9}$$

Since the source point is not really a point but an area with a real diameter m as shown in Fig. 18, in order for the beams to touch each other without overlapping, the slit width B must be

$$B = \frac{a}{3}\left(1 - \frac{m}{s}\right) \tag{10}$$

The shadow mask in addition to its slit usually has tie bars as shown in Fig. 19. These tie bars are required to keep the shadow mask rigid after it is

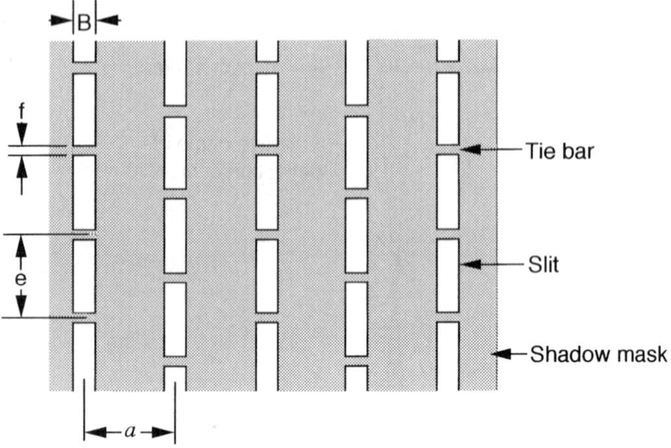

FIGURE 19. Details of slit mask with tie bar.

formed or shaped into a curved surface. When tie bars of pitch e and width f are taken into account, the mask transmission factor T becomes

$$T = \frac{BC}{a} = \frac{C}{3}\left(1 - \frac{m}{s}\right) \tag{11}$$

where

$$C = 1 - \frac{f}{e} \tag{12}$$

The value of T as a function of m/s assuming $C = 0.9$ is shown as the solid "Slit" line in Fig. 20.

If the shadow mask is projected onto the phosphor screen using the beam from the on-axis source point (normally the green beam), we will have a projection on the screen as illustrated in Fig. 21. It is assumed here that between each pair of phosphor stripes a black nonluminescent stripe is provided. This so-called "black matrix" arrangement is discussed further in Section X. With the width of the opening configured by the black matrix (corresponding to the width of the phosphor stripe) defined as M, and the matrix opening factor defined as T_B for each color, then

$$M = E - 2g \tag{13}$$

where g is the guard-band width and

$$T_B = \frac{MC}{D} = \frac{MC}{\lambda a} = \frac{C}{3}\left(1 - \frac{2g(3s - a)}{as}\right) \tag{14}$$

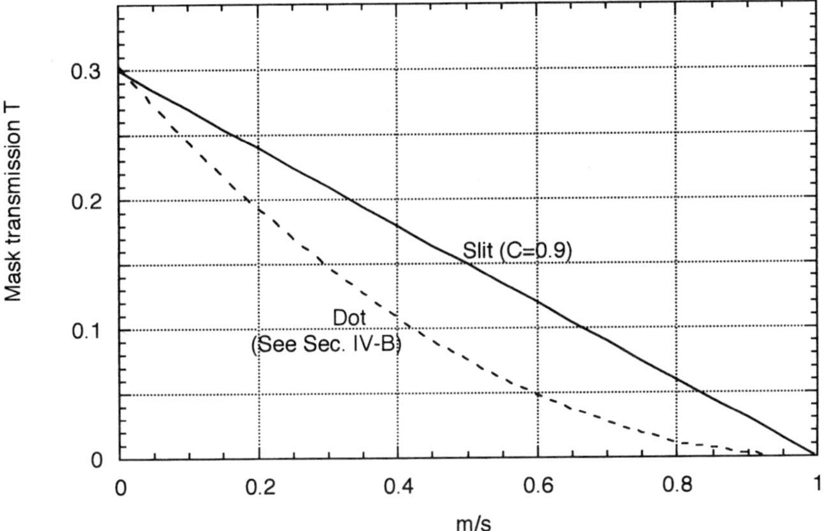

FIGURE 20. Mask transmission T has to decrease in accordance with the increase of m/s (ratio of beam diameter to beam separation at deflection center; see Fig. 18).

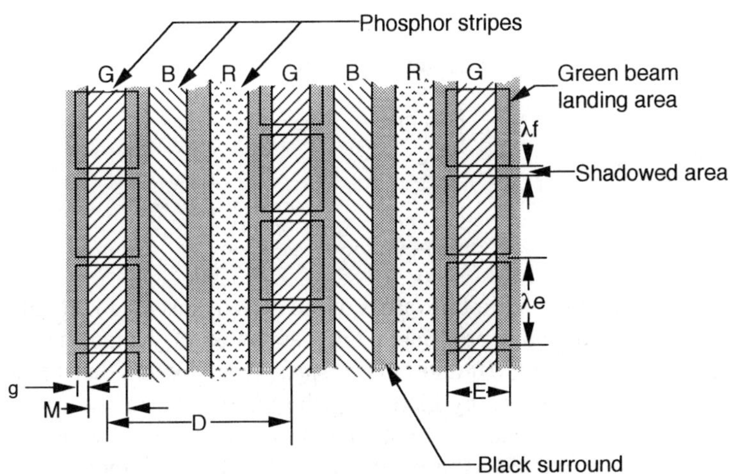

FIGURE 21. Effect of tie bars on the landing areas of green beam for stripe screen.

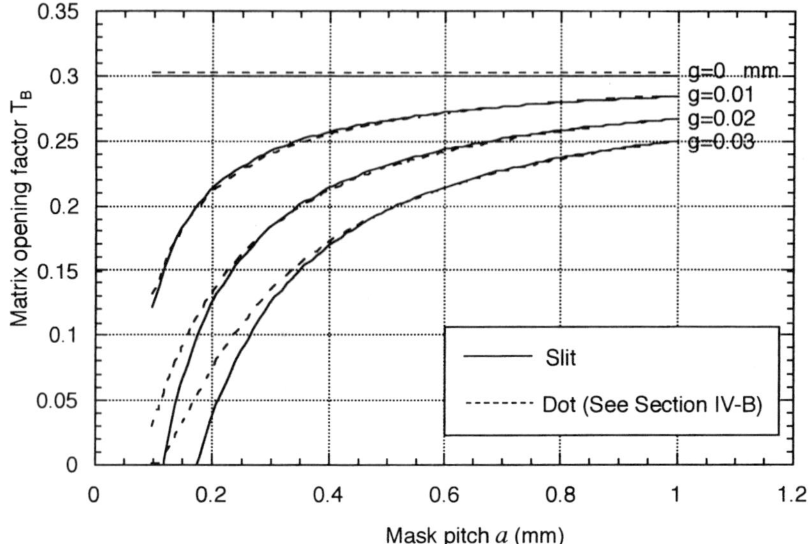

FIGURE 22. Matrix opening factor T_B as a function of mask pitch a for various values of guard band g.

T_B as a function of a is shown in Fig. 22 by the solid lines for various values of g.

Since the beam is not a point source but has an actual size m as already mentioned here, there is a fall-off of beam current at the edges of the landing area as shown in Fig. 23(a). This penumbra effect is shown in greater detail in Fig. 23(b). Here, h is defined as the width of the penumbra and F is the width of the full brightness area. They are expressed as

$$h = m\frac{q}{p} = \frac{am}{3s - a} \tag{15}$$

and

$$F = E - 2h = \frac{a(s - 2m)}{3s - a} \tag{16}$$

Determination of the optimum values of the dimensions M, T_B, B and T can be made by assuming an appropriate pitch a and necessary guard-band width g using Eqs. (13), (14), (10) and (11), respectively. We see that the screen brightness can be increased by increasing the matrix transmission factor T_B but at the same time we should notice that it is also affected by the shadow-mask transmission T due to the penumbra effect if M is larger than F (the actual situation in most cases). Even in the case of a given value of T_B a larger

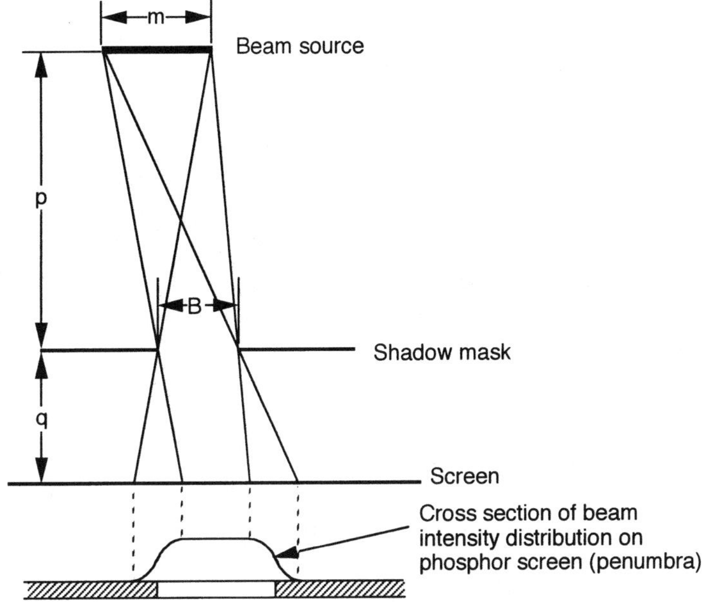

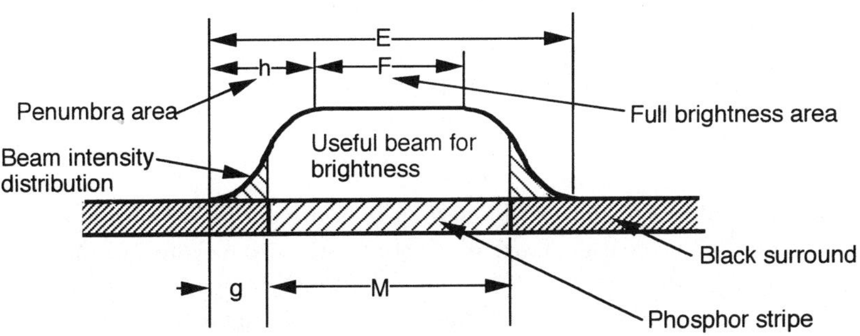

FIGURE 23. Penumbra effect of beam at phosphor screen.

value of m/s gives a smaller T and a larger h gives a smaller F and lower light output accordingly.

B. In-Line Gun with Dot Screen

Figure 24(a) shows a horizontal cross section for this arrangement in which L, p, q and s, and Eqs. (1), (2), and (3) are defined exactly as in Fig. 17. Because the phosphor screen now has a dot structure, the beam triads in a state of optimum nesting are located as shown in Fig. 24(b). If we define the dot pitch of the shadow mask as a, the pitch D of the beams being projected onto the phosphor screen is

$$D = \lambda a = \frac{\sqrt{3}as}{\sqrt{3}s - a} \tag{17}$$

For the landing areas (or triads) of the three beams at the phosphor screen to just touch each other the diameter E of these areas as shown in Fig. 25 must be

$$E = \frac{D}{\sqrt{3}} = \frac{as}{\sqrt{3}s - a} \tag{18}$$

and

$$q = \frac{aL}{\sqrt{3}s} \tag{19}$$

If the diameter m of the source points is considered, the diameter B of the mask holes should be

$$B = \frac{a}{\sqrt{3}}\left(1 - \frac{m}{s}\right) \tag{20}$$

The available mask transmission factor T will then be

$$T = \frac{\pi}{2\sqrt{3}}\left(\frac{B}{a}\right)^2 = \frac{\pi}{6\sqrt{3}}\left(1 - \frac{m}{s}\right)^2 \tag{21}$$

T as a function of m/s is shown as a dashed line in Fig. 20. We should notice that compared with linear decrease of T for the slit case, because of the two-dimensional effect in the case of the dot structure, T decreases much faster.

Figure 26 shows the shadow mask holes projected onto the phosphor screen from the on-axis source point. If we define the window diameter of the black surround, that is, the diameter of the phosphor dot, as M the matrix opening factor T_B for each color will be

$$\begin{aligned} T_B &= \frac{\pi}{2\sqrt{3}}\left(\frac{M}{D}\right)^2 = \frac{\pi}{2\sqrt{3}}\left(\frac{M}{\lambda a}\right)^2 \\ &= \frac{\pi}{2\sqrt{3}}\left(\frac{1}{\sqrt{3}} - \frac{2g(\sqrt{3}s - a)}{\sqrt{3}as}\right)^2 \end{aligned} \tag{22}$$

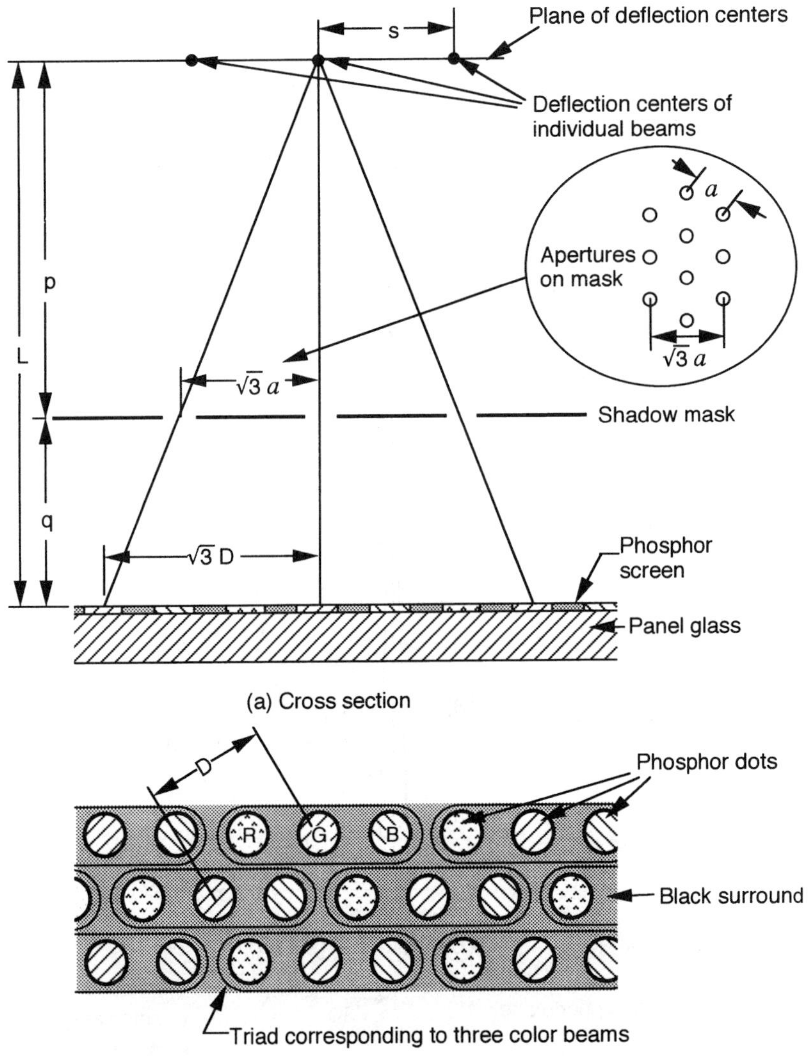

FIGURE 24. Basic geometry of in-line gun with dot screen.

where T_B is a function of a shown by the dashed line in Fig. 22. In the case of a slit structure, although the equations are completely different from the case of a dot structure it is interesting that both curves are fairly close to each other except for small values of a, with the dot structure providing a slightly greater value of T_B.

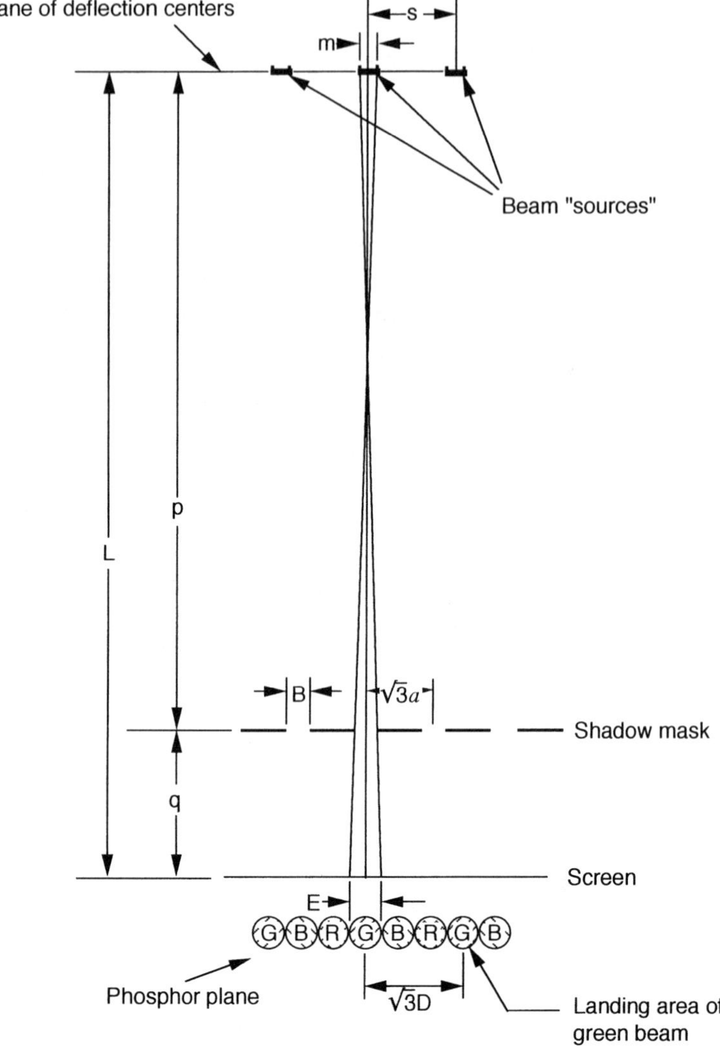

FIGURE 25. Beam landing on phosphor screen (in-line gun with dot screen).

C. Delta Gun with Dot Screen

In the case of the delta-gun system, the electron gun of the blue beam is ordinarily set at the top of the triangular configuration of Fig. 11. Figure 27 shows a cross section of a vertical plane containing both the central axis of the

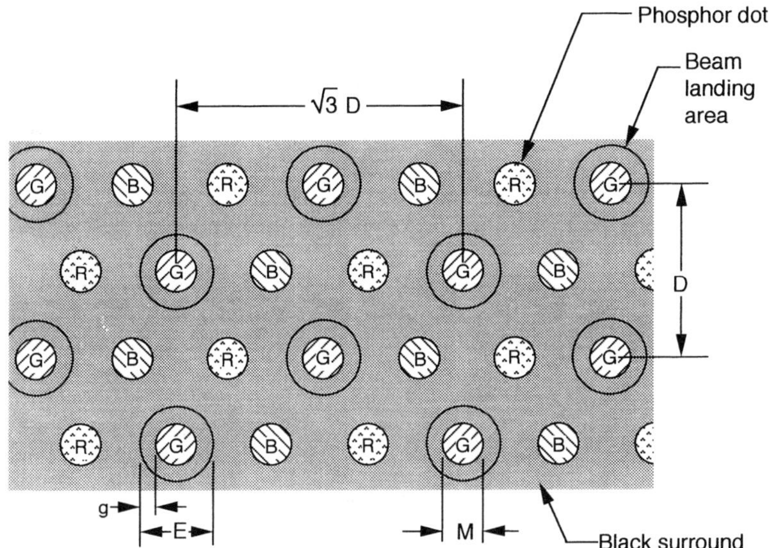

FIGURE 26. Landing area of green beam for dot screen.

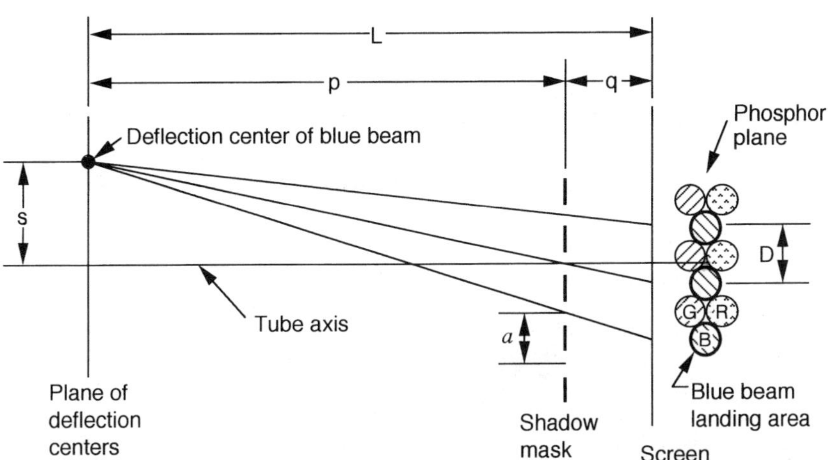

FIGURE 27. Basic geometry of delta gun with dot screen.

blue gun and the axis; L, p and q are defined in the same way as in the previous case, while s is the distance from the tube axis to each source point. Use is also made of the relations indicated by Eqs. (1), (2), and (3).

With the delta-gun arrangement the color triad formed on the phosphor screen has a triangular shape, which is correlated with the dot pitch a of the

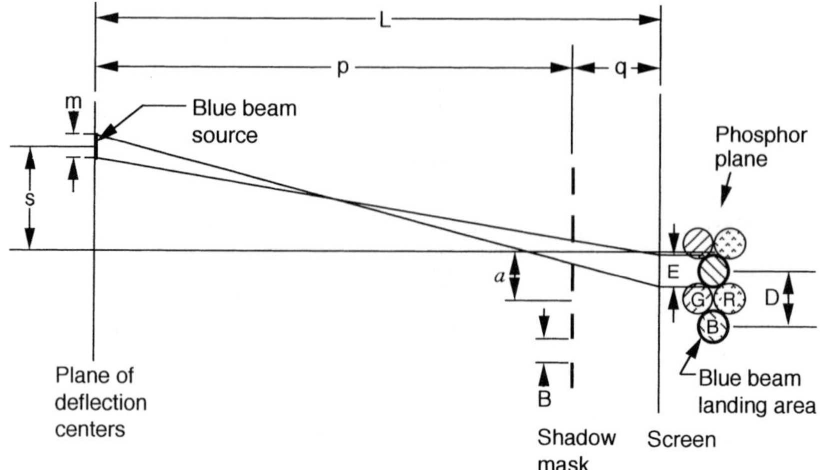

FIGURE 28. Beam landing on phosphor screen (delta gun with dot screen).

shadow mask as follows:

$$q = \frac{aL}{3s} \tag{23}$$

Although this equation is exactly the same as Eq. (9), its physical representation is slightly different. While in the case of Eq. (9) the distance between the source points of adjacent beams was defined as s, in the case of Eq. (23) it is $\sqrt{3}\,s$. Figure 28 shows a cross-sectional view in which the actual width of the source point m is taken into consideration. The diameter of the mask holes and the mask transmission factor will then be (Morrell *et al.*, 1974)

$$B = \frac{a}{\sqrt{3}}\left(1 - \frac{m}{\sqrt{3}s}\right) \tag{24}$$

$$T = \frac{\pi}{2\sqrt{3}}\left(\frac{B}{a}\right)^2 = \frac{\pi}{6\sqrt{3}}\left(1 - \frac{m}{\sqrt{3}s}\right)^2 \tag{25}$$

While the deflection centers of the beams of the delta-gun system are not located on the axis, assuming the shadow mask to be projected onto the phosphor screen from the point where the axis of the picture tube intersects the plane of deflection centers, the projection will appear as illustrated in Fig. 29. The matrix opening factor T_B for each color will be exactly the same as that obtained using Eq. (22).

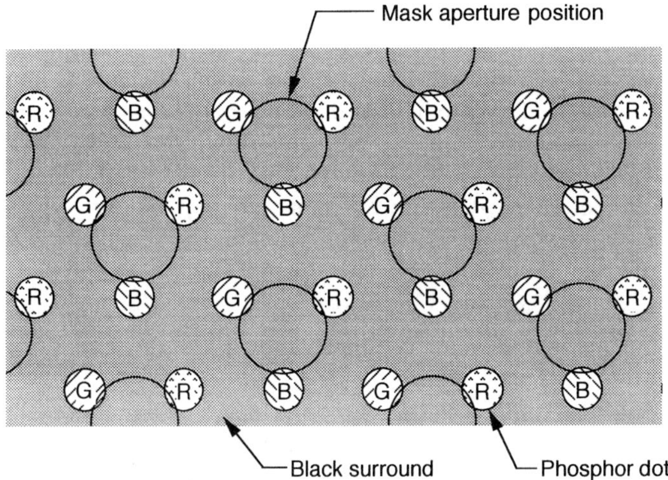

FIGURE 29. Relation between mask aperture and phosphor dots.

V. SHADOW MASKS

A. *Materials*

Since apertures in the shadow mask are formed by etching, as mentioned in what follows, the thickness of the mask material must be very uniform in order to obtain a uniform aperture size over the entire screen area. At the same time each aperture must have a well-defined edge. For this purpose a low carbon content steel is used. Because the initially flat mask material is usually pressed into a spherical shape after etching, another kind of nonuniformity may arise in the apertures known as stretcher strain pattern. This phenomenon can be avoided by applying, prior to press forming, a process referred to as skin pass, in which the material is successively passed through rollers in three different directions resulting in a slight strain to the material surface. However, to avoid the stretcher strain pattern Al-killed steel was developed to eliminate the need for the skin pass process. This material is now commonly used in the industry.

With about 80% of the beam current intercepted by the shadow mask, the thermal expansion caused by the temperature rise results in beam landing errors on the phosphor screen because of the fairly high thermal expansion coefficient (e.g., 1.2×10^{-5}) of the steel. This phenomenon is referred to as doming because the expansion results in an increase in curvature of part or all of the mask. Although landing errors may be reduced by using mask supports designed with compensating thermal shifts, the most effective measure is to reduce the thermal expansion coefficient itself. For this purpose invar material is sometimes used for the mask, because its thermal expansion coefficient (e.g., 1×10^{-6}) is 1/10 of that of steel. Unfortunately, the landing error or color shift is reduced only to one-third or one-fourth that of steel masks since the thermal conductivity of invar is only one-fifth that of steel, thus causing an increased temperature rise. Because invar is expensive and difficult to press into the correct shape it is limited in use to large screen tubes or high-resolution tubes where it is essential (Hamano *et al.*, 1984; Okada and Ikegaki, 1983).

B. Fabrication

To create the apertures a flat sheet of the mask material is coated with a photoresist layer, which is first exposed to a suitable optical pattern and then developed. The bare area is then etched to form apertures or slits through it. The minimum practical aperture size or slit that can be produced is about the thickness of the mask material. For self-supporting masks (whose edges are attached to a surrounding rigid frame) the thinnest material now used is about 0.1 mm, making the diameter of the smallest circular apertures or width of the smallest slits also about 0.1 mm. In the special case of masks held under tension (see Section IX) thinner material can be used, allowing smaller apertures or slits to be produced.

A similar etching process is normally performed from both sides of the material in order to create a knife-edge within the apertures as shown in

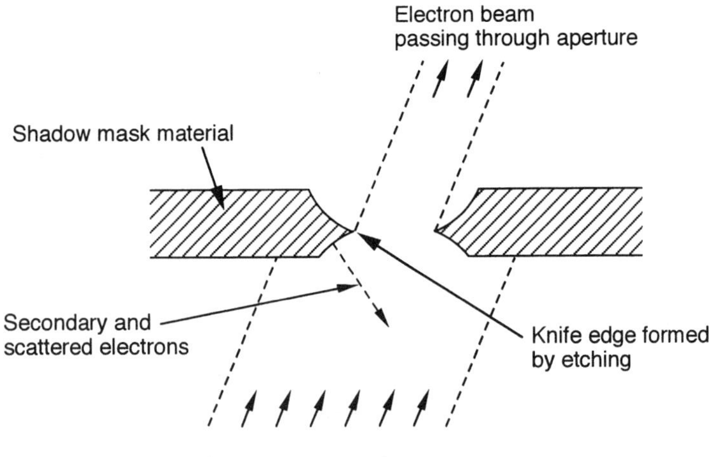

FIGURE 30. Cross section of shadow-mask aperture.

Fig. 30. This minimizes the wall area of the aperture from which scattered or reflected secondary electrons can reach the phosphor screen. The apertured masks are then press formed to a specified curvature. At the same time a shoulder is created at the edge. In the case of invar material press forming requires that the mask and dies be heated to about 200°C.

After press forming, the surface of the mask is exposed to a decomposed LP (liquid propane) gas at about 600°C. The resulting black Fe_3O_4 film produced on the surfaces increases the radiation coefficient of the mask and reduces its temperature rise in tube operation. The film also acts as a protective layer, preventing rust (Fe_2O_3) in the tube-manufacturing process.

A similar process is also used to blacken the frame to which the mask will be attached. This also reduces residual strains in the frame material and is very important to stabilize and minimize subsequent mechanical deformation of the assembled mask and frame unit in the processes that follow.

1. *Positioning of Mask*

In the next step the mask is welded to the frame. In this case support springs are first welded to the outside of the frame. Holes provided in the support springs are then engaged with the three support pins or studs provided on a welding jig fixture (Fig. 31). The location of the pins or studs of the jig must precisely simulate the position of the pins on the panel glass. After the mask is inserted into the jig and positioned at a specified distance from the "reference plane" determined by the studs it is welded to the frame at the point shown in Fig. 31. Of course in the manufacturing process of the panel glass the pins

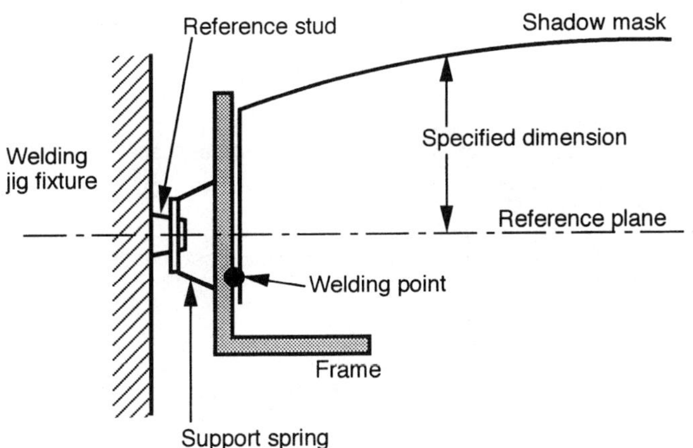

FIGURE 31. Shadow mask assembly using pin plane as reference and assembled with mask welded inside frame (MIFA).

or studs are also precisely positioned in the "reference plane," which is determined by the inside surface curvature of the glass panel. Thus the q spacing is secured through the "reference plane" when the mask is inserted into the panel glass.

In another method called direct q setting, the mask is first welded to the frame. A q set spacer is then placed between the mask and the inside surface of the panel directly to determine the spacing between the mask and the panel. This q spacing (discussed in Section IV) is shown in Fig. 32. With this spacing maintained the support springs are then welded to the frame. Because direct welding is difficult in this arrangement base metal elements are welded to the outside of the frame beforehand and the support springs are then welded to the base metal. In this method, once a panel is mated with a mask, this combination must be maintained throughout all of the following processes. Although in the direct q setting method any variation of the pin position of the panel is automatically compensated theoretically, due to many other errors introduced in this system the total q setting accuracy is almost the same as in a properly designed normal q setting ("reference plane") method.

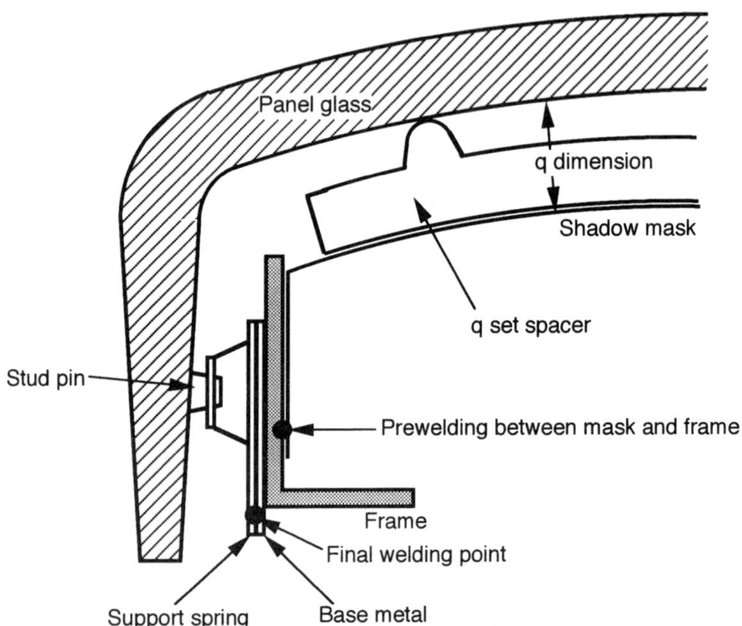

FIGURE 32. Shadow mask assembled by means of q set spacer.

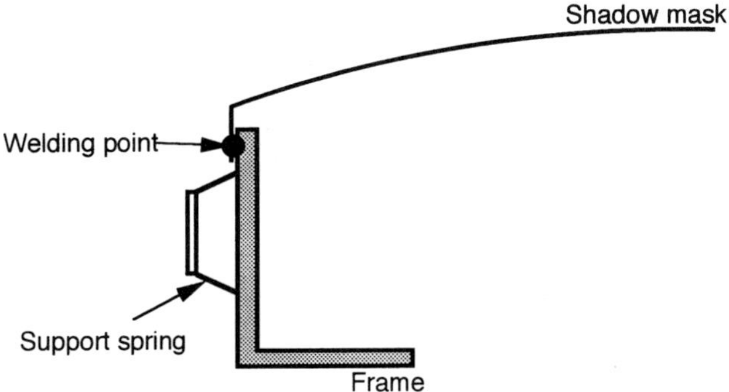

FIGURE 33. Mask assembled outside of frame (MOFA).

2. *Mask and Frame Assembling*

It should be noted that there are two ways of coupling the mask to the frame. The mask skirt may be inserted inside the frame as shown in Figs. 31 and 32 or it may extend around the outside of the frame as shown in Fig. 33. In most cases the former system is currently used because it enables the frame height to be greater, giving it greater strength. Also the protrusion of the edge of the frame above the mask protects the mask from deformation during handling.

C. Pin Systems for Support of Shadow-Mask Structure

The different arrangements of the stud pins and their number are shown in Fig. 34. The 3-pin system shown in Fig. 34a and 34b is the most simple and basic arrangement. This is employed in smaller tube sizes. The off-axis system of Fig. 34b is better in weight distribution of the mask assembly but the center of thermal expansion is shifted from the tube center to some point near A in the figure. In the case of a striped screen this shift does not affect the beam landing but in case of a dot screen it may disturb the beam landing in the vertical direction.

The 4-pin system shown in Fig. 34c is preferable for larger tube sizes because the mask weight is distributed equally among the pins and the thermal expansion of the mask is centered on the tube axis. As in the case of the q setting system using the "reference plane," which is defined by only 3 pins, the mask is fixed to the glass panel using 3 reference pins first. The 4th spring is then welded to the frame engaging it on the 4th pin. Here, welding can be done via a base metal element as mentioned in the direct q setting method of the previous Section or, in the cae of the direct q setting system, all 4 pins are welded via base metals as already mentioned here. To prevent the

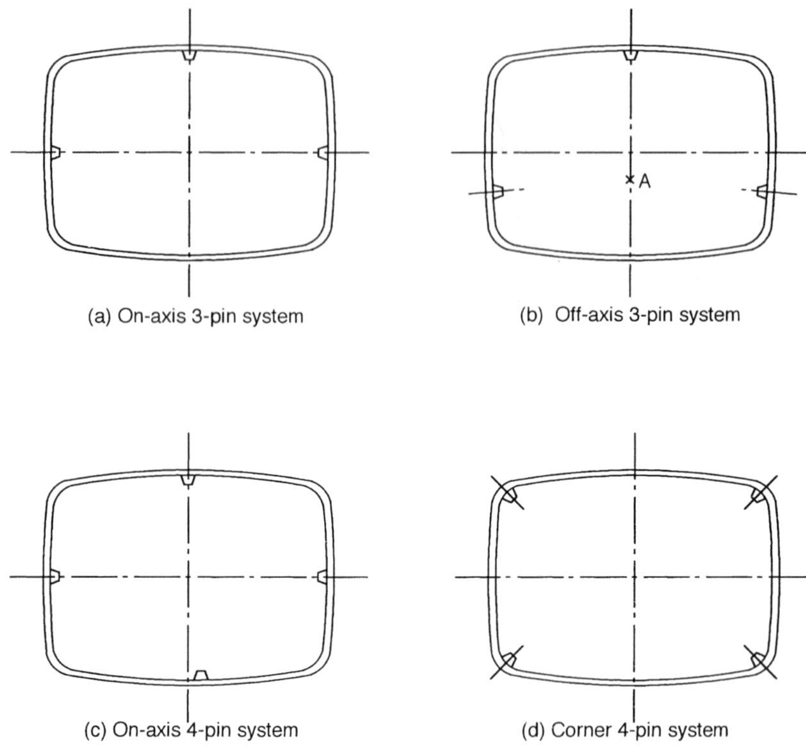

FIGURE 34. Positioning of pin systems.

mask from being inserted in the opposite direction, one of the 4 pins is intentionally positioned slightly off axis.

Another 4-pin system is the corner pin arrangement shown in Fig. 34d. This prevents the twist-like deformation of the mask that can occur in the arrangement of Fig. 34c and is preferred for large size tubes. As there is insufficient space for a long support spring extending along the mask edge shorter springs are used as shown in Fig. 35 (Bauder and Ragland, 1990; Donofrio, 1995).

D. *Compensation for the Thermal Expansion of the Mask*

As previously mentioned, the temperature rise of the mask causes its expansion and results in misregistration of the beam-landing triads on the phosphor screen. In general, when the mask is heated it expands to a dome-like shape. When it is heated over the entire area it is called general doming or simply doming; when it is heated at some local area it is called local doming.

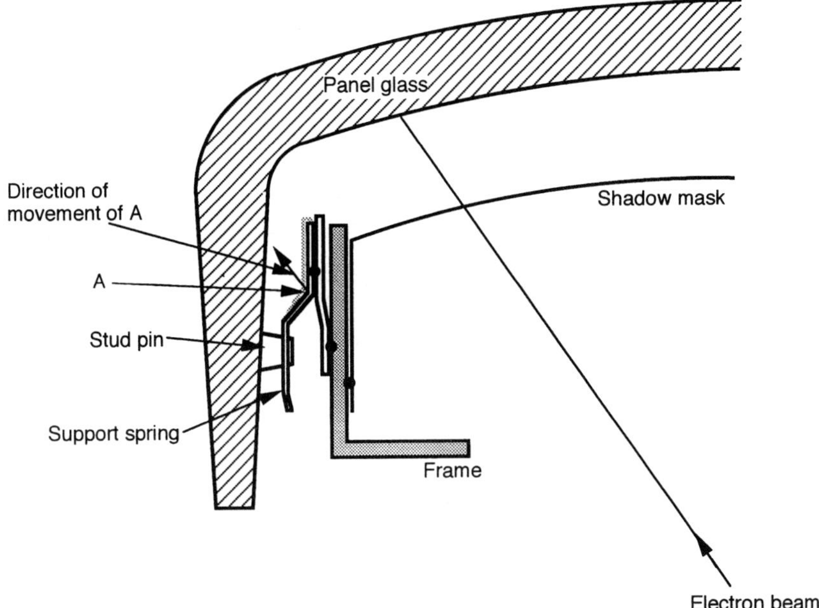

FIGURE 35. Spring system for corner support. When the mask and the frame are heated and expand towards the panel skirt point, A moves in the direction shown by the arrow, that is, nearly parallel with the incident angle of the electron beam at the corner.

To reduce the doming many measures have been proposed and applied. As already mentioned, the mask is blackened to increase its heat loss by radiation. In some cases a Bi coating is applied to the inner surface (electron beam side) of the mask to increase its secondary emission, thus reducing the net power dissipated in the mask by the primary electrons. This coating is referred to as an antidoming coating.

Another method is to blacken the surface of the evaporated aluminum film on the back side of the phosphor screen so that it absorbs more of the heat radiated from the mask. This blackening can be produced by either evaporating the aluminum under poor vacuum conditions or simply spraying graphite on the film.

The general doming can be mechanically compensated by use of a bimetal support spring, an example of which is shown in Fig. 36. Its compensation mechanism is explained in Fig. 37. With the temperature rise of the mask and frame they will expand laterally and at the same time they will be shifted towards the glass panel by the bimetal mechanism. The resultant movement of mask aperture will then be in the direction of the deflected electron beam, ensuring that electrons passing through a mask hole will continue to land on the correct area of the phosphor screen.

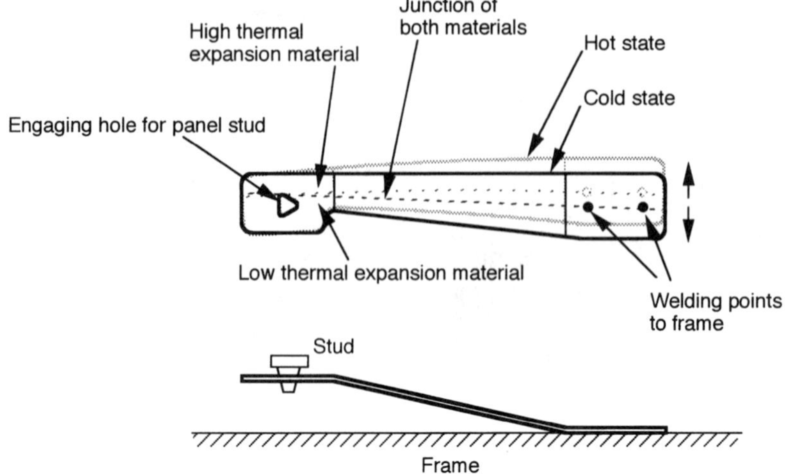

FIGURE 36. Bimetal support spring.

Although this compensation mechanism works well for general doming, it is not effective for local doming. In this case the temperature rise must be minimized by blackening and Bi coating of the mask as well as blackening the aluminum film on the backside of the phosphor screen. In addition, an appropriate mask curvature design may be used to reduce the doming as explained in Section IX.

Another limitation of the bimetallic compensation mechanism is the effect of external temperature on it. Although there is no problem when the tube is operated at normal room temperature, at an abnormal temperature, such as $-20°C$, the bimetallic element causes an undesired change in the landing position of the electron beam.

This problem is avoided by the support spring arrangement shown in Fig. 35, which uses a corner pin system without bimetallic compensation. When the mask expands in this structure an aperture on the mask moves both in the lateral and vertical directions. By correct design of the spring it is possible to make the direction of this movement parallel to the deflected electron beam.

E. Magnetic Shielding

As a result of the earth's magnetic field or other external magnetic field the paths of the electron beams can be affected, causing a shift in raster position or a beam landing misregistration on the phosphor elements as explained

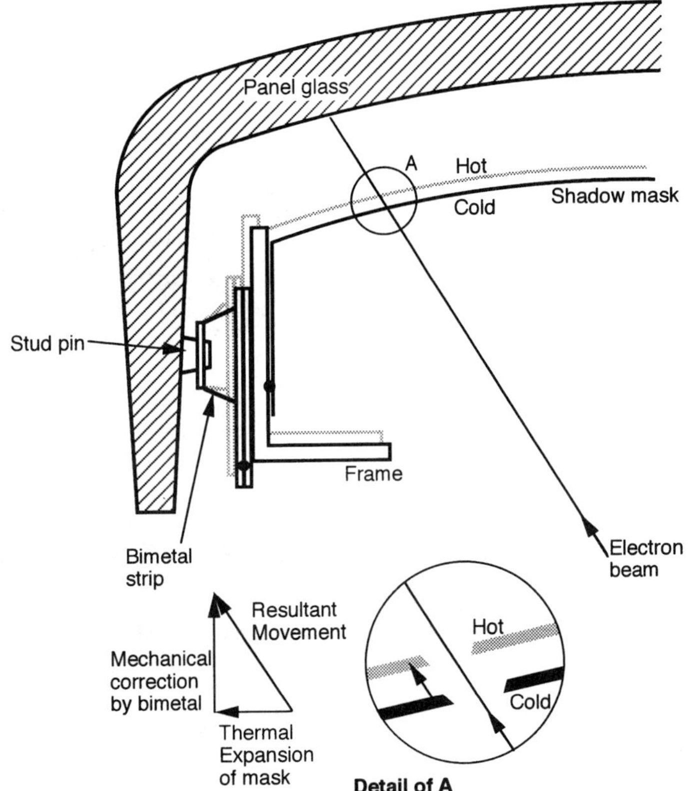

FIGURE 37. Mask frame support system using bimetal structure.

in Section VI.B.2.*a*.ii (Nakanishi, 1990). Although the raster shift can be neglected in most cases, the beam landing errors may cause color shifts. To minimize the effect of these fields appropriate magnetic shielding of the electron beams is thus necessary. Fortunately, the steel used in shadow masks is magnetizable and can be used as a part of the magnetic shielding. To complete the shielding in earlier tubes, an outer steel sheet as shown in Fig. 38 was used; it extended over the funnel portion of the tube. This was magnetically coupled to the periphery of the shadow mask through the skirt of the panel glass.

In recent years an inner shield as shown in Fig. 39 has been commonly used. In this case the shield, inside the funnel portion, is attached to the mask frame directly and can be much smaller than the external shield. This also allows better shielding to be obtained since the shield is more tightly coupled to the mask (Alig and D'Amato, 1989).

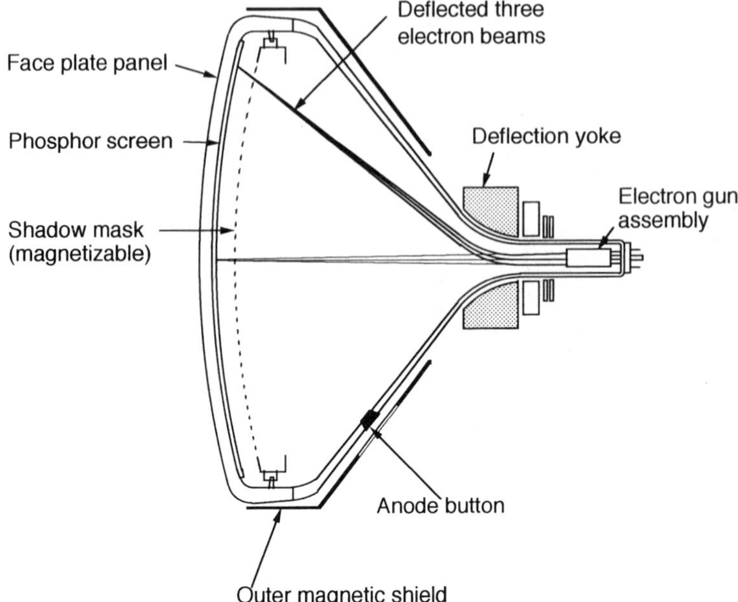

FIGURE 38. Position of outer magnetic shield.

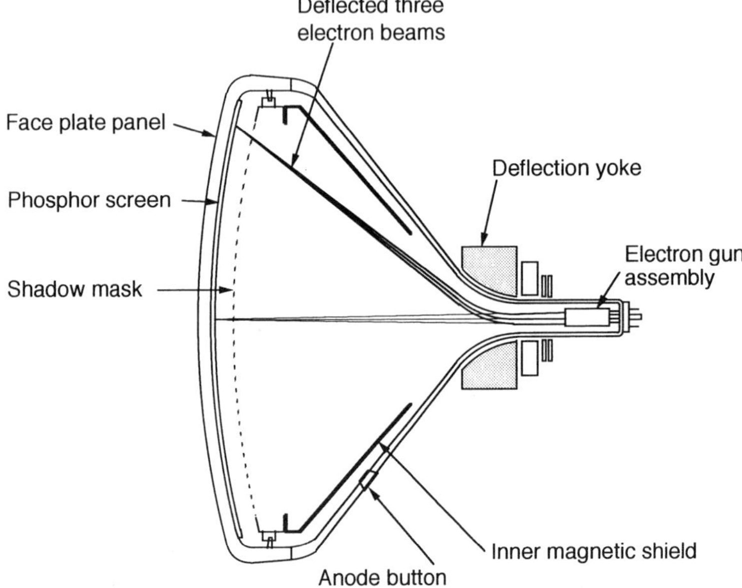

FIGURE 39. Position of inner magnetic shield.

In the case of a dot screen the magnetic shielding should be equally effective both for the horizontal and vertical directions. However, because in the case of a stripe screen only horizontal beam displacements are of concern, stress can be placed on shielding against vertical magnetic field.

In order to reduce the weight of the shield, holes can be cut into it. If not too large the holes will not disturb the magnetic flux paths. These holes are also sometimes useful for allowing better activation as well as radiant heat dissipation from the mask. Normally the shields are blackened as are the shadow masks for both better heat transfer and rustproofing. They are usually attached without welding to the mask frame by small clips inserted in holes provided on both the frame and shield, thus avoiding extraneous forces from being applied to the mask.

To be effective the shield must be properly magnetized in the presence of the external magnetic field. In this process a gradually decaying alternating magnetic field is created by means of a transient ac current passed through a degaussing coil external to the tube and shield. As a result of this procedure the shield retains a magnetization that cancels the external field, thus minimizing the field inside the tube. [Although this process is called "degaussing," it should not be confused with the same term sometimes used for the process of demagnetizing or nullifying the magnetization of a material (Ikeda *et al.*, 1995).]

VI. Phosphor Materials and Screen Fabrication

A. *Phosphor Materials*

1. *Spectral Response of Selected Materials*

Color television images are reproduced by additive mixing of light of the three primary colors: blue, green, and red. The primary colors used by the National Television System Commitee (NTSC) system are defined by the three points indicated by the open circles in the International Commission on Ellumination (CIE) chromaticity diagram shown in Fig. 40 where the wavelengths in nm of pure colors are indicated along the outer horseshoe edge (Morrell *et al.*, 1974). However, the primary colors of the phosphors currently used (not pure colors) are located at the positions shown by the darkened circles in the figure since more efficient phosphors are available at these color points. Although the color gamut that can be obtained with these primary colors, represented by the area of the triangle defined by these points, is smaller than that of the NTSC primaries, the human eye is not very sensitive to this difference in the primaries, with the brighter and more vivid images

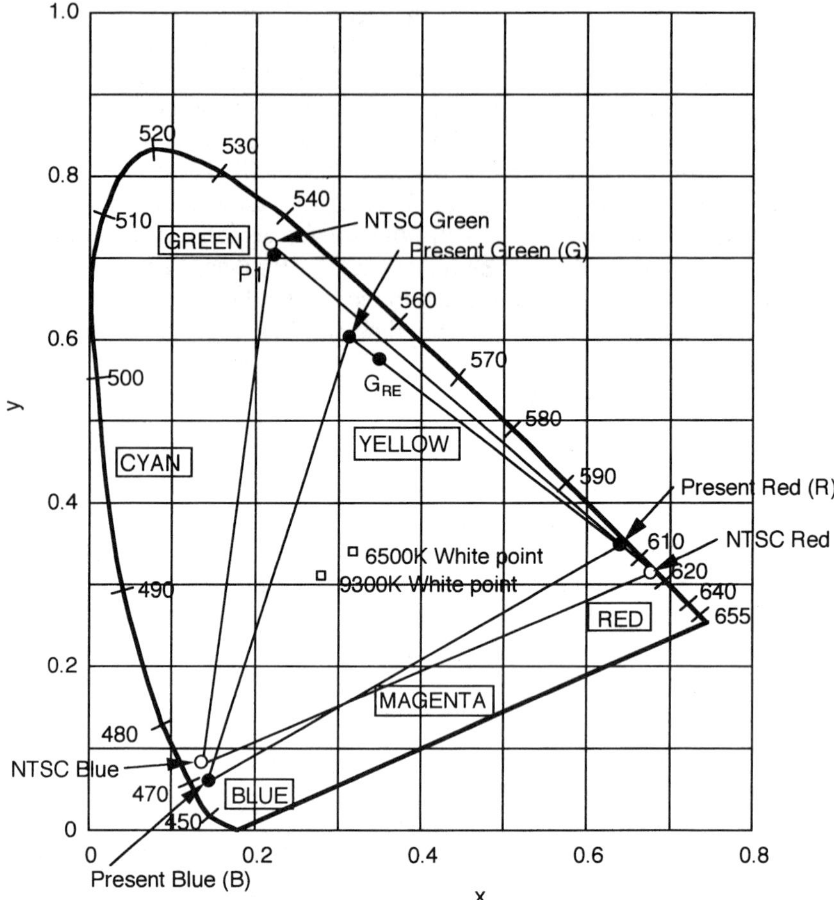

FIGURE 40. The CIE chromaticity diagram showing NTSC color primaries and currently used phosphor points. [Reprinted by permission from Morrell, A. M. *et al.* (1974). *Color Television Picture Tubes*, Academic Press, New York, p. 8.]

obtained by these phosphors being preferred. Typical phosphors used in CRTs are shown in Table I.

a. Blue Phosphors. Because of its saturated blue color and high efficiency, silver-doped zinc sulfide (denoted as ZnS:Ag) powder in its cubic crystal form has been widely used for CRTs since about 1920. Shown as curve B in Fig. 41, the luminescence spectrum of this phosphor has a bell-shaped form that peaked at about 450 nm.

b. Green Phosphors. Zinc sulfide doped with copper and aluminum (ZnS:Cu, Al) in its cubic crystal form has been frequently used for green-

TABLE I

TYPICAL PHOSPHORS USED FOR SHADOW-MASK TUBES [ADAPTED BY PERMISSION FROM HASE, T. *et al.* (1990). PHOSPHOR MATERIALS FOR CATHODE-RAY TUBES, *ADVANCES IN ELECTRONICS AND ELECTRON PHYSICS*, **79**, 271–273.]

Chemical composition	Emission color	Emission peak (nm)	Color point		Decay time[a] (%)	Lumen equivalent[b] (lm/W)	Energy efficiency[c] (%)	Designation		Relative brightness (%)[e]	Points shown in Fig. 40
			x	y				EIA	WTDS		
Normal phosphors for color picture tubes and data display tubes											
ZnS:Ag	Blue	450	0.146	0.061	S	140	21	P22	X	50	(B)
ZnS:Au, Al	Yellowish green	530	0.282	0.620	S	494	17, 23	P22	X	190	
ZnS:Au, Cu, Al	Yellowish green	535	0.306	0.602	S	483	16	P22	X	185	(G)
$Y_2O_2S:Eu$[d]	Red	626	0.640	0.352	M	215	13	P22	X	65	(R)
$YVO_4:Eu$	Red	619	0.664	0.331	M	217	7.1	P22	X	30	
Long persistence phosphors for special data display tubes											
$Zn_2SiO_4:Mn, As$	Green	525	0.265	0.558	L		23	P31	GR	94	
g-$Zn_3(PO_4)_2:Mn$	Red	636	0.655	0.343	L	163	6.7	P27	RE	23	
ZnS:Ag, Ga, Cl	Blue	450	0.147	0.052	L					30	

[a] Decay time is measured from 100% to 1% emission level. S (1 μs–1 ms), M (1–30 ms), L (30 ms–1 s).
[b] Lumen equivalent is the number of lumens contained in 1 W of radiation of the given spectral distribution.
[c] Energy efficiency is defined as the ratio of the energy of the radiation emitted to the energy of the incident electrons.
[d] Measured on pigmented phosphors. See Section X.E for the pigmented phosphors.
[e] Relative brightness at same power input.

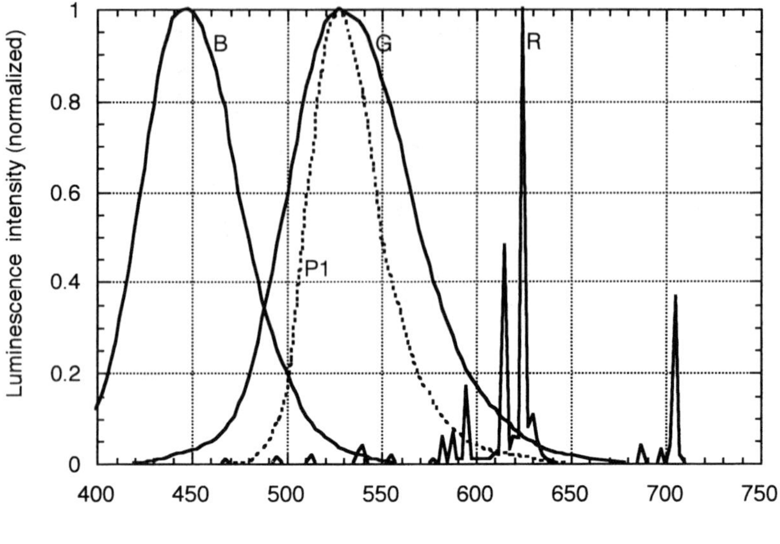

FIGURE 41. Spectral distribution curves of typical phosphors.

emitting phosphors. Its luminescent spectrum, shown as curve G in Fig. 41, is also bell-shaped.

The green color of this material can be shifted to a more yellowish hue either by replacing some of the Zn with Cd or by incorporating Au in addition to Cu as an acceptor. This shift reduces the color gamut of the CRT because of the increase in its red component (Tamatani, 1980). The beam current required to excite the red phosphor must thus be reduced accordingly.

The phosphor producing the most saturated green color is $Zn_2SiO_4:Mn^{2+}$. Its luminescence spectrum is shown as curve P1 in Fig. 41 and its color coordinates plotted in the CIE color diagram (Fig. 40) are designated as P1. The persistence of this phosphor (about 30 ms), however, is too long for conventional TV since it leaves a greenish trace behind the path of a moving image. However, in data display tubes operated at a reduced scan rate where even longer persistence is desired, a green phosphor with the modified composition, $Zn_2SiO_4:Mn^{2+}$, As, is used. This phosphor, known as P39, has the same luminescence color as P1 but a longer persistence of 150 ms (See Section 4 here).

Another large family of green phosphors consists of rare-earth host materials activated with terbium (Tb^{3+}). These phosphors are yellowish, having coordinates as shown in Fig. 40 by point G_{RE}. Although, at best they are only 60–70% as bright as ZnS:Cu,A1 or ZnS:Cu,Au,A1, they exhibit much less loss in efficiency at high brightness (see brightness saturation, Section 3, here) than ZnS-based phosphors. In the hope of reducing the brightness saturation of ZnS:Cu,A1 an attempt was therefore made to mix it with a Tb^{3+}-activated phosphor ($Y_2O_2S:Tb^{3+}$). A mixture of 50% of each phosphor, however, results in a 10% reduction of the green luminance at high beam currents (e.g., 0.5–1.0 mA on a 51 cm diagonal raster area). This drop in luminance is considered too large for such mixed phosphors to be widely used.

c. *Red Phosphor.* In the earliest days of color television $Zn_3(PO_4)_2:Mn^{2+}$ was used for producing the red emission. Later, in the early 1960s "all-sulfide tubes" were developed using the higher brightness $Zn_{0.175}Cd_{0.825}S:Ag$ phosphor as the red primary. As early as 1955, it was recognized that an improved primary should have a narrow emission band around 610 nm and an especially sharp cutoff towards the shorter wavelengths (Bril and Klasens, 1955). Somewhat dramatically, this requirement was actually met first by $YVO_4:Eu^{3+}$ (Levine and Palilla, 1964) and later by $Y_2O_3:Eu^{3+}$ as well as by $Y_2O_2S:Eu^{3+}$ (Royce and Smith, 1968) whose efficiencies are comparable. Because of their higher brightness the latter two phosphors have superseded the former. At the present time almost all shadow-mask tubes use $Y_2O_2S:Eu^{3+}$ as the red primary. This is twice as bright as $Zn_{0.175}Cd_{0.825}S:Ag$ and chemically more stable than $Y_2O_3:Eu^{3+}$. The emission spectrum of $Y_2O_2S:Eu^{3+}$ is shown by the group of peaked curves R in Fig. 41.

2. *Efficiencies*

The energy efficiencies of important commercial color phosphors are given in Table I. (The luminous efficiency of a particular phosphor, given by the product of its energy efficiency and the lumen equivalent of its radiant light (lm/W), is also shown in Table I.)

From theoretical calculations (Klein, 1968; Meyer, 1970), the overall energy efficiency of ZnS:Ag,Cl phosphor (frequently referred to simply as ZnS:Ag) is found to be 0.23. This value agrees well with the observed energy efficiency given in Table I, suggesting that the efficiency of the present ZnS:Ag,Cl phosphor is close to the theoretical limit. On the same basis ZnS:Cu,Al and $Y_2O_2S:Eu^{3+}$ should also have high efficiencies. As the forementioned calculations are rough estimates, however, they do not rule out the possibility that the efficiency of phosphor screens may still be improved by a factor of up to about 20%.

The efficiency of all phosphors is dependent on the accelerating voltage of the electron beam due to the variation in penetration depth of the electron beam. However, over the voltage range in which CRTs are normally operated (e.g., 7–32 kV) the phosphor efficiency is nearly independent of the accelerating voltage of the electron beam. At a very high voltage, for example, 100 kV, most of the primary electrons may completely penetrate the layer of phosphor particles (depending on the thickness of the layer), dissipating almost all their energy in the glass substrate and resulting in an apparent drop in efficiency of the phosphor screen. At low voltages, there is a threshold (sometimes called "the dead voltage") usually in the range of 1–2 kV, below which the efficiency drops rapidly. This is attributed to the electrons penetrating only the surface layer of the phosphor grains where the luminescent efficiency is low because of the large concentration of crystal defects in this region.

The grain size of the phosphors used in normal color CRTs is about 6–8 μm. For high-resolution data display application a slightly smaller grain size of 4–6 μm is normally used. It is well known that a reduction in particle size, for example, below 1 μm, results in a large drop of efficiency even at high voltage. This is attributed to the increased ratio of surface area to volume of such particles, resulting in a greater fraction of beam energy being dissipated in the dead layer. This is clearly observed in the case of ZnS phosphors, in which free electrons have a relatively long diffusion length, allowing them to reach the dead layer before recombining. It is also found that mechanical means used to reduce the particle size can cause a reduction in efficiency by creating cracks and lattice defects in the particles.

3. *Brightness Saturation*

Below room temperature the efficiency variation of most phosphors as a function of temperature is usually small. Although efficiency generally falls with a temperature increase above that of room temperature, under typical operating conditions of shadow-mask tubes the electron-beam power dissipated at the screen is relatively small and its time averaged temperature rises only several degrees. The resulting "thermal quenching" is thus not as serious as in projection tubes where the beam power dissipated per unit area reaches about 100 times that of shadow-mask tubes and the screen temperature can rise as high as 100°C.

At high levels of excitation, the luminescence efficiency tends to decrease at high current densities, aside from any thermal quenching, because all of the available activator centers may become excited. This phenomenon, called "brightness saturation," is especially noticeable in green- and blue-emitting ZnS phosphors but is almost negligible in Y_2O_2S:Eu^{3+} (red) phosphor. Because of the resulting difference in phosphor efficiency the white emission at bright areas may become pinkish.

The different degree of saturation of each color phosphor at high brightness levels also causes a nonuniformity in a white field on the phosphor screen. Because the beam spot is generally smallest at the screen center and becomes somewhat defocused at the screen corners, the local current density at the center may be 4–5 times higher than at the periphery. This results in a much greater brightness saturation at the center, causing a more severe white color shift in this area.

4. *Persistence*

For TV application the persistence of an image should be long enough not to create flicker. At the same time it must not be too long, otherwise it will result in afterimages and image smear of moving objects. This persistence requirement is met by present commercial television phosphors whose decay time (from 100% to 1%) is less than the frame time (16.7 ms for a 60 Hz frame rate) but still long enough to minimize flicker.

In high-resolution data display tubes, because of the increased number of scanning lines and the noninterlaced (or progressive) scanning normally used, the flicker is more pronounced. In this case phosphors with a longer decay time may be useful to reduce the flicker. Some examples of longer persistence phosphors employed for this purpose are shown in Table I. Unfortunately their luminous efficiencies are generally about half that of the conventional phosphors and they produce smear when viewing moving images. In recent years, however, as a result of the progress in drive circuitry, higher line-scanning frequencies and considerably higher frame rates are commonly employed, making the need for such long-persistence phosphors smaller.

5. *Life*

In general, bombarded areas of the CRT phosphor screen tend to become darkened and degraded in efficiency, usually turning brown or black. This phenomenon is called "burning" when it occurs within a short time (e.g., minutes) and "aging" when it occurs over a long period (e.g., many hours). Usually, both cases are called "aging" for simplicity, unless the two need to be distinguished.

In addition to aging of the phosphor, coloration of the glass substrate also occurs when electrons have sufficient energy to penetrate the phosphor into the glass. This phenomenon is called "browning" (See Section *b* here and Section VIII.A.).

a. Aging of Phosphors. Aging has become more and more serious with the increasing operation of CRTs at high current density to obtain high brightness. In some cases, the darkening can be thermally bleached or reversed by annealing the phosphor materials at several hundred degrees Celsius. In practice the phosphor may be bleached by leaving the tube at room temperature for many days. Such observations suggest that the darkening orginates from the generation of many kinds of complex color centers.

It is observed that the aging is roughly correlated with the total amount of charge per unit area deposited on the phosphor. For many commercial phosphors, Pfahnl (1961) and Infante (1985) have given the half-lives of phosphor efficiency in terms of the charge deposited per unit area in C/cm^2. Such data show that the most durable phosphor is P53($Y_3Al_5O_{12}$:Tb) whose half-life occurs at 200 C/cm^2. In a 38 cm diagonal tube operated at a beam current of 0.3 mA this would correspond to an operating time of 5.67×10^4 hours.

The total charge deposited per unit phosphor area, however, is not the only factor controlling the aging. For example, in the case of $InBO_3:Tb^{3+}$ (Yamamoto *et al.*, 1987) and $LaOBr:T^{3+}$ (Raue *et al.*, 1989), a higher current density induces faster aging even with the same total charge deposited. On the other hand, at the low anode voltages used in vacuum fluorescent displays (below 50 V), the half-life of ZnO expressed in C/cm^2 is found to be about 100 times longer (Kazan, 1985) than that of the same material at high voltages (20 kV).

In general, chemically stable phosphor materials that consist of hard crystallites with a high-melting point are resistant to damage by electron irradiation (Leverenz, 1950). Consequently modification of the composition may greatly influence the aging. Partial substitution of Al in $Y_3Al_5O_{12}:Tb^{3+}$ with Ga, for example, improves the luminescence efficiency but shortens the life.

Impurities in phosphor crystals can also have a remarkable effect on the life. For example, a trace of F^- ions in $Y_3Al_5O_{12}:Tb^{3+}$ or $Y_3(Al,Ga)_5O_{12}:Tb^{3+}$ contaminated with the BaF_2 flux used for grain growth induces rapid degradation. On the other hand, Si^{4+} or Yb^{3+} impurities in the same materials increase the life (Yamamoto and Matsukiyo, 1991).

In many cases the aging is accelerated by an increase in phosphor operating temperature. In this sense the screen thickness, packing of phosphor particles, and the amount of binder left in a screen may affect the aging because of their effect on heat dissipation. In this connection the aging depends on various factors that influence particle packing such as particle shape, particle size distribution, and surface coating. Materials that tend to deteriorate are particularly sensitive to these factors. One example is the decreased aging of $InBO_3:Tb^{3+}$ by an improved deposition process (Raue *et al.*, 1989) or by surface coating (Yamamoto *et al.*, 1987), which results in more closely packed screens.

b. Browning of Glass. As mentioned, when exposed to high-energy electrons the faceplate glass turns brown. This is caused by irradiation by X-rays generated in the glass and shadow mask as well as the direct effect of electrons penetrating the glass. As expected, the browning of the glass is increased when the phosphor screen has many pinholes; however, in commercial tubes with a controlled screen weight and deposition process, the loss in image brightness by glass browning is much less (about 10% or less) than that caused by phosphor aging.

The X-ray browning results from color centers formed by exposure to the X-rays (for details, see Hase *et al.*, 1990). It is found, however, that the light absorption resulting from the X-ray browning can be suppressed by the addition of CeO_2 as a dopant to the glass (Levy, 1960). The faceplate glasses of CRTs thus usually contain 0.1 to 0.2 at % Ce.

On the other hand, the electron browning, which is caused by direct electron bombardment of the glass, results in the formation of very fine grains of alkali metals within the glass (Ishiyama *et al.*, 1975), a process that is unaffected by the incorporation of CeO_2.

It should be noted that the glass browning produced by either X-rays or electron bombardment can be thermally or optically bleached more readily than the reversal of phosphor aging (Hase *et al.*, 1990).

B. Screen Deposition

1. *Phosphor Screening Process*

In the fabrication of shadow-mask tubes, deposition of the pattern of phosphor stripes or dots as shown in Figs. 3, 5 and 11 is the most laborious and critical process. Commercially this is accomplished by a process in which ultraviolet (UV) light rays are used to simulate the electron beam trajectories in exposing a photoresist. The black-matrix pattern surrounding the phosphor is also deposited by a photoresist printing technology that is carried out before deposition of the phosphors.

Figure 42 shows the process flow for printing and depositing the black matrix and phosphor. For printing the black matrix a very thin film (1–2 μm) of photoresist is first applied to the inside of a washed face-plate panel. After drying, the film is exposed to a UV light source through the correctly positioned shadow mask associated with the particular tube being fabricated. In this process the UV light source is placed successively at the positions corresponding to each of the beam deflection centers (green, blue, and red). This causes the exposed portions of the photoresist to become hard and insoluble to water. During development the unexposed portions are washed away, leaving the hardened resist pattern behind. A slurry containing a black

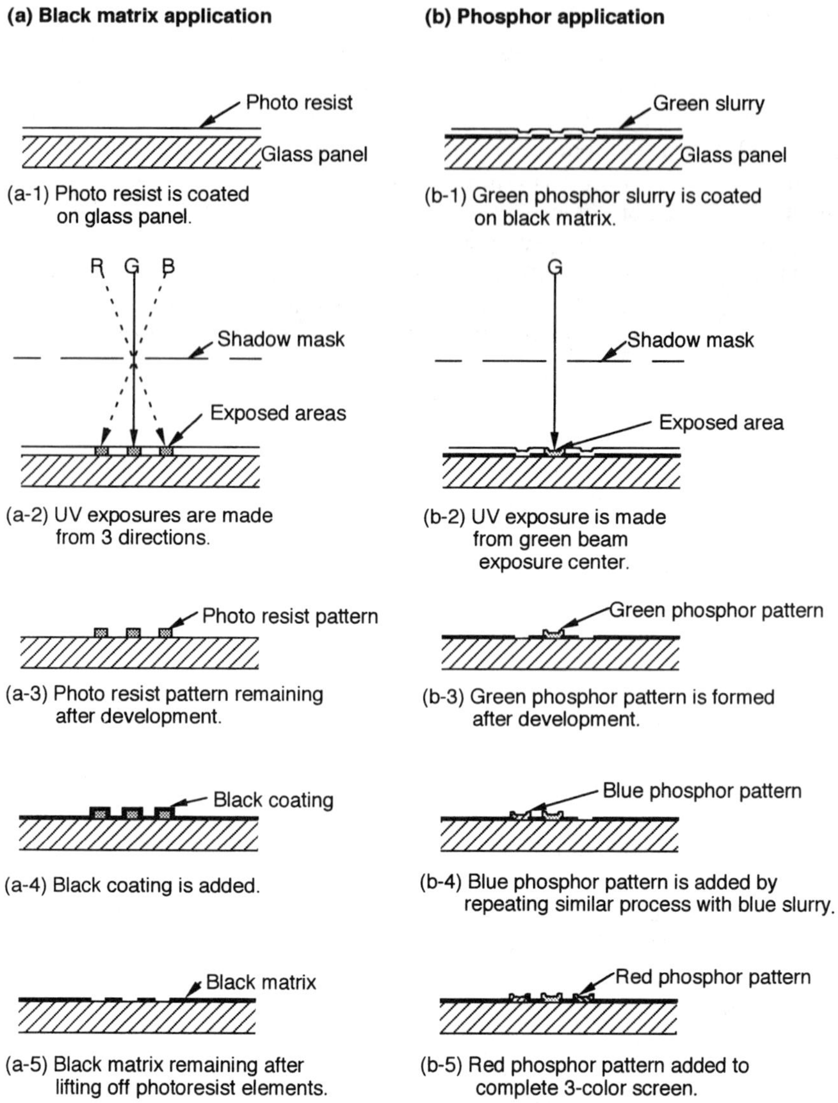

(a) Black matrix application

Photo resist
Glass panel

(a-1) Photo resist is coated on glass panel.

R G B
Shadow mask
Exposed areas

(a-2) UV exposures are made from 3 directions.

Photo resist pattern

(a-3) Photo resist pattern remaining after development.

Black coating

(a-4) Black coating is added.

Black matrix

(a-5) Black matrix remaining after lifting off photoresist elements.

(b) Phosphor application

Green slurry
Glass panel

(b-1) Green phosphor slurry is coated on black matrix.

G
Shadow mask
Exposed area

(b-2) UV exposure is made from green beam exposure center.

Green phosphor pattern

(b-3) Green phosphor pattern is formed after development.

Blue phosphor pattern

(b-4) Blue phosphor pattern is added by repeating similar process with blue slurry.

Red phosphor pattern

(b-5) Red phosphor pattern added to complete 3-color screen.

FIGURE 42. Matrix and phosphor printing process flow.

substance such as graphite is then coated over the entire area. The dried black coating is then immersed in a solution of H_2O_2, which causes the hardened resist pattern to swell and lift off. This carries away the black coating on the photoresist areas, leaving behind a black layer with suitably positioned aperture windows.

Printing of the phosphor patterns is carried out in a somewhat similar manner. Green phosphor material is mixed with a suitable amount of photoresist material to produce a quantity of slurry, which will result in a layer of phosphor 20–30 μm thick. This slurry is deposited on the face-panel that was previously coated with its black-matrix pattern while, at the same time, the panel is constantly rotated and gradually tilted from a horizontal to vertical position, spreading the slurry into a uniform layer. The panel is then exposed through the shadow mask to a UV light source located at the green deflection center. In the subsequent development process a pattern of green phosphor stripes or dots is left behind. The same process is repeated for the blue and red phosphors, with the location of the UV light source shifted to the blue and red deflection centers, respectively.

Finally, a thin film of lacquer or emulsion is deposited on the phosphor screen to produce a smooth surface on which a mirror-like aluminum film about 200–300 nm thick is evaporated. The lacquer or emulsion film is then decomposed during a baking process leaving the shiny aluminum film intact and resting on the high points of the phosphor grains.

2. *Optical Exposure*

As already mentioned, UV light is used in the phosphor printing process to simulate the electron-beam trajectories. This is possible because the electron beams beyond the deflection region travel in a field-free space and move in straight lines like light rays. During exposure the ultraviolet light source is placed, respectively, at the position of the point sources shown in Figs. 17, 24 and 27. However, the use of an UV optical beam in place of the electron beam leads to various kinds of errors as will be described here. To reduce these errors, a special correction lens is placed between the UV light source and the mask as shown in Fig. 43. Before discussing the characteristics of this correction lens, however, the errors resulting from the use of an optical beam without a correction lens are discussed first. (Errors caused by manufacturing variations are not considered here because they cannot be corrected with a lens.)

a. Errors Affecting Beam Landing Characteristic

i. Error Stemming from Linear Extension of the Deflection Field. In an actual tube, as the electron beam passes through the field of the deflection yoke its deflection angle gradually increases. However, as an approximation the resultant trajectory is usually represented by a straight line with a single virtual turning point, such as that shown in Fig. 44. This point is called the effective deflection center. Actually, however, this deflection center is not fixed, shifting toward the phosphor screen as the deflection angle increases.

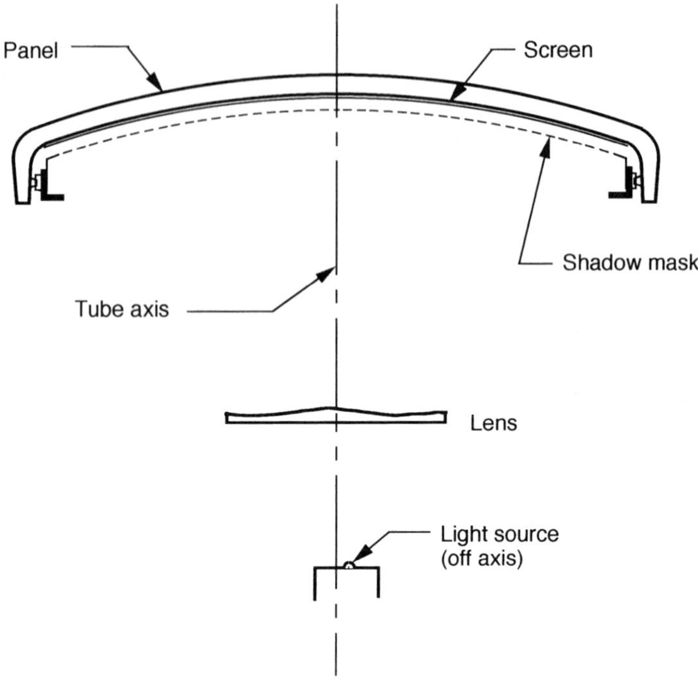

FIGURE 43. Arrangement used for exposure of one color of phosphor screen.

Because of this, placing the UV light source at a fixed "deflection center" would not expose the phosphor screen satisfactorily over the entire area. The plotted points in Fig. 45 show measured values of this movement as a function of deflection angle. Because this movement corresponds to a reduction in the dimension p (mentioned in Section IV.A), it is called Δp, the curve being referred to as the Δp characteristic. Misregistration on the phosphor screen caused by this Δp is called Δr (Janes *et al.*, 1956); Δp can be approximated by the following equation, which is obtained by assuming a uniform deflection yoke field with an effective length of l_y as shown in Fig. 46:

$$\Delta p = \frac{l_y}{2} \tan^2\left(\frac{\theta}{2}\right) \tag{26}$$

where θ is a half deflection angle. This approximation is also shown in Fig. 45 by the dashed line.

In addition to moving axially the deflection center also moves perpendicular to the axial direction, that is, in the s direction. This variation is called Δs and is shown in Fig. 47. With the delta gun arrangement in particular,

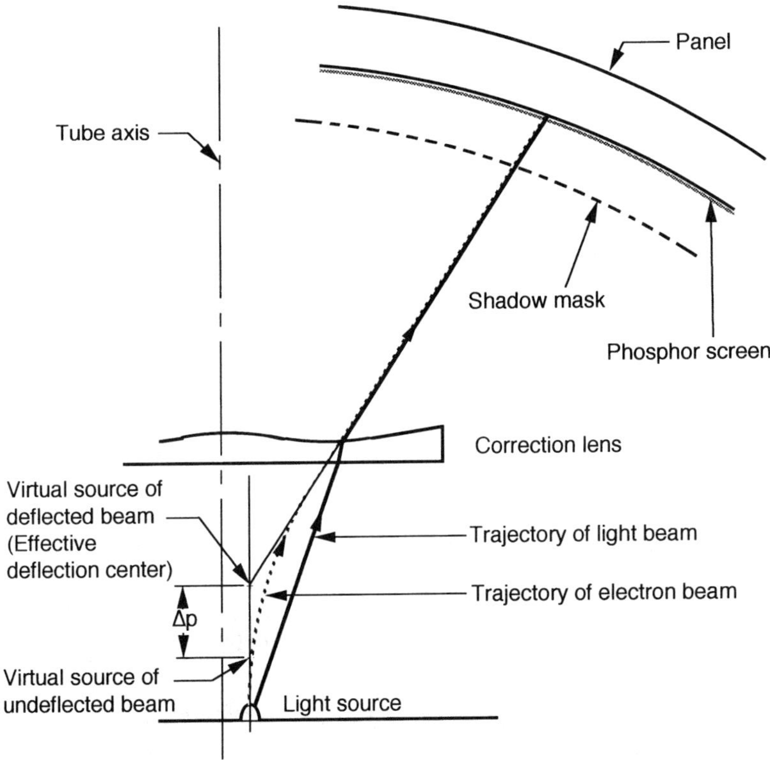

FIGURE 44. Use of correction lens to optically simulate trajectory of electron beam.

dynamic convergence causes an initial deflection of the beam before it reaches the deflection center, resulting in a change as large as 20–30% of the original value of s. Because the dot triangle at the phosphor screen corresponding to the landing of the three electron beams that pass through a mask hole is larger than the triangle corresponding to the phosphor dots, the Δs characteristic is also called the degrouping error (Janes *et al.*, 1956).

The Δs error is also present in the in-line gun system, but it is not as serious as with the delta gun because in the in-line system the magnetic field is intentionally distorted to ensure self-convergence. This results in different Δs characteristics in the horizontal and vertical directions.

ii. Error Stemming from External Magnetic Field. An external magnetic field, such as the earth's field of about 50 μT (0.5 gauss), can cause the beam trajectory to curve instead of being a straight line, also resulting in a beam landing error. Although a magnetic shield is usually provided, as discussed in

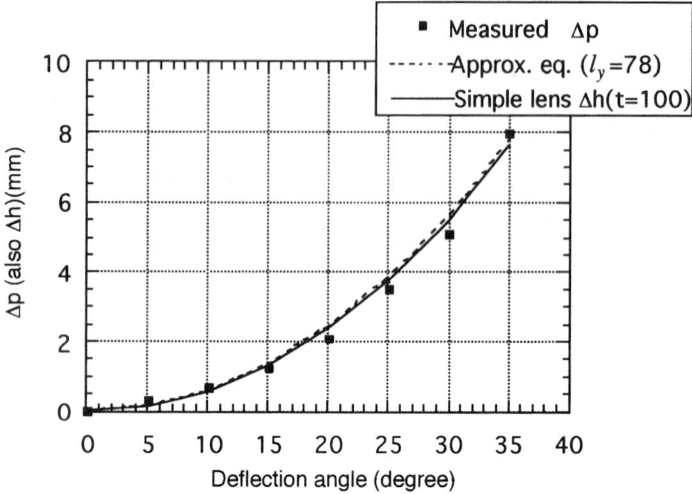

FIGURE 45. Shift in axial position Δp of virtual source of electron beam as a function of deflection angle (see Fig. 49 for explanation of Δh).

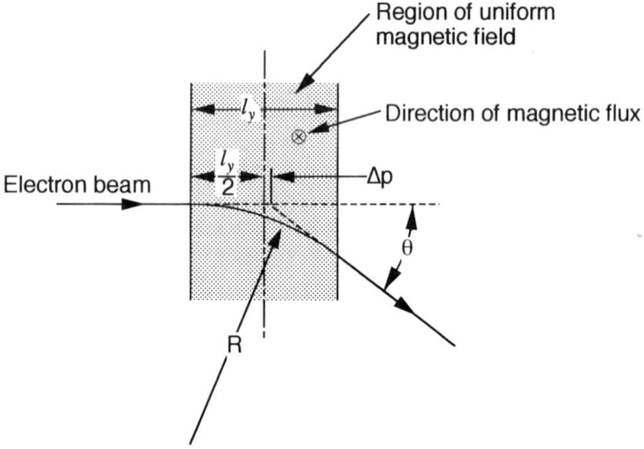

FIGURE 46. Explanation of Δp in the case of a uniform magnetic field.

Section V.E, some residual field still remains. This field can be resolved into horizontal and vertical components. The effect of the horizontal component on the electron beam trajectories depends on the orientation of the TV set. On the other hand, as the vertical component of the earth's magnetic field varies relatively little over a wide range of latitudes from place to place, its effect is independent of the TV set's orientation.

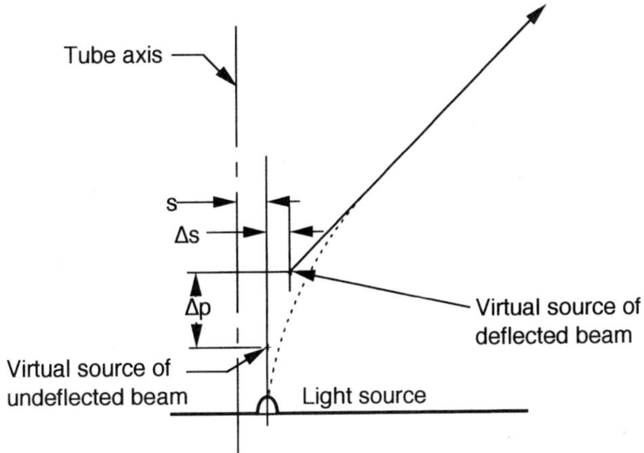

FIGURE 47. Displacement of virtual source Δs.

Figure 48 shows the trajectory of a beam affected by the vertical component of the residual external magnetic field. Assuming that this field is uniform within the tube, the radius of curvature R' of the beam trajectory is given by

$$R' = \frac{1}{B}\sqrt{\frac{mV}{2e}} \tag{27}$$

where V is the anode voltage and e/m is the electron charge to mass ratio. Because the actual residual magnetic field within the tube is not uniform, depending on the shape of the shield, the actual effect on the beam trajectory must be generally determined by actual measurement. To compensate for the effect of the vertical field a correction can be made by slightly shifting the position of the light source during fabrication of the phosphor screen. However, correcting for this kind of error limits the use of the tube to either the northern or the southern hemisphere depending on the correction. For further details on the effects of ambient magnetic field on the beam trajectory, see Morrell *et al.*, 1974.

iii. Error Stemming from Glass Deformation. If the front glass panel with its black matrix and phosphor elements already deposited is heated during tube bakeout and then cooled, some dimensional change generally results in the glass. (For example, a contraction of about 0.06% may result after a heat treatment of 450°C.) This deformation causes beam landing errors.

When the tube is evacuated in the exhaust process, the atmospheric pressure on its outer surface causes further deformation of the front panel to a degree that depends on the panel size, thickness, material and processing

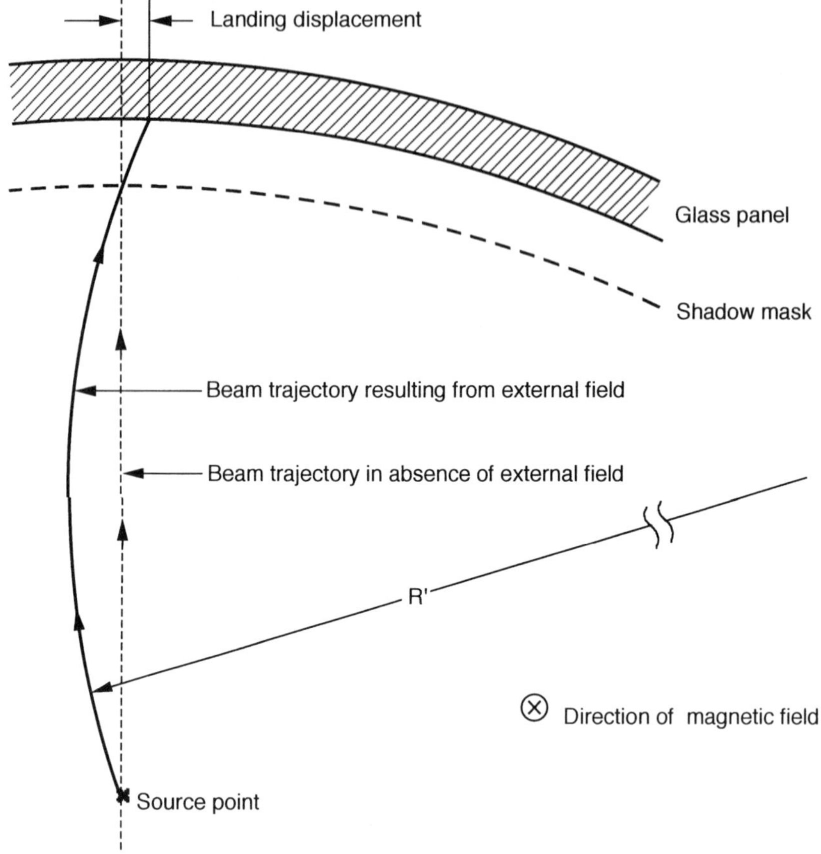

FIGURE 48. Beam trajectory in magnetic field (R′, radius of curvature of beam).

conditions. For example, the panel of a 68 cm diagonal tube may be depressed about 0.6 mm at the center. However, this may be slightly reduced by the steel band tightly wrapped around the panel for implosion-protection (see Section VIII.B).

iv. Error Stemming from Thermal Expansion of Shadow Mask Relative to Phosphor Screen. As already mentioned, during tube operation the scanning electron beams dissipate about 80% of their power in the shadow mask. The resulting thermal expansion is quite complicated and is governed by many factors. Aside from the beam current levels used, these include the material used for the shadow mask (steel or invar) and its thickness, the thickness of the support frame around the mask, the method of assembling the shadow mask and its frame (number of weld points, etc.), and the surface treatment of the shadow mask (in so far as it determines the radiation coefficient of its surface), and whether or not an antidoming coating is added. The result of this expansion depends on the shadow-mask profile, which in turn is related to the glass panel profile.

b. Error Correction by Means of Optical Lens Used in Screen Fabrication. With the exception of the variable component of the errors such as those thermally induced, the effect of the forementioned errors can be compensated during fabrication of the phosphor screen by using a suitably designed optical lens for error correction.

Although the magnitude of each of the errors can be determined either by computation or simulation, these methods are time-consuming and very laborious and may not result in satisfactory accuracy. In practice, the degree of optical correction required is determined by actual measurements of beam landing error using an experimental test tube fabricated for this purpose. Preferably, this tube is fabricated using a correction lens whose characteristics are approximately correct. Information from the measured errors is then used to further improve the lens design. By successive iteration of this process, the lens error can be gradually reduced.

i. Simple Lens. Before explaining the design of complex lenses, the results obtained with extremely simple glass structures are discussed. Figure 49 indicates a simple glass plate with a refractive index n and thickness t. Light emitted from the point light source A at an angle α, when refracted by the glass plate, will appear as if it were emitted from point A' on the axis. In accordance with Snell's Law, the distance h between A and A' will be

$$h = t\left(1 - \sqrt{\frac{1 - \sin^2\alpha}{n^2 - \sin^2\alpha}}\right) \tag{28}$$

If we define $h = h_0$ when $\alpha = 0$, we obtain

$$h_0 = t\left(1 - \frac{1}{n}\right) \tag{29}$$

The virtual shift Δh is then

$$\begin{aligned}\Delta h &= h - h_0 \\ &= t\left(\frac{1}{n} - \sqrt{\frac{1 - \sin^2\alpha}{n^2 - \sin^2\alpha}}\right)\end{aligned} \tag{30}$$

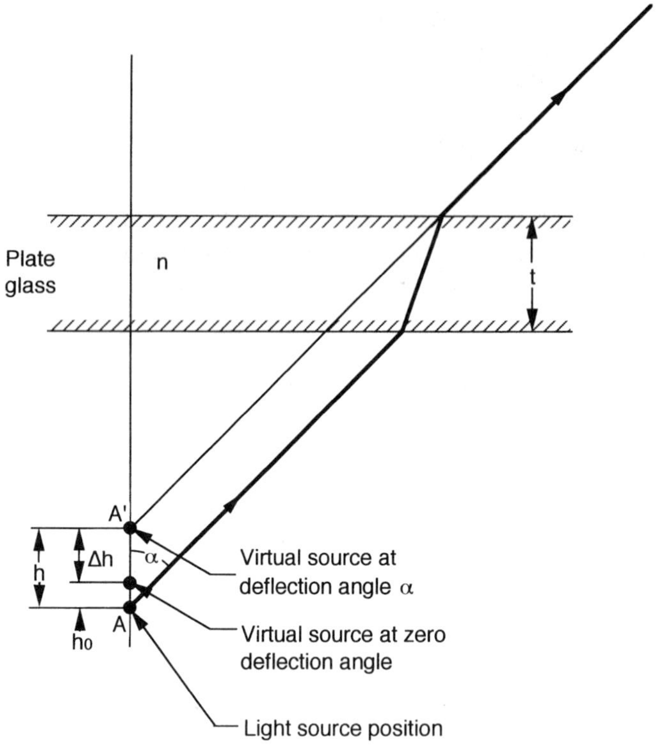

FIGURE 49. Axial shift Δh of virtual source point produced by flat glass plate.

For example, employing BK7 type glass (often used for optical purposes) with an index of refraction n of 1.530 at a UV wavelength of 365 μm, and a thickness t of 100 mm, the value of Δh obtained from Eq. (30) as a function of α is shown by the solid curve of Fig. 45. As indicated, this coincides with the Δp characteristic mentioned earlier, making such a thick plate glass by itself useful for producing the Δp correction. By adjusting the glass thickness t, a fairly good match can be obtained to the Δp characteristic as a function of angle.

If such a thick glass plate is inclined or tilted with respect to the tube axis, some shift in the direction perpendicular to Δp results corresponding to the Δs shift. Although detailed calculations are omitted, Fig. 50 shows an example of what occurs as a function of azimuthal angle θ. The Δs shift caused by lens inclination will be largest for the inclination directions ($\theta = 0°$ and 180°) and smallest for the perpendicular directions ($\theta = 90°$ and 270°). Needless to say, there is a limit to the correction accuracy. However, such thick plate lenses are valid for rough corrections. In addition, it can be used as

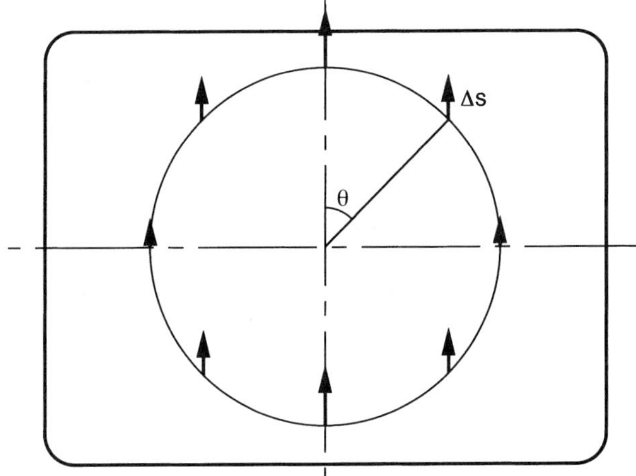

FIGURE 50. Change in radial position Δs of virtual source as a function of azimuth angle θ.

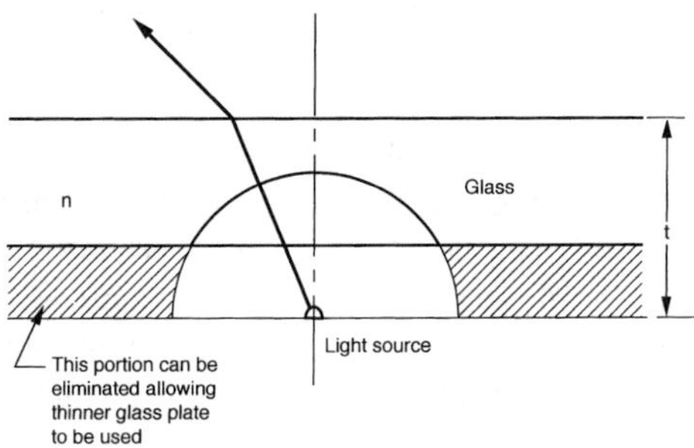

FIGURE 51. Axial shift correction equivalent to correction of Fig. 49 produced by concave spherical surface.

the primary approximate lens when designing completely new lenses. (Because it has a simple structure, without inherent error further errors are not introduced in developing a more accurate lens.)

It is of interest that the axial position of these thick glass plates is arbitrary, allowing them to be placed anywhere between the light source and mask. As an extreme example, assume that the plate is placed so that its bottom side comes in contact with the light source as indicated in Fig. 51 and that it has a spherical surface whose center is located at the light source. Whether or not the glass inside this spherical surface is scooped away the light path will not be affected. The resultant concave (spherical) lens thus used is thinner and lighter than the thick glass plate. However, in this case, it is necessary to

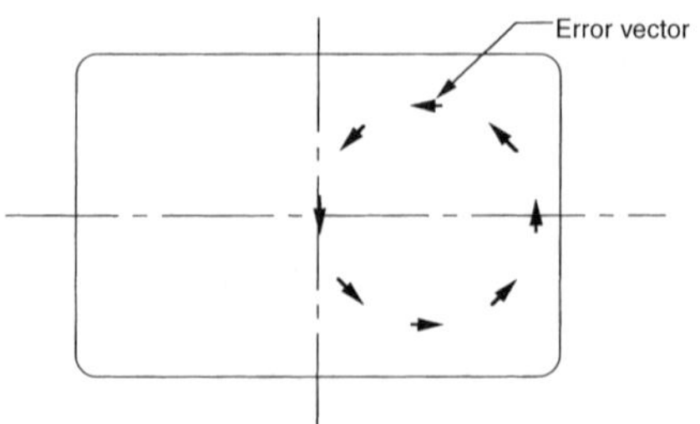

FIGURE 52. An example of rotating correction vector.

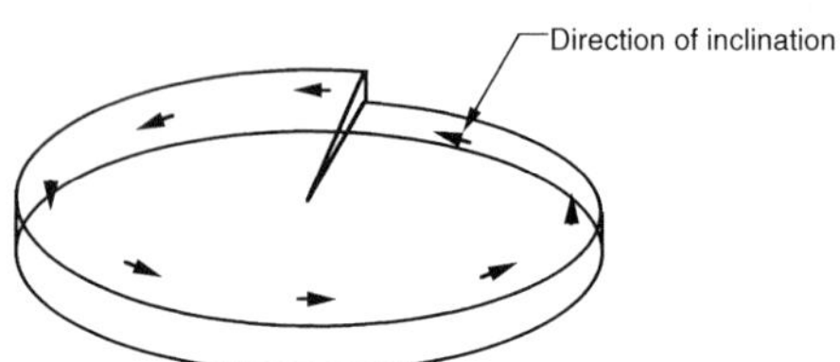

FIGURE 53. Spiral-like lens surface showing discontinuity. [Reprinted by permission from Yamazaki, E. *et al.* (1973). A segmented lens for improving color television dot pattern, *SMPTE*, **82**(3).]

ensure that the center of curvature of the spherical surface and the light source coincide. In order to compensate for the Δs shift, the spherical lens can be inclined as in the case of the thick glass plate provided that the center of curvature of the spherical surface is maintained coincident with the light source. As the light source diameter may appear smaller due to the effects of the strong concave lens, it is necessary to employ a larger light source to compensate for this.

ii. Continuous Surface Lens. To make accurate corrections over the entire phosphor screen it is necessary to employ a lens with a curvature that varies over its surface. However, such a lens has its limitations and regardless of the curvature chosen, there will be some errors that cannot be corrected. This fact can be explained qualitatively in the following way. If landing error vectors on the phosphor screen happen to fall in a circular pattern in some local area as shown in Fig. 52, in order to compensate for this kind of error, the lens surface for this area would have to have a spiral inclination as shown in Fig. 53. However, because this would cause a discontinuity at some point, errors such as those shown in Fig. 52 cannot be corrected with a conventional lens.

Mathematical analyses of this situation have been carried out by Ogura (1966), Morrell *et al.* (1974), and Fujimura (1988), each in a different way. However, since such analyses are beyond the scope of this article, in order to provide a better understanding of the problem, paragraphs from the *Principles of Optics* by Max Born and Emil Wolf are quoted here[1]:

> *When* n *is defined as an index of refraction, and* s *as a unit vector along the ray trajectory then*
>
> $$rot^2\, n\mathbf{s} = 0 \tag{31}$$
>
> *characterizes all the ray systems which can be realized in an isotropic medium and distinguishes them from more general families of curves. In a homogeneous isotropic medium* n *is constant, and Eq. (31) then reduces to*
>
> $$rot\, \mathbf{s} = 0 \tag{32}$$
>
> *Rays in a heterogeneous isotropic medium can also be characterized by a relation independent of* n*. System rays in any isotropic medium must satisfy the relation*
>
> $$\mathbf{s}\, rot\, \mathbf{s} = 0 \tag{33}$$
>
> *A system of curves which fills a portion of space in such a way that in general a single curved line passes through each point of the region is called a congruence. If there exists a family of surfaces which cut each of the curves orthogonally the congruence is said to be normal; if there is no such family, it is said to be skew. For ordinary geometrical optics (light propagation) only normal congruence is of interest, but in electron optics skew congruences also play an important part.*
>
> *If each curve of the congruence is a straight line the congruence is said to be rectilinear. Equations (33) and (32) are the necessary and sufficient condition that the curves should represent a normal and a normal rectilinear congruence, respectively.*

As quoted here, electron optics has an intrinsically different nature than light optics. Consequently, an electron trajectory cannot always be replaced by a light trajectory.

Needless to say, these considerations also apply when several continuous surface lenses are used in combination. For easier determination of Δp and Δs (mentioned earlier) as well as the other errors on the phosphor screen, these must be converted to errors on the light source plane S_0. The Δp components generally do not include the "rot." However, Δs contains the "rot" in general, making correction difficult.

Such problems do not exist at all for striped phosphor screens because only the horizontal correction must be considered and the error along the vertical

[1] From Born, M., and Wolf, E. (1997). *Principles of Optics*, sixth edition, Cambridge University Press, Cambridge, p. 126. Reproduced by permission of Prof. Emil Wolf.

[2] rot **ns** is sometimes expressed as curl $n\mathbf{s}$.

axis can be ignored. This makes it possible to obtain complete correction using a smooth continuous surface lens.

Two specific procedures can be used to design such a lens. The first involves choosing the lens curvature that corresponds to the correctable components, thereby eliminating the noncorrectable components (rot) from the error data that was measured. In the second method the curvature minimizing the residual errors is chosen by changing the polynomial coefficients representing the smooth continuous surface. The latter is more practical and widely used because the noncorrectable components can be eliminated and the residual errors minimized at the same time.

Various types of polynomials for representing the lens surface can be considered but the simplest can be represented as

$$Z = \sum \gamma_{ij} x^i y^j \tag{34}$$

where Z, the height of the surface of the lens, is a function of the x-y coordinates, i and j are the powers of x and y, respectively, and γ_{ij} is the coefficient of $x^i y^j$. In this case γ_{ij} can be determined by making the slopes at each point of the lens surface as close as possible to the required slopes $(\partial z/\partial x)$ and $(\partial z/\partial y)$ at each point according to the measured data. More accurately stated, coefficients with the minimal mislanding error at each point on the phosphor screen should be determined. Methods minimizing the error include the method of least squares, the min-max method that minimizes the maximum value of the error, the method of minimization by placing weight on the central and peripheral portions of the phosphor screen, and other similar methods, depending on the criteria selected.

It is possible to design a lens without residual error with respect to all measured points if sufficiently high powers (i and j) are assigned to the measured points. However, this does not always result in the design of a good lens. If the power is higher than necessary, a fine lenticulation will be generated over the lens surface and, regardless of the slope matching at the measured points, there is a possibility that a large error will be generated between measured points. With the values of i and j usually around 5 to 6, it is necessary to limit these values so that unnatural lenticulation is not generated. When there is an insufficient number of measured points then other points can be interpolated between the measured points to increase the number of data points without resulting in unnatural lenticulation.

As mentioned earlier, in the case of striped phosphor screens the error in the vertical direction can be ignored (assuming that the stripe is continuous in the vertical direction and the effect of the tie bars of the mask can be neglected), making total correction possible. In this case Eq. (34) can be employed to determine the coefficients for the lens curvature, considering only the horizontal error. However, a more practical method would be to determine the surface inclination by considering the horizontal error from the lens center toward the left and right directions, and then in a step-by-step procedure determining the thickness through integration.

When employing the delta gun arrangement with a dot screen, the noncorrectable rot components of the continuous surface lens increase. However, the "second-order printing method" (see Morrell *et al.*, 1974) can be used to minimize such rot components. This method results in exposure of the phosphor screen through the adjacent mask aperture instead of the mask aperture that the electron beam will actually pass through. Although the rot components can effectively be minimized in this case, it is necessary to set the q value slightly higher than the optimal ideal value (preventing optimal nesting of the dots). Furthermore, when the q value varies slightly during assembly, the direction of movement of the beam spot and phosphor dot occur in opposite directions, easily generating landing error. Because of such reasons and the fact that the delta gun system itself is no longer employed, the "second-order printing method" is not used in practice.

iii. Discontinuous Surface Lens. The basic concept of the discontinuous surface lens is to avoid the limitations of lenses with a smooth, continuous surface by breaking the surface into several hundred small segments, each of which has the optimal surface inclination. Figure 54 indicates an example of such a lens. A problem occurring with this lens, however, is the optical effect at the screen caused by the abrupt steps between the neighboring segments. To alleviate this problem, the entire lens can be oscillated in the x and y directions during exposure, allowing slight differences in correction between

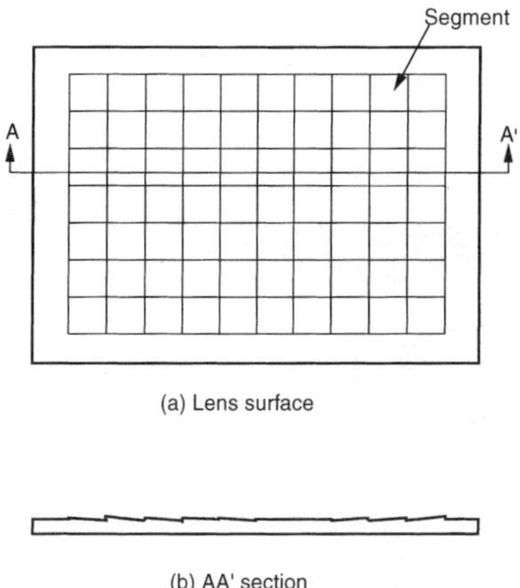

FIGURE 54. Fresnel-like discontinuous surface lens for phosphor exposure. (a) Lens surface divided into segments, each with its own specified inclination; (b) cross section at AA′.

neighboring segments to be blended into each other. In this oscillatory process the amplitude of the oscillation should be confined to the size of one segment. Also it is desirable for the periods of the horizontal and vertical oscillations to be adjusted so that they are not in synchronism. In any case, as much correction as possible should be made with a continuous-surface lens, correcting only the residual error with a discontinuous lens (Yamazaki *et al.*, 1971 and 1973).

iv. Sequential Exposure Method. This method avoids the need for special lenses such as the discontinuous surface type. The phosphor screen in this case is divided into several regions and the position of the light source for each region is sequentially shifted during exposure. The smaller each region is made, the smaller the error will be. However the exposure time will be correspondingly longer, thus posing problems of practicability (Fujimura, 1970).

v. Lens Manufacturing. The continuous surface lens can generally be prepared in one of two ways. The first method involves cutting and grinding the glass with a computer-controlled machine commonly known as an NC (numerically controlled) machine. Although this allows the desired dimensions to be accurately obtained, it involves a high manufacturing cost. In the second method, referred to as the sag method, a flat glass plate of uniform thickness is placed on a metal mold as shown in Fig. 55 that was cut and processed with the NC machine. When the temperature is elevated to the point where the glass softens, it sags into the mold, leaving the upper surface shaped like the mold after the glass plate is cooled. To complete the lens, the bottom surface of the glass that was in contact with the metal mold is ground

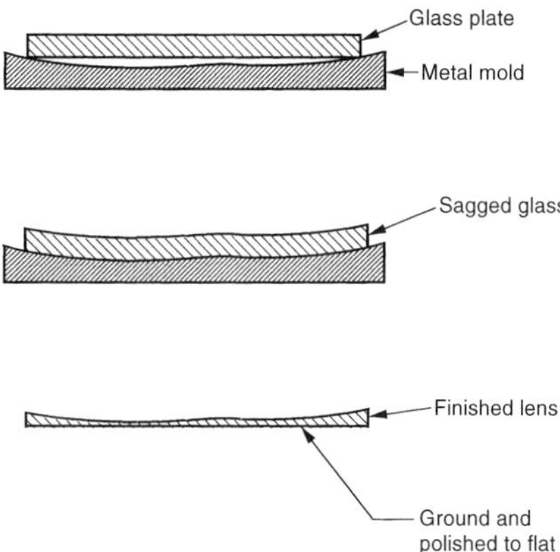

FIGURE 55. Steps in fabrication of correction lens by sagging heated glass plate onto mold. [Reprinted by permission from Morrell, A. M. *et al.* (1974). *Color Television Picture Tubes*, Academic Press, New York, p. 86.]

flat. Although the accuracy obtained with this method is slightly inferior to that obtained with the former method, by careful attention to the processing procedure a level of accuracy can be obtained that is sufficient for practical use. Once the mold is made, the later processing steps are relatively simple, making the process ideal for the mass production of similar lenses at a fairly low cost (Morrell *et al.*, 1974).

Discontinuous surface lenses are preferably made of acrylic resin. First, several hundred metal blocks (each, e.g., 8 mm × 8 mm in area) are precisely ground and polished to match the slope of the corresponding segment of the discontinuous surface lens. The blocks are then secured in a frame that will be used as a mold. Following this, a 3 mm thick acrylic sheet, for example, at elevated temperature is pressed against the mold, transferring the slope of the mold surface to the acrylic surface. Although processing the metallic block may consume time and money, the resulting acrylic lenses can be produced relatively easily (Yamazaki *et al.*, 1971).

VII. Deflection

As shown in Fig. 1 the electron beams emanating from the electron gun of a color CRT are deflected by the common magnetic field of the deflection yoke positioned around the tube neck. A sectional view of the deflection yoke is shown in Fig. 56 together with the magnetic field lines for producing horizontal deflection. The electron gun is normally designed so that, in the absence of any deflection field the three beams are converged at the center of the screen. As already mentioned, in the case of an in-line gun system the deflected beams can be maintained converged if a specially designed

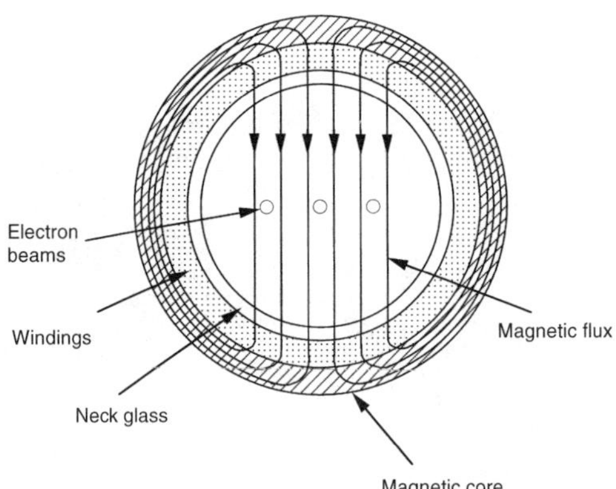

Figure 56. Cross section of a deflection yoke showing outer magnetic "core."

deflection yoke with an appropriate field distribution is used (Haantjes and Lubben, 1959; Morrell *et al.*, 1974). This self-convergence system is commonly used in the industry, eliminating the complicated dynamic convergence correction circuitry previously used with the delta gun system (which is now obsolete). The discussion that follows is thus concerned entirely with the self-convergence system.

A. Astigmatic Convergence Error

We assume first a combination of a horizontally arranged in-line gun system and a deflection yoke with uniform field distribution. In this case the raster patterns of the individual beams will appear as shown in Fig. 57 (Takano, 1983). Although the raster pattern scanned by the center beam is symmetrical with respect to the vertical axis the rasters created by the side beams are asymmetrically deformed as shown in the figure due to the tilt angle of their electron-gun axes with respect to the tube axis. Assuming the undeflected beams were converged at the screen center, they will thus become misconverged when they are deflected. This error is referred to as the astigmatic convergence error.

This convergence error can be corrected by modifying the magnetic field distribution of the deflection yoke, in particular by using a barrel-shaped distribution of the vertical deflection field as shown in Fig. 58 and a pincushion-shaped distribution of the horizontal deflection field as shown in Fig. 59. In Fig. 58 when the beams are deflected vertically the curvature of the field lines produces a component of force in the x-direction on the outer beams, reducing their overconvergence angle and enabling them to converge

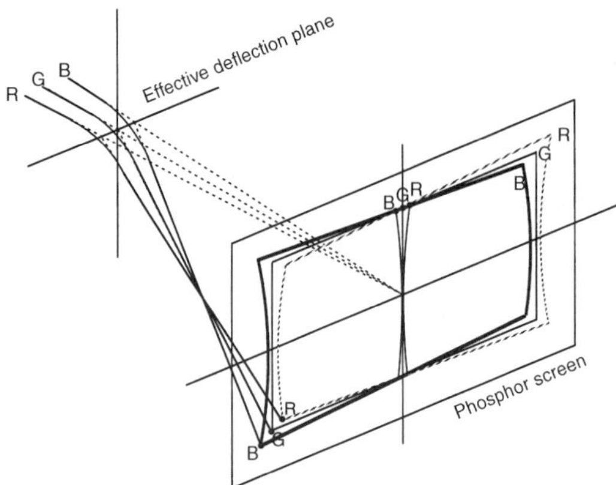

FIGURE 57. Astigmatic convergence error. [Reprinted by permission from Takano, Y. (1981). Recent advances in color picture tubes, *Advances in Image Pickup and Display*, **6**, 7.]

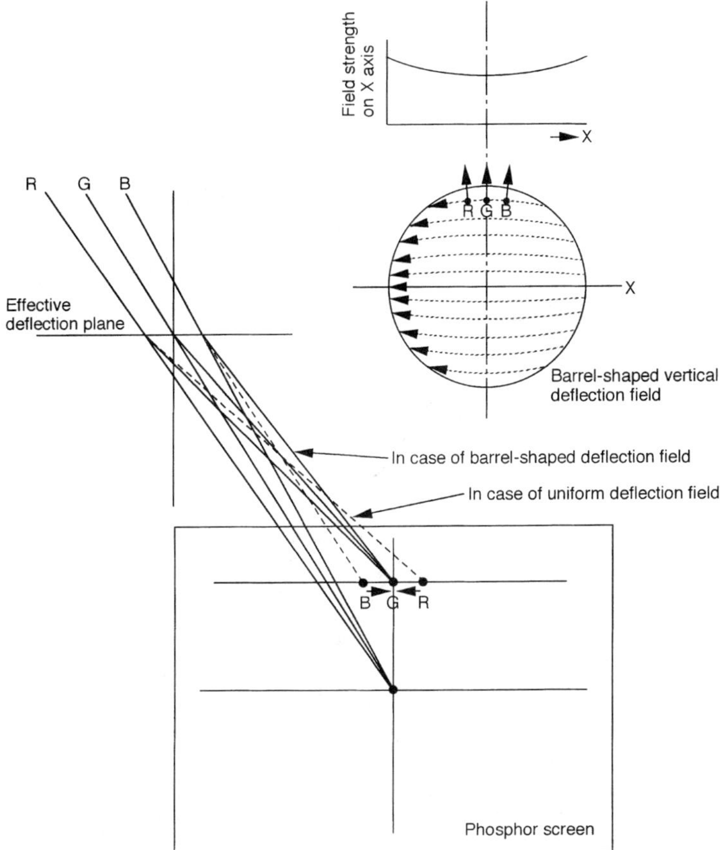

FIGURE 58. Astigmatic error correction by barrel-shaped vertical deflection field. When beams are deflected vertically curvature of field gives divergent force to the beams, canceling overconvergence and making them converge on the screen.

at the screen. In Fig. 59, where the beams are deflected horizontally by the pincushion-shaped field, they experience a field that is increasingly stronger in the x-direction from the center to the edge. The resulting variation in deflection angle also reduces the convergence angle of the beams, enabling them to converge at the screen.

B. Coma Error

As shown in Fig. 59, however, because the strength of the horizontal deflection field does not vary linearly but quadratically, the convergence error corrections on both side beams are not balanced as shown in the circular inset

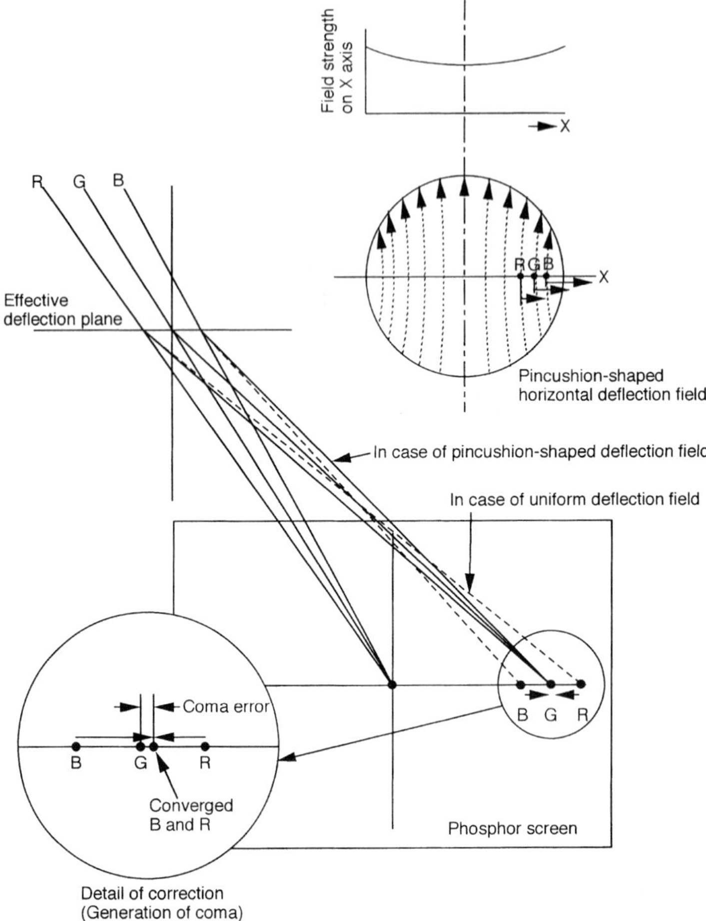

FIGURE 59. Astigmatic error correction by pincushion-shaped horizontal deflection field. When beams are deflected horizontally difference of field (stronger toward outside) gives divergent force to the beams, canceling overconvergence and making them converge on the screen.

at the lower left of Fig. 59. Beam B at the outer side of the deflection field in this example is corrected more than the beam R, which is closer to the center of the deflection field. This results in the two outer beams converging to a common point BR on the X-axis of the screen and the center beam G falling slightly to the left as shown in Figs. 59 and 60. This type of convergence error is referred to as coma error. In a somewhat similar manner coma error is introduced when the beams are deflected by the vertical deflection field of

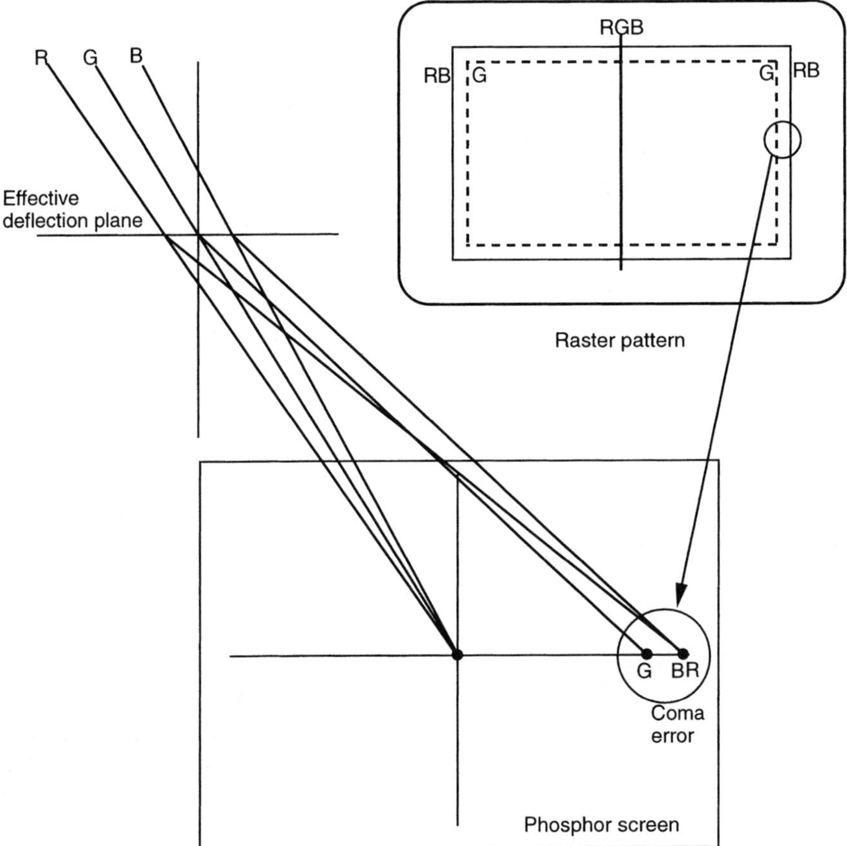

FIGURE 60. Coma error.

Fig. 58 where the two outer beams experience a somewhat stronger field than the center beam.

There are basically two different ways to eliminate the coma convergence error. One involves attaching appropriately shaped magnetically permeable field controller pieces at the exit end of the electron guns. Such an arrangement, introduced by Barbin and Hughes, (1972) is shown in Fig. 61. These magnetic pieces, exposed to the fringing field extending beyond the entrance side of the yoke, modify the strength of both the horizontal and vertical deflection fields in the neighborhood of each of the three beams before they enter the yoke. As shown, the deflection field of the outer beams, which are shielded by the circular shunts, is reduced while the field of the

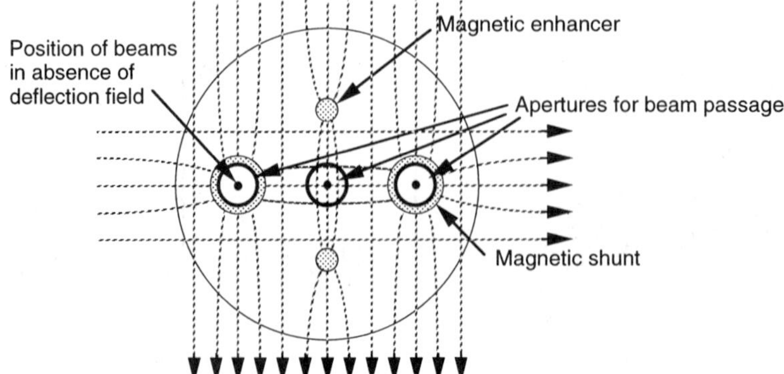

FIGURE 61. Magnetic field controller. [Reprinted by permission from Barbin, R. L., and Hughes, R. H. (1972). New color picture tube system for portable TV receivers, *IEEE, Broadcast Television Receivers*, **18**, 193–200 © 1972 IEEE.]

central beam is increased by the presence of the field-enhancing magnetic elements above and below the beam. This method for coma reduction is often used in television applications because it permits the use of a simple and low-cost deflection yoke.

The other way of minimizing the coma error involves modifying the distribution of the winding of the deflection yoke itself so that at the electron-gun end of the yoke the field shape is made opposite to the field shape used for the astigmatic correction mentioned in Section A. For example, if a barrel-shaped field is used for astigmatic correction of the vertical deflection a pincushion shape is used for the portion of the yoke near the gun. Such a field distribution can be achieved either by properly varying the angular distribution of the windings along successive parts of the yoke in the axial direction. Alternatively, magnetic pieces can be added around the deflection yoke as will be explained in more detail in Section D.

C. *Raster Distortion*

In an actual tube the raster pattern of the center beam will not be rectangular as shown in idealized form in Fig. 57 but will have a pincushion shape as shown in Fig. 62. (The raster patterns of the side beams will also be more pincushion-shaped than shown in Fig. 57.) This is due to the fact that the distance the beam is deflected at the phosphor screen increases more than linearly with the increase in deflection field as a result of the increasing distance between the spot at the screen and the deflection center of the yoke with increasing deflection angle (a fact that was neglected in Fig. 57). Also, a strongly deflected beam will have a longer and more curved trajectory

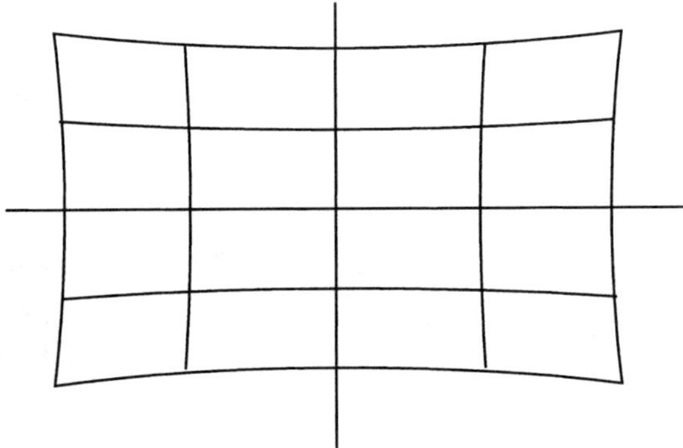

FIGURE 62. Pincushion-shaped raster distortion.

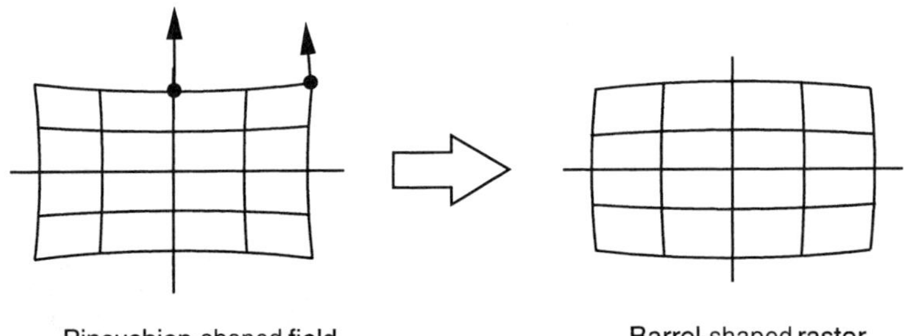

FIGURE 63. Raster correction. In pincushion-shaped field, denser field line means stronger field on axes, causing larger deflection which makes raster barrel-shaped.

through the yoke field (and a correspondingly longer transit time in the magnetic field) than a less deflected beam.

This raster distortion can also be corrected by a proper field distribution of the deflection yoke. For example, a pincushion-shaped field for vertical and horizontal deflection creates a barrel-shaped raster distortion as shown at the right in Fig. 63. By modifying the deflection field in this way the pincushion raster distortion can be cancelled (Heijnemans *et al.*, 1980).

D. Correction of Convergence Error and Raster Distortion

The objective in deflection yoke design is to achieve several goals simultaneously, namely obtaining beam convergence (i.e., avoiding astigmatic error), minimizing coma error, and reducing raster distortion. However,

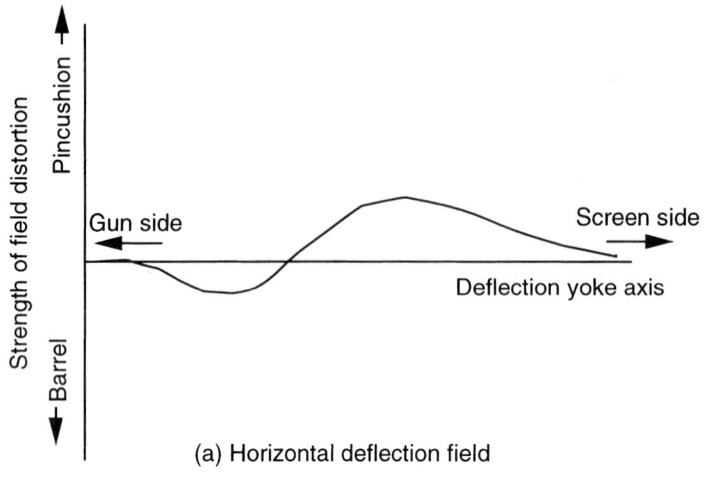

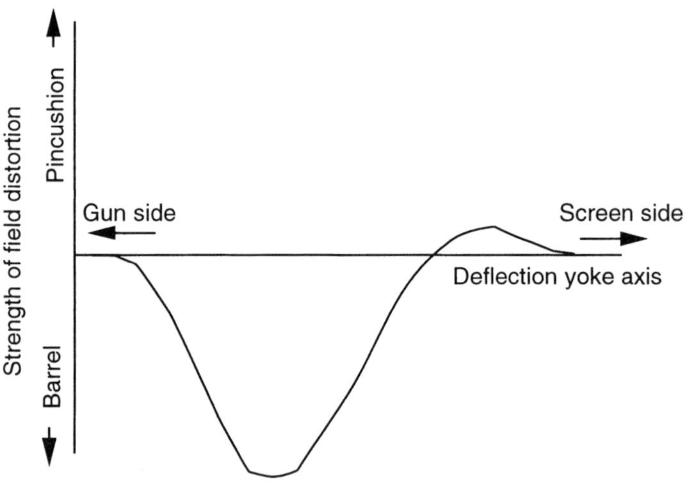

(b) Vertical deflection field

FIGURE 64. Axial distribution of deflection yoke.

since raster distortion is affected mainly by the field on the screen side and coma more affected by the field on the electron gun side, astigmatic error can be dealt with by the field at the center portion of the yoke (Takano, 1983). Thus by optimizing the axial distribution of the deflection field the convergence error and raster distortion can be minimized simultaneously, achieving self-convergence and raster correction-free characteristics.

Figure 64 shows an example of the variation in the type of field (pincushion or barrel shape) with position along the axis of the deflection yoke in order to simultaneously achieve self-convergence and correction of the raster shape. In addition to varying the winding distribution, one or more of the following techniques (Shimoma *et al.*, 1978) can be used to obtain the desired field distributions:

1. attaching permanent magnets to the open end of the deflection yoke (at the screen side);
2. using auxiliary coils in order to supplement the field distribution;
3. embedding magnetically permeable metal plates in, or attaching them to, the open end (screen side) of the deflection yoke.

E. Data Display Application

For data display the performance requirements of the deflection yoke are far more stringent than for TV applications, making it necessary to impose much tighter manufacturing tolerances. In particular, it is necessary for the winding distribution to be very precise so as to achieve the necessary field distribution. To obtain the necessary winding-distribution accuracy slits are provided in the magnetic core of the yoke in which the wires are held precisely in place. To minimize any residual convergence error auxiliary coils or nonlinear reactors may also be connected in series with the vertical deflection yoke windings.

Although effective for the low line-scanning frequencies used in TV, at the higher line-scanning frequencies such as 63, 90, and 130 kHz used in high-resolution displays, the hysteresis of any magnetic pieces added to the yoke will cause asymmetrical convergence errors along the left and right sides of the screen as shown in Fig. 65 making complete convergence unobtainable. For operation at high frequencies it is thus necessary to dispense with such pieces and employ a deflection yoke specially designed to minimize coma error. Because it is sometimes difficult in this case to eliminate all of the errors completely, the yoke is designed with some coma error remaining in the vertical direction as shown by the beam positions in Fig. 66. Since this error can then be corrected by adding a special convergence yoke (such as shown in Fig. 67), it allows the yoke designer to concentrate on correcting the horizontal convergence errors. As indicated, the coils of the convergence yoke, mounted close to the gun-side of the deflection yoke, are connected in series with the vertical deflection coils. Because the frequency of the vertical deflection current is relatively low this presents no problem.

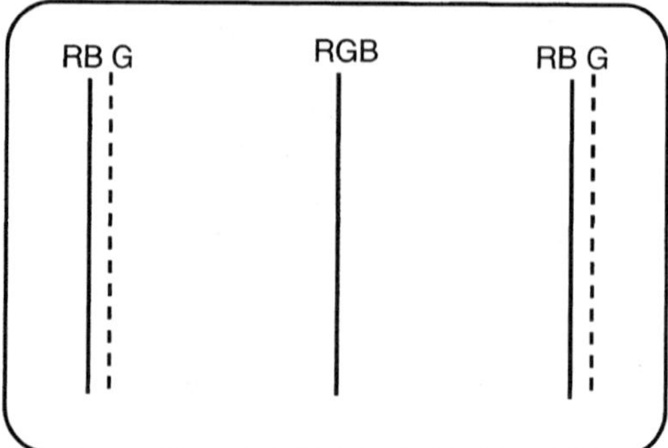

FIGURE 65. Asymmetrical convergence errors.

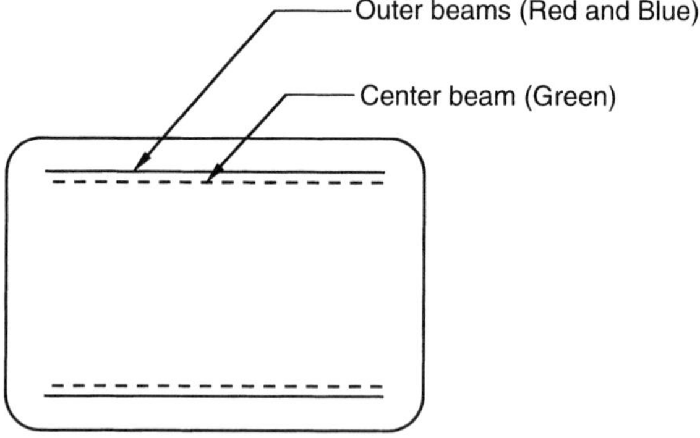

FIGURE 66. Coma error (vertical).

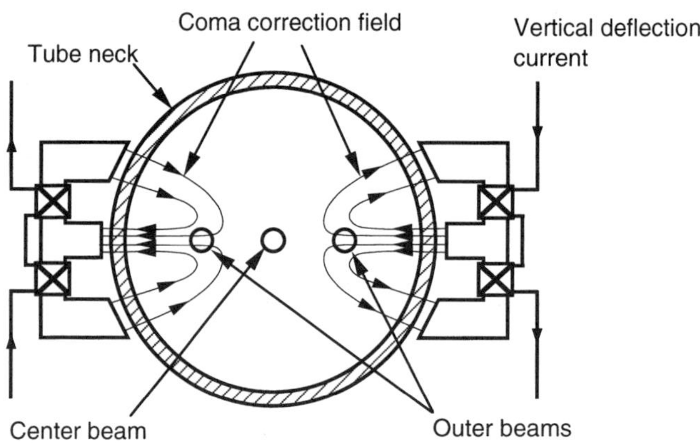

FIGURE 67. Coma correction system.

In the case of CRT monitors capable of operating at multiple vertical scanning frequencies, the correction current waveforms applied to the convergence yoke are usually synthesized in an analog manner. However, where extremely high precision is required they are synthesized digitally after digitally recording the convergence error for many representative positions on the screen. The residual error in this way can be reduced virtually to zero.

F. Deflection Angle

Instead of the 110° deflection angle generally employed in conventional television tubes, a deflection angle of 90° is generally used for data display tubes. Although this results in a small increase in overall tube length, it reduces the spot distortion at the edges of the screen as well as the misconvergence and raster distortion.

Unlike systems with 110° deflection, which normally require an external correction circuit for raster distortion, systems with a 90° deflection angle usually do not require such a correction circuit. However, if necessary, fairly simple circuits for modulating the amplitude of the line-scanning yoke currents can be used in 90° tubes to correct E-W distortion (curvature of raster sides). This also provides an advantage if multiple scanning frequencies are used. Another advantage of a 90° deflection angle is the reduced deflection power required at the higher horizontal scanning frequencies.

G. Temperature Rise

Because of the increased number of scan lines in a high-resolution display and the associated higher horizontal scanning frequency, copper losses in the windings of the horizontal deflection yoke and eddy current losses in the core material are increased, leading to excessive heat generation in the deflection yoke. Figure 68 indicates the relationship between the horizontal scan frequency and the temperature rise for different types of windings and cores. Because the resistivity of glass falls very sharply with increasing temperature, in order to prevent electrical breakdown of the neck glass, the temperature rise of the deflection yoke is normally limited to 40°C as shown in the figure. Since, as shown in the figure, the temperature of a normal deflection yoke operated at 42 kHz scan rate exceeds the allowable limit, Litz wire is used to reduce the effective resistance for the windings and low-loss magnetic materials are used for the core to reduce the temperature rise. If these measures are insufficient, it may also be necessary to employ forced air cooling.

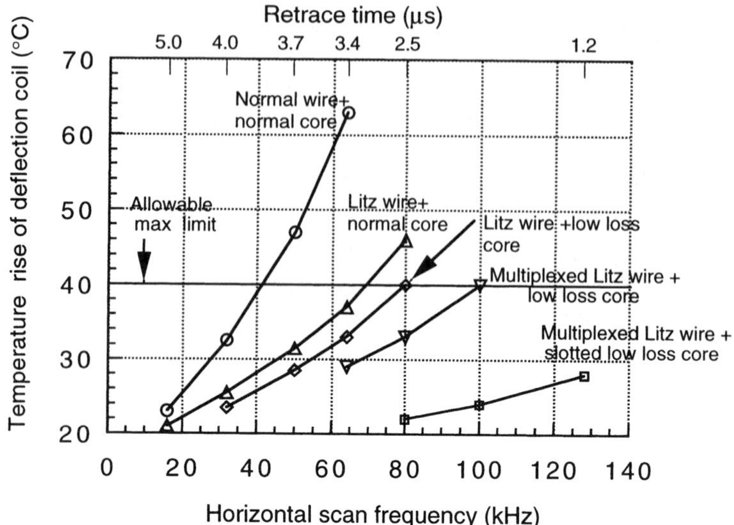

FIGURE 68. Temperature rise of deflection yoke components.

H. Leakage Flux Suppression

Instead of the saddle-toroidal (S/T) deflection yoke winding commonly used in television, saddle-saddle (S/S) windings are increasingly being used for data display tubes (Yamazaki, 1993). These produce far less external leakage of the vertical-deflection magnetic field because the magnetic core in this case surrounds both the horizontal and vertical windings whereas in the S/T yoke the magnetic core surrounds only the horizontal winding. The reduced magnetic leakage of the S/S yoke minimizes interference with neighboring monitors and avoids possible health problems (see Section XIII). If a further reduction in leakage is required compensating coils can be added just outside the deflection yoke.

VIII. GLASS BULBS

A. Glass Bulb Design

Glass is generally used for the CRT envelope because of its transparency, its ability to withstand a high vaccum, and its capability of absorbing a high fraction of the internally generated X-rays (McLellan and Shand, 1984; *New Glass Handbook*, 1991; Sakka, 1985). However, as will be discussed, there

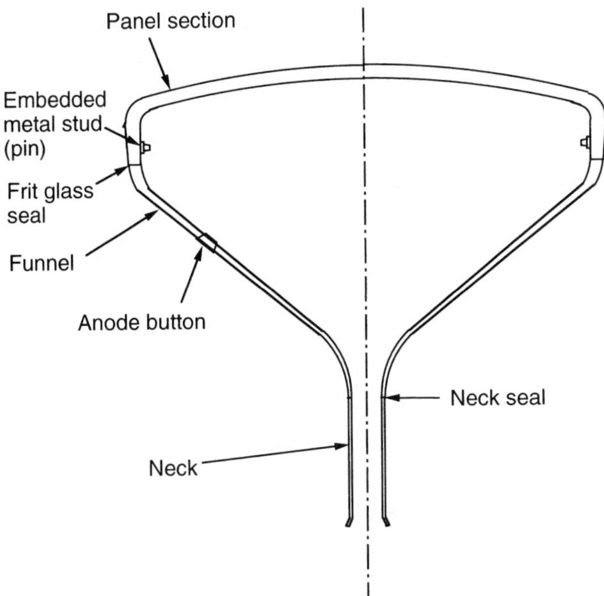

FIGURE 69. Glass bulb parts.

are a number of technical problems associated with the use of glass. As shown in the cross section of Fig. 69, a typical glass bulb consists basically of three parts: the panel section, the funnel, and the neck, sealed together at the points indicated.

1. *X-Ray Absorption*

To absorb X-rays generated at the shadow mask and screen, lead glasses containing about 23 and 35% of PbO are used for the funnel and neck, respectively. However, this glass can not be used for the panel section because during electron bombardment the PbO is converted to metallic colloidal particles of lead that appear as a brownish coloration referred to as the "browning phenomenon." To avoid this, barium and strontium are substituted for the lead components of the glass. To further reduce the browning a small amount of CeO_2 (normally about 0.3%) is added to the barium-strontium glass.

For a layer of glass the X-ray absorption can be expressed by Eq. (35).

$$\frac{I}{I_0} = e^{-\mu t} \tag{35}$$

where I_0 and I are the initial and transmitted X-ray intensity, respectively, μ is the absorption coefficient, and t is the thickness of the glass. The values of m for the glasses currently used in the color CRT industry are $28\,\mathrm{cm}^{-1}$ for the panel, $62\,\mathrm{cm}^{-1}$ for the funnel and $90\,\mathrm{cm}^{-1}$ for the neck glass. In order to achieve the necessary X-ray absorption the minimum glass thicknesses are determined, assuming the preceding absorption coefficients.

2. *Optical Absorption*

Optical absorption of the glass panel is an important factor in determining the contrast of an image. This can be controlled by adding metal oxides such as NiO and Co_3O_4 to produce a neutral gray and by adding Nd_2O_3 for selective spectral absorption.

For a given type of glass the percent transmission T is given by

$$T = (1 - R)^2 e^{-Kt} \times 100\,(\%) \tag{36}$$

where R is the reflectance of the front surface (normally about 0.04) and K is the absorption coefficient (mm^{-1}). Glass panels are categorized into clear ($T \geq 75$), gray ($60 \leq T < 75$), tinted ($50 \leq T < 60$) and darkly tinted ($T < 50$) according to their level of transmission. Figure 70 shows the relationship between the glass thickness and the optical transmission for the different types of glasses. After choosing the appropriate thickness of the

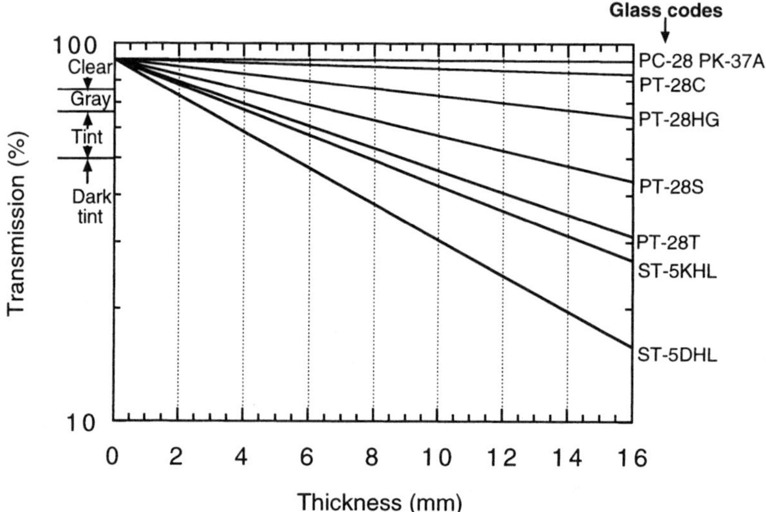

FIGURE 70. Examples of transmission versus thickness of some commercial types of glass (manufactured by Nippon Electric Glass) used for CRT faceplates.

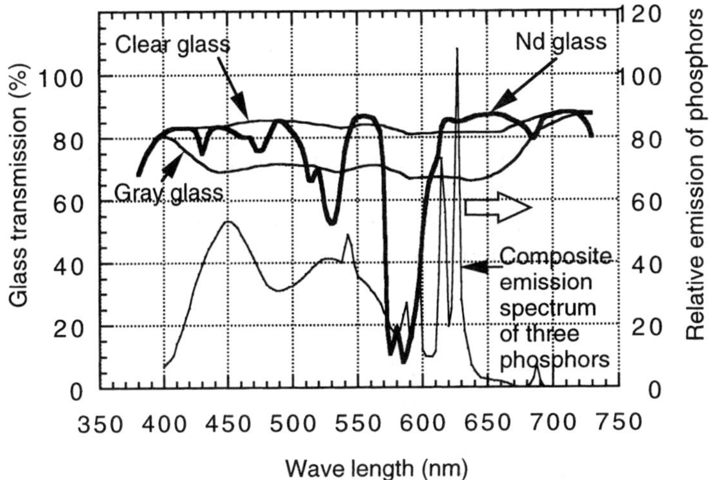

FIGURE 71. Spectral absorption curves for different types of glass panels.

panel based on mechanical strength considerations, glass with the desired absorption coefficient can be selected from the curves of Fig. 70 to obtain the desired transmission.

Figure 71 provides an example of the change in transmission of glass obtained over a range of wavelengths by adding Nd. To some degree the transmission spectrum of this glass is opposite to the composite emission spectrum of the phosphors, absorbing ambient light more effectively at areas between peak emission bands and reducing the attenuation of emitted light from the phosphors, thus further increasing the contrast ratio (Shimizu *et al.*, 1981). However, such glass may cause the color of darkened areas of the tube face to vary significantly according to the ambient light conditions.

3. *Sealing of Glass Components*

The thermal expansion coefficients of these glasses and the frit glass mentioned here for sealing the components together are chosen so that their values are close to each other, with a value of about 98×10^{-7}. The temperatures corresponding to the strain point, the annealing point and the softening point of the glass, which define the glass characteristics, must also have specific values because they are important in preventing glass deformation during the bakeout of the tube (about 430°C) during manufacture.

After the phosphor screen is deposited and the shadow-mask is mounted, the funnel, which is sealed with the neck beforehand, and panel sections are joined together with a frit (powder) of a PbO-based glass. This has a very low softening temperature and devitrifies in the frit sealing process.

4. *Envelope Strength*

Since the completed CRTs are evacuated to about 10^{-4} Pa (7.5×10^{-7} Torr) the glass envelope has to withstand full atmospheric pressure. However, thermal stresses induced in the tube-manufacturing process and surface scratches created during normal tube operation or handling may weaken the

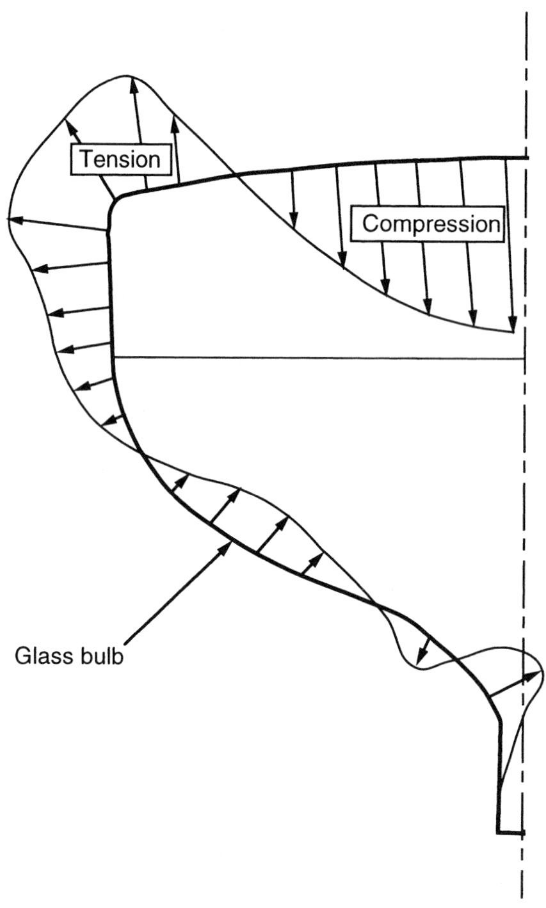

FIGURE 72. An example of results obtained utilizing FEA model. Although not shown, the tension at the shoulder occurs at the region where the minor axis of the tube face joins the skirt.

glass and lead to the breakage of the bulb. In the design of the glass bulb these factors have to be carefully considered (Gulati, 1993; Ghosh and Gulati, 1995).

In general, glass is fairly strong under compressive stress but weak under tensile stress. The current practice of glass design utilizes finite element analysis (FEA) technology to minimize the tensile stress in regions where it is greatest. Figure 72 shows qualitatively the results of such an analysis. As indicated, the weakest point (where the tension is highest) appears normally around the outer shoulder portion joining the panel face and the skirt on the minor axis. Simulation is iterated with the design parameters and panel thickness being successively modified until the tensile stress is sufficiently lowered below the maximum allowable tensile stress of 7.9 MPa (Enstrom *et al.*, 1978; Elst and Wielenga, 1977; Itoh *et al.*, 1987).

After fabrication of the glass bulb is completed its strength is checked by subjecting it to an air pressure difference of 3 atm (no evacuation). Before this test the outer surface of the bulb is abraded to simulate normal scratches that might occur.

5. *Thickness and Mass of Glass Bulb*

The thickest part of the glass bulb is the faceplate panel in the region where curvature of its surface is smallest. The thickness of a panel required to

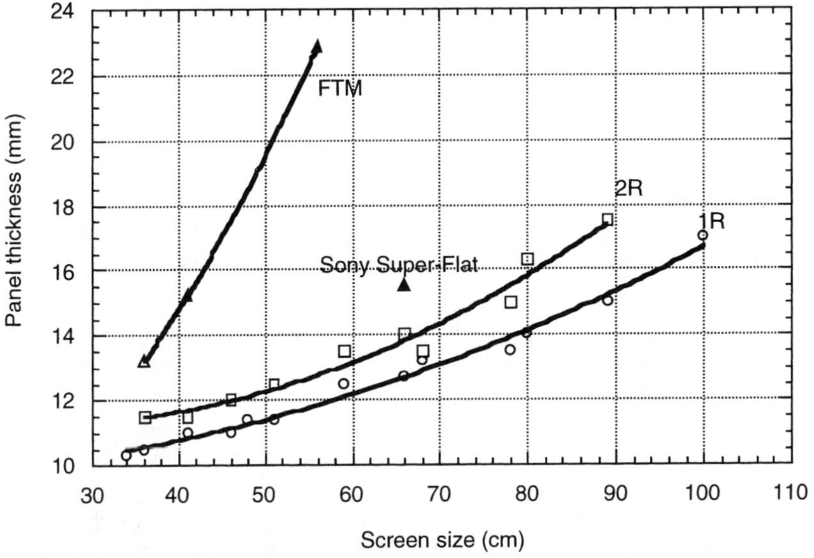

FIGURE 73. Panel thickness versus tube size.

achieve a given mechanical strength depends upon the curvature and diagonal size of the faceplate. The curves of Fig. 73 show examples for current designs of panel thickness versus diagonal size of the faceplate for different curvatures. As indicated the panel thickness increases with an increase of screen size and the flatness of the screen. In the case of the flat tension mask (FTM) tube, in particular, with its flat panel design (see Section IX) the thickness of the panel must be considerably increased.

Although the panel of the FTM tube was originally fairly thick, recent tubes with flat panels, such as the Sony Super-Flat, indicated in Fig. 73, have a much thinner panel (see Section IX.C for more details).

An increase of the panel thickness also causes an increase in total weight of the glass bulb. Aside from the higher cost of a bulb with a thick panel, because of the greater amount of glass used more power is consumed and more time required in processing the heavier bulbs.

The thickness of the funnel portion of the bulb is about 3–5 mm, becoming increasingly thicker toward the screen end. The neck glass is normally 2.5 mm thick but this may be increased to 4.2 mm for larger-sized tubes in order to withstand the high voltage (exceeding 30 kV) applied between the inner graphite coating of the neck and its outside (which is grounded).

B. Implosion Protection

In an accidental breakage of a tube it may collapse very violently (or implode) causing dangerous glass pieces to fly out. In order to prevent such hazardous implosions, two basic systems have been developed. In one a separate glass panel is laminated onto the CRT face plate with a resin layer in between. This is an expensive system used for special applications such as high resolution monitors where an antireflection (AR) coating is needed. As explained in Section X.B.2.*a* the AR coating is usually applied to the separate glass panel before lamination because the high temperature required in tube processing cannot be applied directly to the anitreflection panel of the finished tube.

In the other system a metal band is provided around the skirt of the CRT faceplate as shown in Fig. 74. By applying strong tension to the band compressive stress is applied to the CRT faceplate, thus reducing the tensile stress in this area and preventing violent implosion of the tube. For smaller tube sizes the metal band is wrapped around the faceplate panel and its ends joined by welding or crimping. For larger tube sizes a heart-shrink method is generally used in which the completed band is made slightly smaller than the circumference of the faceplate. After preheating the band to expand it, it is placed around the panel and then allowed to shrink as it cools down (Itoh *et al.*, 1992; Hondoh *et al.*, 1978; Nakamichi, 1979). In many cases mounting lugs are attached to the band to support the tube.

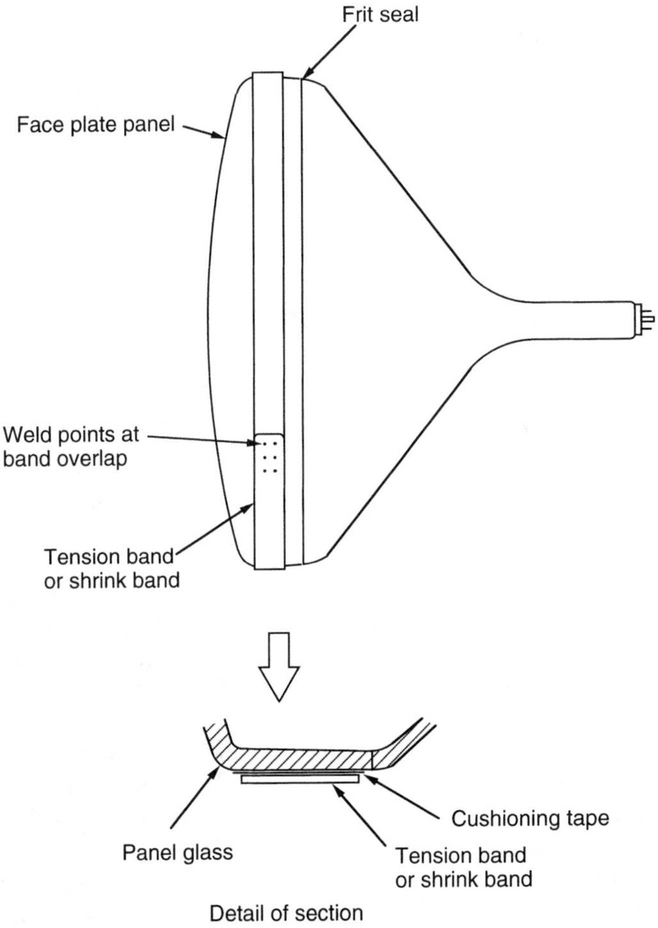

FIGURE 74. Tension band used for implosion protection.

IX. REDUCTION OF SCREEN CURVATURE AND USE OF FLAT TENSION MASKS

A. Panel Curvature

In the discussions of Section IV it was assumed that both the phosphor screen and shadow mask have flat surfaces. Except in special cases, however, the tube faceplate (phosphor substrate) is usually given some curvature to enable it to withstand the atmospheric pressure as discussed in Section VIII. Additionally, the use of a curved faceplate results in a number of supple-

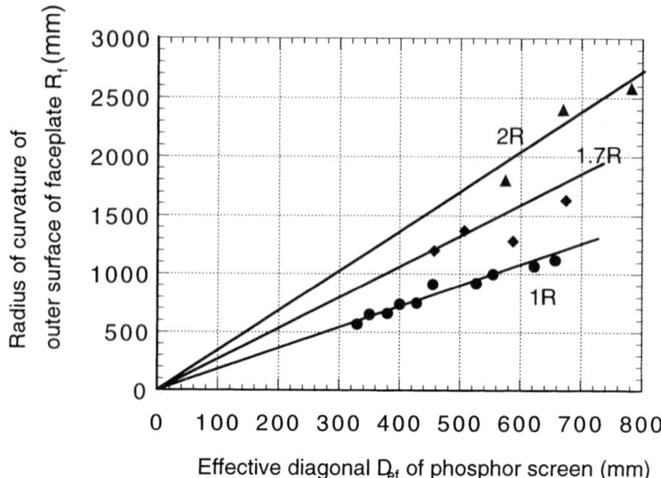

FIGURE 75. Relation of screen size to faceplate curvature.

mentary geometrical advantages, including better beam convergence and focus in areas near the picture periphery as well as less raster distortion, thus improving the overall performance. From the viewpoint of the CRT user, a phosphor screen as flat as possible is preferred for good viewability and thus some compromise is required.

For a long time tubes were designed in which the relation between the radius of curvature R_f of the panel outer surface and the effective diameter D_{ef} of the phosphor screen corresponded to the round dots of Fig. 75. As indicated, these dots fall almost along the line labeled 1 R. Tube designs with this relation have long been considered as the standard. In recent years, however, panels with a surface much flatter relative to their size have come into use (Tokita *et al.*, 1984). The relations in such tubes are indicated by the diamonds and triangles of Fig. 75, which fall approximately along the lines 1.7 R and 2 R, respectively.

The use of a flatter panel, however, entails a number of technical difficulties. As already mentioned in Section VIII, to withstand atmospheric pressure, it is necessary to make the glass thicker, thereby increasing not only its weight but also substantially increasing the processing time required for heating and cooling during tube bakeout. Both of these factors affect manufacturing cost. In addition, technical difficulties involving convergence, focusing, and raster distortion also inevitably increase, requiring more sophistiated electron gun structures for compensation or the use of additional compensating circuits. The most serious problem is the loss of color purity caused by a local increase of mask curvature resulting from its thermal expansion. (This phenomenon is explained in more detail in Section V.A.) Although many measures are being developed to overcome this, there are few that are effective. As already mentioned the simplest, yet most effective solution, is to replace the steel currently used for the shadow mask with invar, which has a lower coefficient of thermal expansion. Invar, however, provides only a partial solution, reducing the doming effect to only one-third or one-fourth of the previous level. In addition it is both extremely expensive and difficult to process.

This situation has led to the idea of making the faceplate aspherical with an overall profile, which is relatively flat over the central area but which has a strong curvature where mask doming is more likely to occur. In an actual design, such as that shown in Fig. 76, the areas at the boundary of the useful

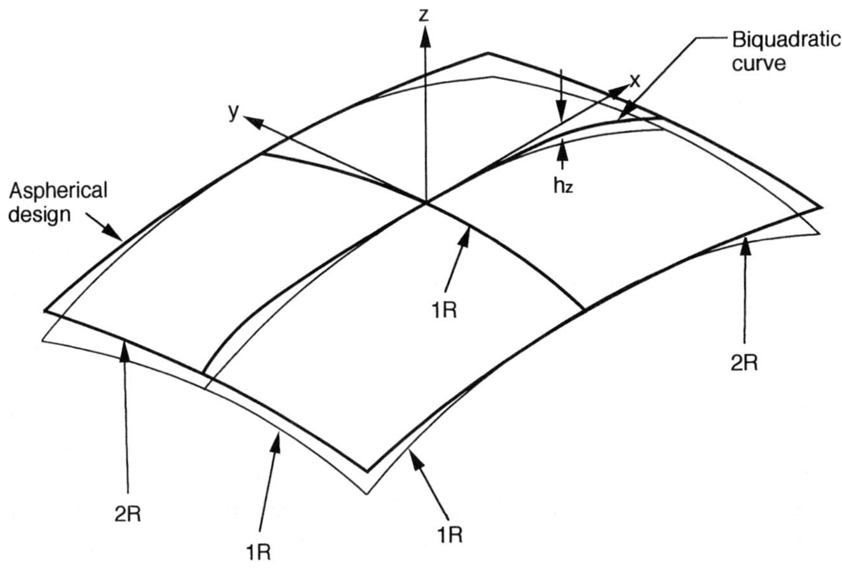

FIGURE 76. Comparison of shapes of aspherical panel and conventional spherical faceplate panel.

screen where it is closest to the escutcheon of the TV cabinet, are made relatively flat, equivalent to 2 R, for example (Suzuki *et al.*, 1989; Yonai *et al.*, 1986). At the sides, a little farther from the edges, the screen is given a greater curvature (defined by the biquadratic relation $h_z = -ax^2 - bx^4$, for example) as shown in Fig. 76, thus alleviating the adverse effects where the doming may most severely affect the color purity.

In spite of its advantages the aspherical surface panel has a number of drawbacks. First of all a higher-order raster distortion is produced. Though this distortion can, to a small degree, be remedied by the design of the deflection yoke it becomes increasingly difficult to correct as the distortion increases. Furthermore, such higher-order distortion may be much more visible to an observer even though only minimal distortion remains. Another problem results from the distortion of reflected images. Although the reflected image is in itself quite unrelated to the displayed image on the CRT screen and may be ignored, its distorted image seems to disturb some viewers and requires careful consideration. Finally, although an aspherical surface can easily be formed in a mold, much more complex polishing procedures are required than in the case of spherical surfaces, especially if the surface is strongly aspherical.

B. Mask Curvature

In general, the curvature of the shadow mask is chosen so that its surface is approximately parallel to the inner surface of the panel. However, the spacing q between the panel and mask is not set at a constant value over the entire screen but is set so as to realize optimum nesting of the beamlet triads over the entire screen area. In the case of the delta gun system, for example, the value of the "s" (distance from the tube axis to center of each electron beam at the deflection center) will become larger around the peripheral portion of the screen caused by the beam shift resulting from the dynamic convergence adjustment. The value of "q" should, therefore, be made proportionately smaller. As a result the shadow mask will have to be slightly flatter than the inner surface of the panel. In the case of an in-line gun system, where a self-converging deflection yoke is normally used, the modification of the magnetic deflection field requires that the "q" value near the peripheral portion of the mask in the horizontal direction be a little larger than the "q" values near the peripheral portion in the vertical direction. As a result the mask will be somewhat more cylindrical in shape, even though the inner surface of the panel will still be spherical.

To reduce doming a "variable pitch" shadow mask is sometimes employed, whose pitch near the right and left edges is greater by about 20 to 30% than the pitch at the center. Because it is necessary to increase the spacing "q" between the mask and the panel in proportion to the pitch, the radius of curvature of the mask in the horizontal direction is made smaller than that of the panel's inner surface. The smaller radius of curvature at the edges of the mask as well as its greater pitch near the screen's periphery also serve to alleviate the doming effect considerably. A mask constructed in this way is referred to as a super arch mask (SAM).

C. Flat Tension Mask Tubes

The first experimental models of shadow-mask tubes were made using a flat tension mask (Law, 1951) in the form of a 76 mm × 76 mm thin copper sheet in which holes with a pitch of 0.76 mm were etched. It was then stretched and welded to a steel frame while the mask was heated. Tension was thus applied to the mask when it was cooled. An assembly consisting of this mask and a flat phosphor plate was then mounted in a tube with a transparent curved face plate. In later tubes this approach was abandoned and the phosphor was deposited directly on the curved faceplate of the tube as shown in Fig. 1 to simplify tube construction.

Tubes with a flat face, however, are ideal for data display because of their reduced image distortion with viewing angle as well as a reduction in disturbing surface reflections from overhead lights. Also, images viewed on such tubes appear more similar to those on paper. Unfortunatley, such tubes present additional technical difficulties such as obtaining overall focus uniformity over the screen and minimizing geometric distortions.

In recent years there has been a growing interest in color tubes with a completely flat faceplate on which the phosphor dot pattern is directly deposited. To fabricate such structures the use of a flat tension mask was reintroduced by Zenith (Dietch *et al.*, 1986; Strauss *et al.*, 1986). A cross section of the flat-screen structure used by Zenith is shown in Fig. 77. The panel of the tube consists of a flat plate glass without any skirt. The mask, consisting of a very thin sheet of steel about 0.025 mm thick, is stretched at an elevated temperature and welded onto elongated metal rails attached by glass frit to the inner surface of the panel. Because the mask is maintained under tension, local heating by the electron beam causes a relaxation of the stress rather than a deformation (doming) of the mask. Compared with a conventional curved shadow mask more than 8 times the power per unit area can therefore be absorbed before comparable doming effects are observed. For high-resolution data display applications holes with a pitch of about 0.2 mm or smaller are required. This can only be achieved with the thin material used in the tension mask.

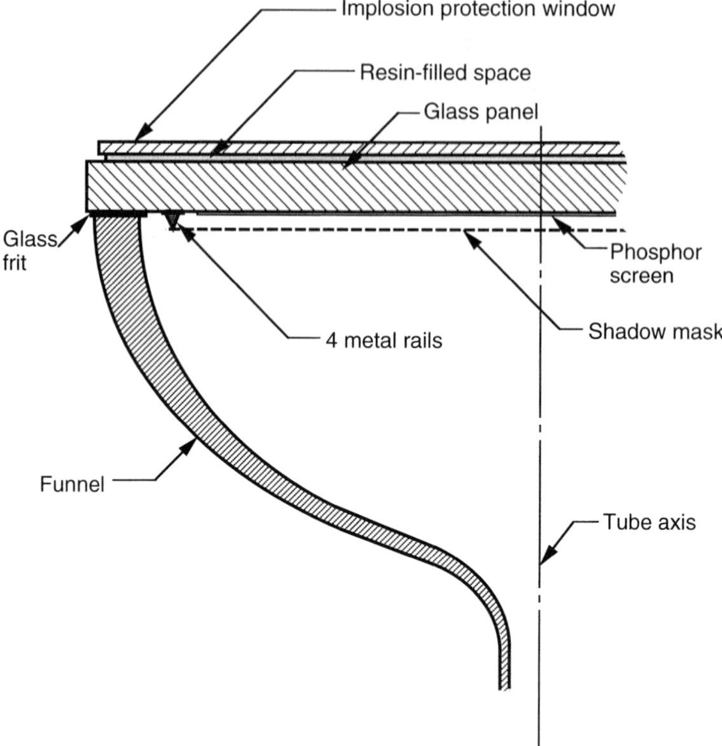

FIGURE 77. Zenith flat tension mask suspension system.

The use of a flat glass plate for the panel also makes it possible to cement a plate of ordinary window glass to the panel for implosion protection. A major problem with such tubes, however, is the need to use a faceplate thick enough to withstand the atmospheric pressure as mentioned in Section VIII.A.5. In the case of a 41 cm tube, for example, the panel thickness must be more than 15 mm compared to a thickness of only about 11 mm for a conventional curved panel. As mentioned before, the tube-processing time must be substantially increased to allow proper heating and cooling of the flat faceplate.

Matsushita also introduced a completely flat face tube (Taki and Arimoto, 1996). In this case a glass skirt is attached to the flat glass panel as in a conventional tube. The shadow mask (0.025 mm thick), maintained under tension on its frame, is stretched flat and has slit apertures of 0.24 mm pitch. This arrangement was used in a 41 cm diagonal tube.

Sony also introduced flat face tubes with much larger diagonal screen sizes of 66 and 76 cm for consumer use (Iguchi *et al.*, 1997). In these tubes a slight curvature was initially given to the panel glass in order to obtain an almost completely flat surface after evacuation. In addition, the panel glass was tempered to give it more strength and to minimize the increase of panel thickness and weight. In the case of the tube with the 66 cm diagonal screen, for example, the center thickness of the panel was reduced to 15.5 mm, a 10% reduction compared with nontempered glass.

X. Contrast Enhancement

Along with resolution and brightness, contrast is one of the major factors determining the performance of a CRT. Sufficient contrast allows better readability even if brightness is low. On the other hand, increasing the brightness of a low contrast image will not always improve its readability. Obtaining high contrast has thus always been a major objective in improving the performance of CRTs.

The brightness of the excited screen area is determined by beam power and light-emitting characteristics of the phosphor. When viewing an image in a dark room the phosphor areas not excited by the electron beams generally remain sufficiently dark to yield good contrast images because scattered light from excited phosphor area does not reach them. However, when viewing the same image under normal room lighting the contrast is reduced largely by the reflectance from the dark areas of the screen. The following discussion is concerned with methods of minimizing such undesirable "wash-out" phenomena.

A. Reduction of Optical Transmission of Faceplate

The conventional way of improving the contrast is to reduce the transmission T of the panel glass. The brightness of the display image will thus proportionally drop by this factor. Although ambient light reaching the phosphor will also be lowered by this factor, after reflection from the phosphor, this light will again be reduced by a factor of T.

Mathematically the contrast C (not contrast ratio) of the display exposed to ambient light is expressed as follows:

$$C = \frac{BT + IRT^2}{IRT^2} = \frac{B}{IRT} + 1 \tag{37}$$

In this formula, B refers to the luminance of the phosphor produced by the electron beams, T refers to the panel glass transmission, I refers to the illuminance at the outer surface of the glass panel by the external light and R

refers to the reflectance of the phosphor screen. Thus BT denotes the amount of light reaching the viewer, which was generated by the electron beam, and IRT^2 denotes the reflected light from the tube reaching the viewer. As shown by Eq. (37), the contrast improves as T is decreased.

Recently, thanks to the improved efficiency of phosphor screens and their operation at higher brightness, an industry trend has been to use panel glass with lower transmission. A transmission of 40–50% is now commonly used compared with the high transmission of 85% used in the early black matrix tubes, resulting in much higher contrast (for more details see Sections VIII).

B. Treatment of Outer Surface of Faceplate

A major concern in data display applications is specular reflection. Images specularly reflected from the CRT panel surface often disturb the observation of the displayed image and also reduce its contrast.

The simplest method to reduce specular reflection is to roughen the panel surface, scattering the light and diffusing the reflected images. Because this causes a loss of resolution the degree of roughening must be limited. A more basic approach is to reduce the surface reflectance itself by means of a multilayer optical coating, avoiding the loss in resolution. Although this is a more ideal approach it usually involves more expensive processing. Also, the nonreflecting effect can be easily destroyed by a fingerprint or other surface contamination. A combination of both methods for reducing reflection is also possible, as will be explained in the sections that follow.

In addition to the surface treatment for reducing reflection, a conductive layer is also combined with it in many cases. Capacitive charging of the outer surface of the screen by high voltage applied inside the panel section can be annoying because of the discharge sensation felt when a finger is brought close to the screen. The external potential also attracts dust from surrounding air, which builds up on the screen. To maintain the outer surface at ground potential a slight surface conductivity (about $10^9\ \Omega$/square) is added to the surface. This is often referred to as antistatic coating. Figure 78 shows the drop in surface voltage with time, which is obtained by adding an antistatic coating compared to the case of a bare glass surface. As shown, the surface voltage is reduced by the antistatic coating to less than 5 kV within 10 s, thereby significantly reducing the problem. In recent years, however, more stringent regulations have been instituted specifying the allowable electrostatic radiation from the display monitor, as mentioned in Section XIII. To satisfy this regulation a much lower resistivity, such as $10^3\ \Omega/\text{square}$ is required.

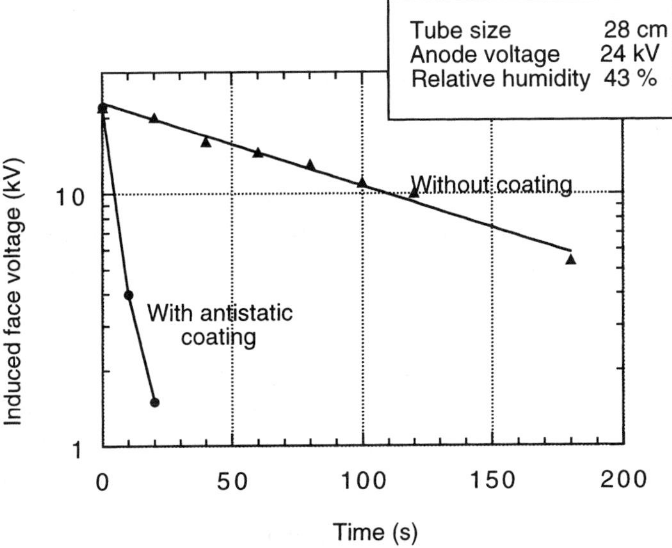

FIGURE 78. Examples of decay of static charge.

TABLE II
PANEL SURFACE TREATMENT SYSTEMS

Type	System	Conductivity (antistatic)	Figure in which shown
1. Nonglare	(a) Direct etching	No	Fig. 79(a)
	(b) Silica coating	No	Fig. 79(b)
	(c) Direct etching with tin oxide coating	Yes	Fig. 79(c)
	(d) Antistatic coating	Yes	Fig. 79(d)
2. Antireflection	(a) Antireflection (AR)	Yes	Fig. 80(a)
	(b) Direct antireflection (Direct AR)	Yes	Fig. 80(b)
	(c) Double-layer AR	Yes	Fig. 80(c)
	(d) Filtered double-layer sol-gel AR	Yes	Fig. 80(d)
3. Nonglare + antireflection	(a) Triple-layered sol-gel AR coating	Yes	Fig. 80(e)

For production, many kinds of surface treatment have been developed. These are summarized and classified in Table II. Brief explanations of each system are given here (Kawamura *et al.*, 1994).

1. *Nonglare Type*

a. Direct Etching. One method for roughening the panel surface is chemical etching (Ishii *et al.*, 1983). This is normally done to a bare panel glass before tube manufacturing. The appearance of the etched surface is shown schematically in Fig. 79(a).

b. Silica Coating. Another method for producing a rough surface is to use a silica coating. A sol solution made from $Si(OR)_4$ (R: alkyl radical), for instance, is sprayed on the finished tube face after CRT manufacture. The sprayed film is then polymerized by heating, being converted to a roughened silica gel film (a procedure referred to as a sol-gel process) which acts as a nonglare coating (Ohta, 1975). A schematic of this surface structure is shown in Fig. 79(b). Compared with direct etching, silica coating is believed to provide slightly better resolution performance for a given level of glare reduction because of the detailed structure of the roughened surface. Because the silica coating can be made at lower cost, this system is widely used in normal display applications. Based on this sol-gel process, many new processes have been developed as will be explained here.

c. Direct etching with tin oxide coating. To provide surface conductivity, a transparent tin oxide film can be coated directly on an etched surface using a chemical vapor deposition (CVD) method (Daiku and Okada, 1988). As shown in Fig. 79(c), the tin oxide film follows the surface contour.

d. Antistatic silica coating. In this case a very thin conductive coating, less than 1 μm thick, is deposited on the tube face, usually by applying an alcohol solution of ethylsilicate mixed with conductive material, such as SnO_2 or In_2O_3. A normal silica coating is then deposited (Tawara, 1987). The resulting structure, shown in Fig. 79(d), is widely used commercially.

2. *Antireflection Type*

a. Antireflection (AR) System. For reducing specular reflection a multilayer antireflecting coating commonly used in the optical industry for lenses and other glass surfaces (Rock, 1969) is applied to the CRT panel. Such coatings typically consist of alternate film of low refractive index material (such as SiO_2) and high refractive index material (such as TiO_2 or SnO_2) deposited by evaporation. By selecting materials with the proper refractive

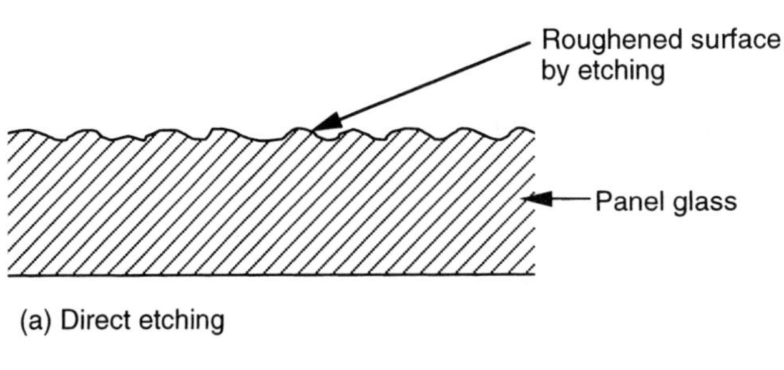

(a) Direct etching

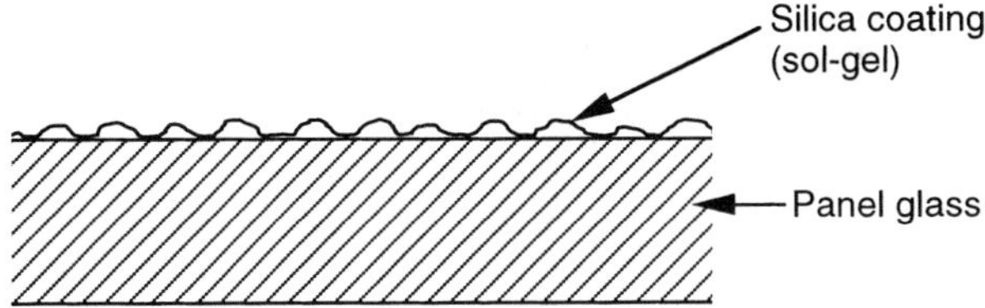

(b) Silica coating

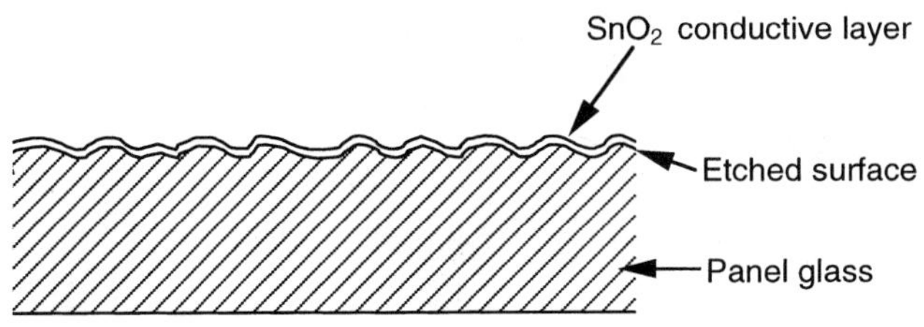

(c) Direct etching with tin oxide coating

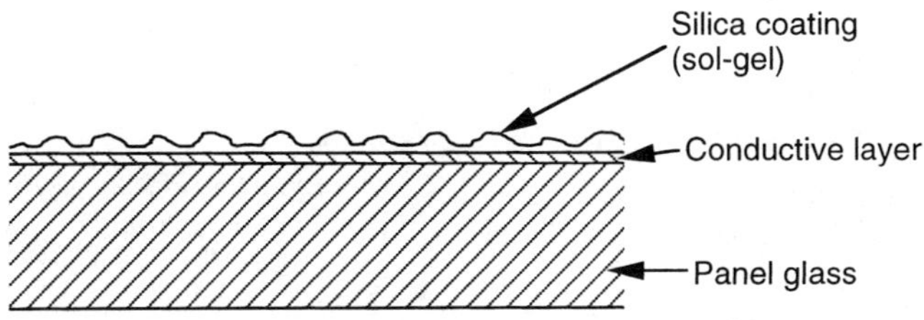

(d) Antistatic coating

FIGURE 79. Forms of nonglare coatings.

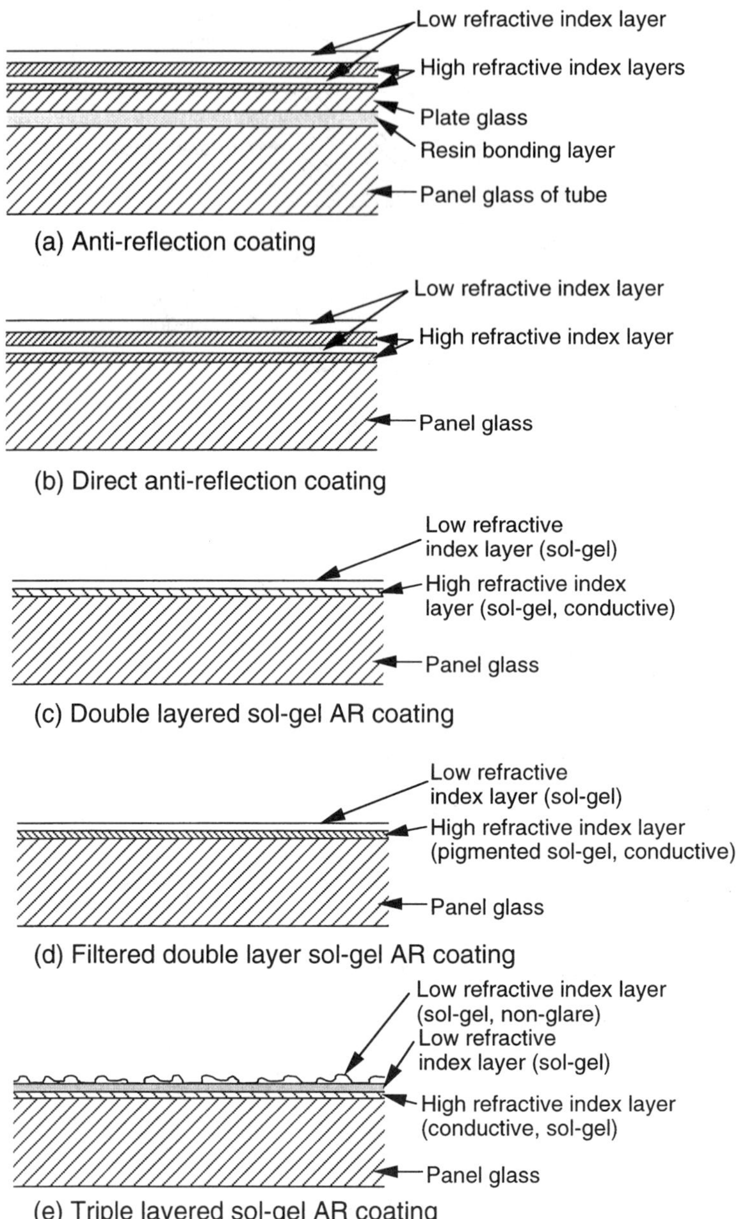

FIGURE 80. Forms of antireflection coatings.

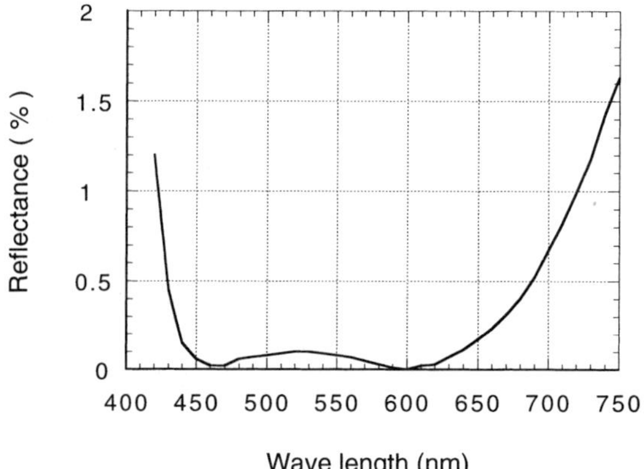

FIGURE 81. Spectral reflectance of antireflection coating of Fig. 80a. [Reprinted by permission from Kawamura, H. *et al.* (1994). Trends in treatment of the viewing surface of color picture tubes, *Display and Imaging*, **3**, 51–57.]

indices and controlling their thickness, optical interference effects greatly reduce the surface reflections. Figure 80(a) shows an example of the entire panel structure and Fig. 81 shows a typical curve of the reflectance as a function of wavelength. As shown the reflectance is less than 0.3% over the visible spectrum. This should be compared with a reflectance of 4.2% for light incident normally on bare glass.

The antireflection coating is first applied to a separate tempered glass plate about 3 mm thick and with a curvature that matches the CRT screen panel. The plate is then bonded to the tube face by a layer of resin about 2.5 mm thick. Because of the very small difference of refractive index between the two, reflection at the bonding surface between the glass and resin can be ignored. This is an ideal system because there is no loss of resolution and almost complete reduction of reflection. However, the manufacturing cost is very high so this approach is usually employed only for high-end high-resolution graphic display monitors. Unfortunately, if the outer surface is touched by a finger, the surface contamination left will disturb the optical condition required for canceling the reflections at the local area, making it necessary to clean the face frequently with an appropriate solvent. In most cases in this system an outer conductive layer is combined for antistatic purposes.

b. Direct Antireflection (Direct AR) System. As already mentioned, in order to eliminate the bonding process required in the antireflection system, the multilayered coating can be deposited directly on the tube face by magnetron sputtering at room temperature (Scobey *et al.*, 1989). The resulting structure is shown in Fig. 80(b) (Kuhlman and Kurman, 1994).

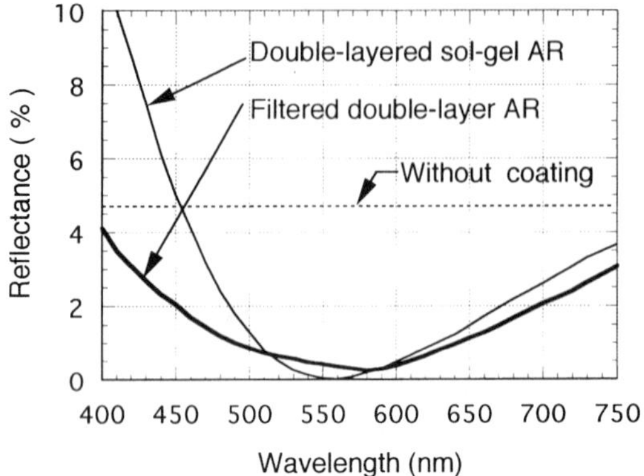

FIGURE 82. Spectral reflectance of double-layered sol-gel AR coating of Fig. 80c and d. [Reprinted by permission from Kawamura, H. *et al.* (1994). Trends in treatment of the viewing surface of color picture tubes, *Display and Imaging*, **3**, 51–57.]

Although satisfactory performance can be achieved with this method, for mass production a large investment is necessary.

c. Double-Layered Sol-Gel AR Coating. By combining two sol-gel layers of low and high refractive indices, respectively, an antireflection effect can also be obtained. An example of the structure used and the performance obtained are shown in Fig. 80(c) and Fig. 82, respectivley. In this case a conductive coating such as antimony tin oxide (ATO), indium tin oxide (ITO) or TiO_2 can be used for the layer of higher index. The finished coating has a smooth surface and high resolution can be obtained. On the other hand, because there is no diffusing effect reflected images are not suppressed. Because of its relatively low cost, however, this system has been applied to TV picture tubes. The possibility of applying it to data display tubes depends on how much further improvement can be achieved (Ono and ohtani, 1992; Onodera and Matsuda, 1993; Hayama *et al.*, 1994).

d. Filtered Double-Layer Sol-Gel AR Coating. In order to improve the contrast ratio, pigment or dye can be added to the conductive (high index) layer of the forementioned double-layer sol-gel AR coating. By using a pigment or dye having a sharp absorption band midway between the peaks of the phosphor emission spectra, effective absorption of ambient light can be obtained with minimum absorption of the phosphor emission, thus improving the contrast. The structure is shown in Fig. 80(d) and the performance is shown in Fig. 82. The body color of an unexcited phosphor screen is usually greenish. If desired, however, this color can be controlled by modifying the absorption characteristics of the pigment or dye. This approach is used mainly for TV picture tubes. To enhance the contrast over the enitre spec-

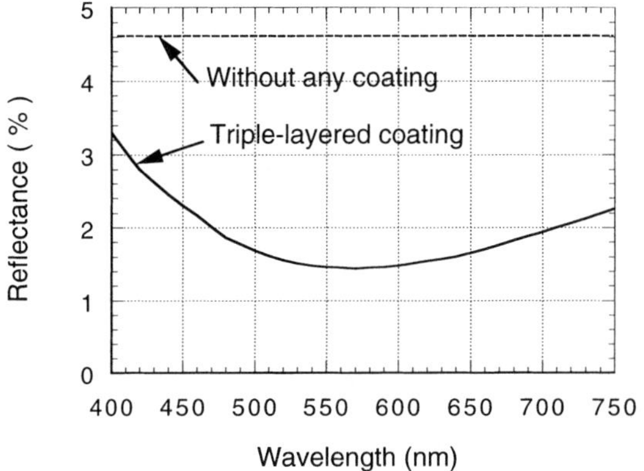

FIGURE 83. Spectral reflectance of triple-layered sol-gel AR coating of Fig. 80e. [Reprinted by permission from Kawamura, H. *et al.* (1994). Trends in treatment of the viewing surface of color picture tubes, *Display and Imaging*, **3**, 51–57.]

trum black pigment or dye can be used for TV tubes as well as data display tubes (Toshiba, 1989).

3. *Nonglare Plus Antireflection Type*

a. Triple-Layered Sol-Gel AR Coating. In order to enhance the antireflection effect of a double-layered sol-gel AR coating, an additional layer can be added in the form of a low index nonglare silica coating, resulting in a combination of antireflection and nonglare technology. By balancing both characteristics, fairly good performance can be achieved at a reasonable cost. The structure of this arrangement and the performance obtained are shown in Fig. 80(e) and Fig. 83, respectively (Tohda *et al.*, 1992).

C. *Treatment of Inner Surface of Faceplate*

The preceding discussions all relate to the outer surface of the tube face. However, some reflection also occurs at the inner surface of the glass panel. This surface is normally very finely roughened (referred to as "stippling") during the process of molding the panel. Although stippling scatters light from reflected images, it causes no loss of resolution, because the scattering surface is directly in contact with phosphor screen.

D. *Black Matrix*

As mentioned in previous sections the black matrix is used in all color CRTs today to reduce the reflected ambient light. Before the black matrix

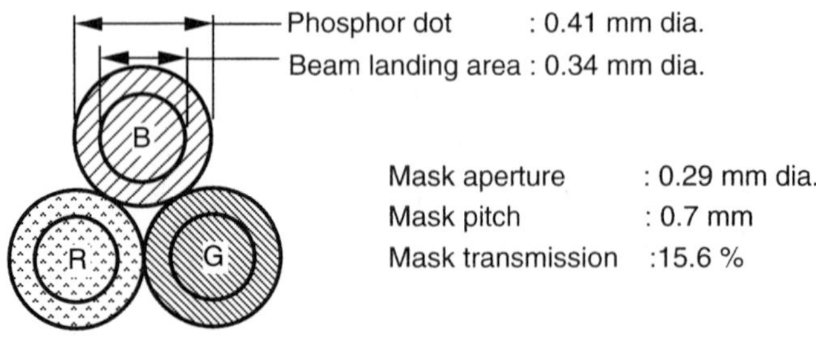

(a) Non black-matrix

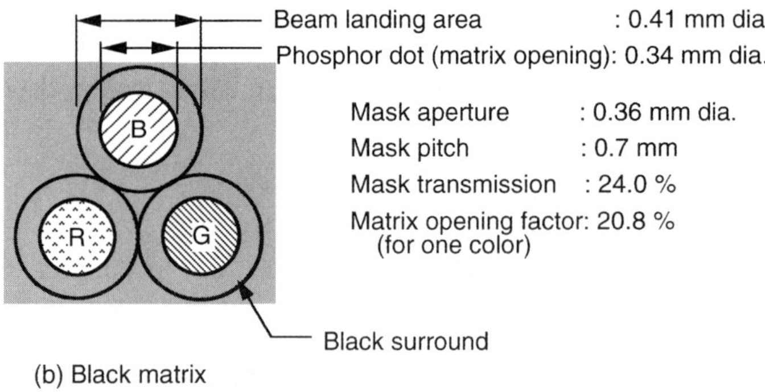

(b) Black matrix

FIGURE 84. Comparison between non black-matrix and black-matrix screen assuming same shadow-mask pitch of 0.7 mm and same light emitting area of 0.34 mm diameter for each color dot.

technology was introduced, the three-color phosphor dots of a dot screen were deposited so that they almost touched each other. To prevent electrons passing through a mask hole from partially landing on the wrong color dot the diameter of the shadow mask apertures was reduced as shown in Fig. 84(a), so that the diameter of the areas on which the electron beams landed was smaller than the diameter of the phosphor dots. One-half of the difference between these diameters is regarded as the landing tolerance. Viewed from the front side, the phosphor screen appears almost entirely covered with the three-color phosphor dots except for the small bare area left between them, which is coated with aluminum. As a result, overall about 87% of the ambient light is diffusely reflected from the phosphor screen.

In the case of the black matrix shown in Fig. 84(b) the diameters of the phosphor dots and the beam landing areas are, respectively, reversed (Fiore and Kaplan, 1964). The phosphor dots are now surrounded by a black nonluminescent film whose area is generally about 40 to 50% of the total area of the phosphor screen, allowing the same landing tolerance to be obtained as in the nonmatrix tube. The overall reflectance of the phosphor screen is thus reduced to about 48%, thus doubling the contrast of an image viewed in ambient light. In the case of a striped phosphor screen a line type of black matrix is similarly applied with comparable increase in contrast (Morrell *et al.*, 1974).

The forementioned discussion assumed that the transmission of the panel glass was constant. However, the image brightness can be improved if, instead of using a panel glass with 40% transmission (as employed with the non black matrix), an almost clear glass of 85% transmission is used with the black matrix. In effect, this doubles the brightness without causing a loss in contrast. In effect the black matrix can be used either to increase contrast or brightness.

The fabrication of the black matrix, however, presents special problems in the optical exposure processes used. In non black-matrix tubes phosphor dot diameters larger than the mask aperture diameters can be obtained by simply selecting a light source with an appropriate diameter. However, in black-matrix tubes openings in the black matrix smaller than the mask apertures must be created. To accomplish this a smaller diameter portion of the penumbra must be used (see Fig. 23) during exposure. This makes it difficult to ensure uniform opening diameters because the opening is not sharply defined by slope of the penumbra and is dependent on the processing conditions. Furthermore, because exposures are separately made for the black matrix openings surrounding the red, green and blue phosphors, it becomes difficult to maintain the aperture diameter of the black matrix constant from one color to another. In particular, if the ratio of aperture diameters of the three colors varies with location, the uniformity of the white color will be degraded. However, currently, the increased precision used in the process and the development of a new photoresist sensitizer with a more nonlinear characteristic have resolved these problems, enabling this process to be used for mass production. It should be noted that in the case of a non black matrix screen, even if the diameter of the phosphor dots is not uniform, the actual light-emitting area of each phosphor dot of a triad will be the same because the size of the beamlets passing through individual mask apertures will be equal.

E. Pigment Coating of Phosphor Grains

An additional method for enhancing contrast is to reduce the reflectance of the phosphor particles themselves. It is obvious from Eq. (37) that the

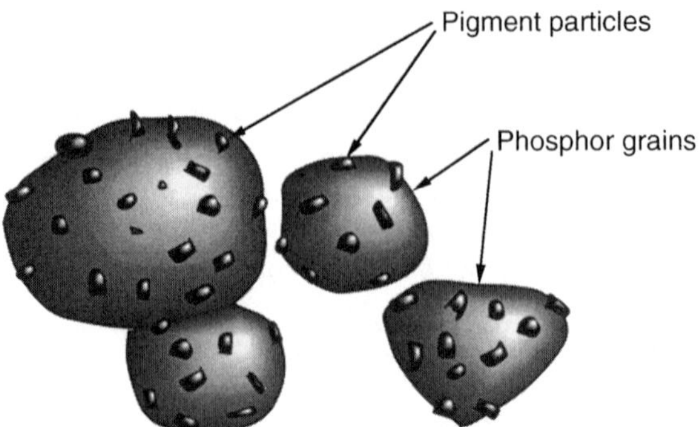

FIGURE 85. Pigmented phosphor.

reduction of R would be effective in increasing contrast if the luminanace of the phosphor was unchanged. One way to achieve this is to add to the surface of the phosphor grains an appropriate bandpass optical filter that does not reduce the light from the phosphor but absorbs the light of other wavelengths.

To accomplish this in practice the phosphor grains are coated with small pigment particles as shown in Fig. 85. For red emitting phosphors a red pigment of ferric oxide (Fe_2O_3) is applied; for blue phosphors a blue pigment of cobalt aluminate ($CoAl_2O_4$) is applied. However, green pigment is not usually applied to the green phosphor because this phosphor normally has a slightly greenish body color, which provides a self-filtering action. Also no suitable pigment for this purpose is available (Trond, 1979). Multiple reflections occur within and between the phosphor grains so a large improvement in reflectance can be obtained with a small pigment coverage of the phosphor grains.

Pigmenting the phosphor provides a significant advantage in contrast improvement with minimum brightness sacrifice. Examples of the reflection of conventional red and blue phosphor screens with and without pigment are shown in Fig. 86(a) and (b), respectively. Examples of the quantitative performance improvement obtained with these phosphors are shown in Table III.

F. Elemental Color Filters

A more effective way to absorb ambient light is to place a separate tiny bandpass filter element between each of the color phosphor elements and the faceplate, allowing the light emitted from each phosphor element to be transmitted while absorbing ambient light outside of the passband of the filter.

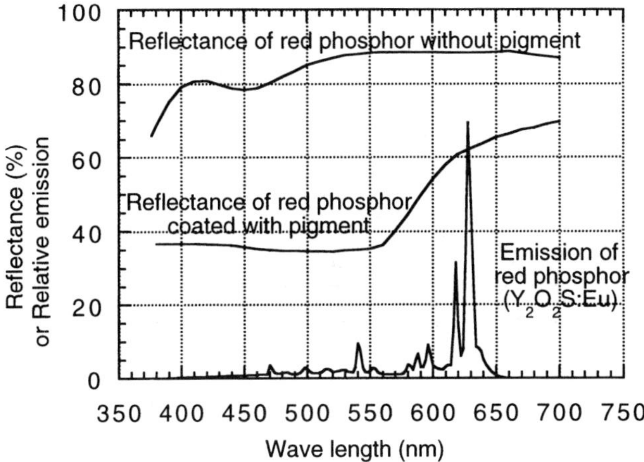

(a) Pigmented red phosphor

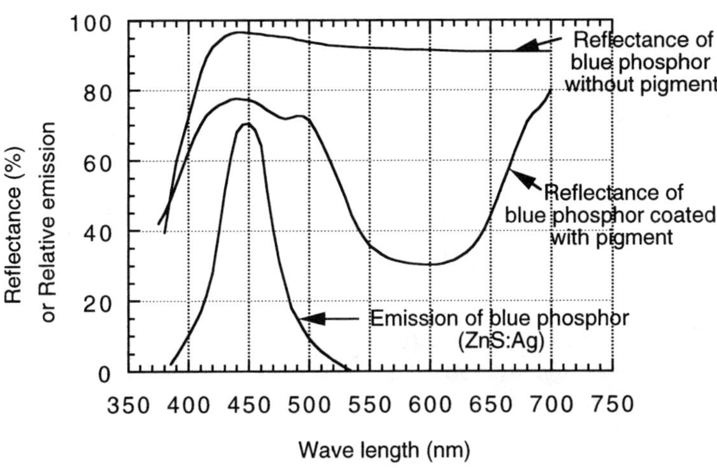

(b) Pigmented blue phosphor

FIGURE 86. Spectral reflectance and emission of red (a) and blue (b) phosphors with and without pigments.

TABLE III
EXAMPLES OF PERCENTAGE CHANGE IN PERFORMANCE OBTAINED BY EMPLOYING PIGMENTED PHOSPHORS RELATIVE TO UNPIGMENTED PHOSPHORS

	Percent pigment	
	Blue 1.5% Red 0.15%	Blue 5% Red 0.5%
Brightness	97%	88%
Contrast	111%	124%
Reflection	88%	70%

TABLE IV
EXAMPLES OF PERCENTAGE CHANGE IN PERFORMANCE OBTAINED BY EMPLOYING FILTERED PHOSPHOR SCREEN RELATIVE TO UNPIGMENTED PHOSPHORS

	Filtered screen	Conventional (unpigmented)
Brightness	120%	100%
Contrast	174%	100%
Reflection	69%	100%

As expected, the results obtained are far better than with the pigmented phosphor technology as shown in Table IV (Itoh *et al.*, 1995). Because of the significant improvement, production of tubes with the new type of filters is increasing even though their fabrication cost is very high due to the additional processing required.

XI. MOIRÉ

In operation a shadow-mask tube sometimes produces a moiré pattern on the screen because of the difference in spatial frequency between the vertical structure of the screen f_1 and the spatial frequency of the horizontal scanning lines f_2. This appears as sets of dark and light horizontal bands whose spatial frequencies are the difference between the two frequencies f_1 and f_2 and their harmonics. The spatial frequency f_1 corresponds to the dot structure of the shadow mask or to the tie bar structure in the case of the striped screen.

If the light and dark bands of the moiré appear too prominent they may interfere with viewing of the images on the screen. There are two ways to reduce the moiré effect. One is to make the moiré frequency high enough so that it is not detected by the eye. The other is to reduce the contrast of the moiré pattern itself.

In a system where the spatial frequency of the horizontal scanning lines is fixed, the screen can be fabricated with a vertical pitch, which causes the spatial frequency of the moiré to be sufficiently high so as not to be disturbing. For this purpose many studies have been conducted (Morrell *et al.*, 1974; Wittke, 1987; Shiramatsu and Inoue, 1994). A useful mathematical formulation giving the spatial period of the moiré produced by different combinations of the vertical shadow-mask period, the spacing (i.e., period) between scan lines and their spatial harmonics was developed by Wittke (see the forementioned reference). This is given in Eq. (38), where M is the moiré pitch, m,n are the index numbers of the Fourier components, h is the vertical screen pitch (i.e., $1/f_1$) and s is the scan line spacing (i.e., $1/f_2$). It should be noted here that the vertical screen pitch is not the pitch of the mask but a slightly enlarged pitch projected to the screen by electron beam

$$M_{mn} = \left|(m/h - n/s)^{-1}\right| \tag{38}$$

Using Eq. (38), a number of curves, normalized with respect to the scan-line spacing, have been plotted in Fig. 87 for the moiré pitch produced by combinations of f_1, f_2 and their first harmonics as a function of the vertical screen pitch. In order to reduce moiré visibility as much as possible the moiré pitch M should be made as small as possible. As shown in Fig. 87 this can be achieved by choosing a normalized vertical screen pitch of 2/3 or 3/2, depending on which combination of fundamental and harmonic is considered, as shown by the abscissa of the downward pointing arrows in the figure.

Recently, multimode scanning systems have been incorporated in most data display monitors and even in some TV receivers. Because these allow the number of the scan lines in a frame to be varied there is no single optimum screen pitch that can be selected from Fig. 87. The only solution is thus to reduce the contrast of the moiré pattern by reducing the amplitude of the brightness variation of f_1 or f_2 spatial frequencies. Because the screen structure is designed to satisfy other aspects of the required tube performance it cannot be modified easily.

The only other alternative is to increase the beam spot size at the phosphor screen so that the space between scan lines is filled as much as possible, thus minimizing the visibility of any line structures. Although the beam spot size is

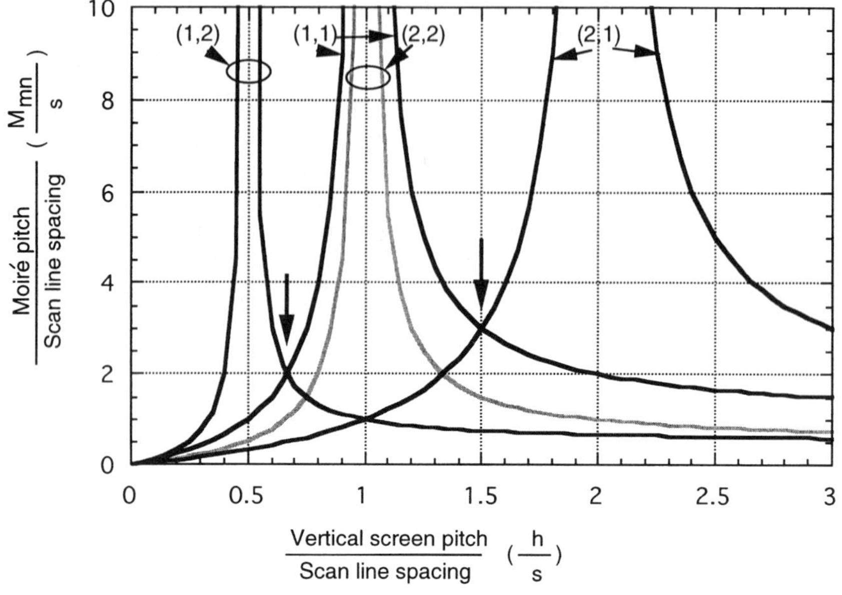

FIGURE 87. Evaluation chart for moiré generation. Minimum moiré period at vertical screen pitch of 2/3 or 3/2 of the scan line spacing (shown by arrows). (Permission for Reprint, courtesy Society for Information Display.)

normally minimized to achieve best focus one possibility is to defocus the spot slightly. Alternatively the beam spot may be elongated only in the vertical direction, thus reducing the moiré without any sacrifice in horizontal resolution. Althoug this can be accomplished by modifying the electron gun design it is very difficult to keep the same spot shape over the entire screen. An alternative method is to slightly deflect the beam at a high frequency in the vertical direction. This is the ideal solution, allowing maximum moiré reduction to be obtained for each scanning mode by adjusting the amplitude of the high-frequency deflection.

It should be added that in some cases moiré produced between the horizontal screen structure and the horizontal dot pattern produced by the video signal can be observed. The moiré in this case can also be minimized by selecting an optimum screen structure (stripe pitch) and reducing the visibility of the dot structure produced by the signal by increasing the width of the dots created by the signal.

XII. Shadow-Mask Tubes for High-Resolution Display Applications

A. *Data-Display Tubes*

Color CRTs employed in computer monitors, medical equipment, industrial instruments, computer-aided design (CAD) and computer-aided manufacturing (CAM) generally have performance requirements that are different and frequently more demanding than those for TV applications. Such CRTs are referred to here as "data display tubes" to distinguish them from the color picture tubes commonly used for TV. The various aspects of such tubes are discussed in what follows.

1. *Resolution*

Generally, the most important factor for data display tubes is the resolution. As discussed in previous sections, this is determined by the spot diameter of the electron beam and the phosphor dot (or stripe) pitch of the screen. A useful way to express the resolution capabilities of a tube is in terms of its modulation transfer function (MTF) (Ohishi, 1970). This is indicated by a curve that shows the magnitude of the brightness variation (response) as a function of the spatial frequency expressed in TV lines when the beam current is 100% modualted at different frequencies. Figure 88(a) shows examples of the MTF curves obtained from a 34 cm tube without a shadow mask for different spot diameters D_S. The spot diameter is defined here as the width between the points where its brightness is 5% of the peak brightness.

Figure 88(b) shows the MTF curve obtained when a tube with a dot screen pitch of 0.28 mm and infinitely small apertures is scanned with an infinitely narrow beam. Depending on the phase relation between the signal, which modulates the beam and the mask structure, two MTFs (a max and a min) result, representing the best and worst conditions. Assuming the min curve and convoluting it with the MTF curves of Fig. 88(a) we can obtain the combined MTF curves shown in Fig. 88(c). For example, if we assume a combined MTF of 30% as a resolution limit for data display purposes, this figure shows that with a beam diameter of 0.5 mm a resolution of 640 TV lines would be obtained. In effect, in order to increase resolution, both the mask pitch and the beam spot must be made smaller.

Sluyterman (1993) reexamined the effect of shadow-mask pitch and concluded that the resolution will not be deteriorated as much as was indicated in the foregoing. This is due to the fact that the mask aperture size was assumed to be infinitely small, resulting in essentially no brightness gradation within a dot. In practice, however, this is not the case. It is thus not

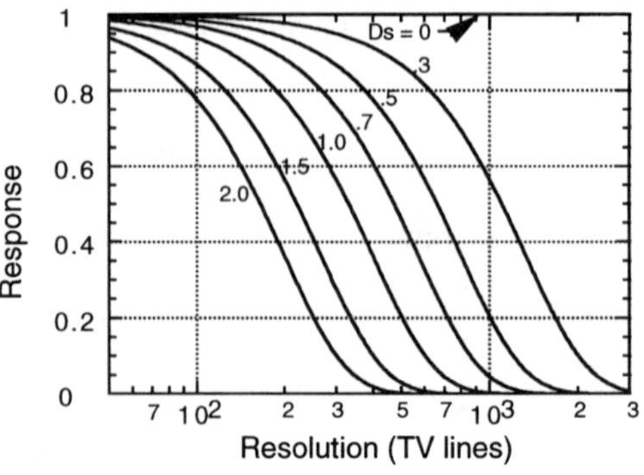

(a) Electron gun MTF

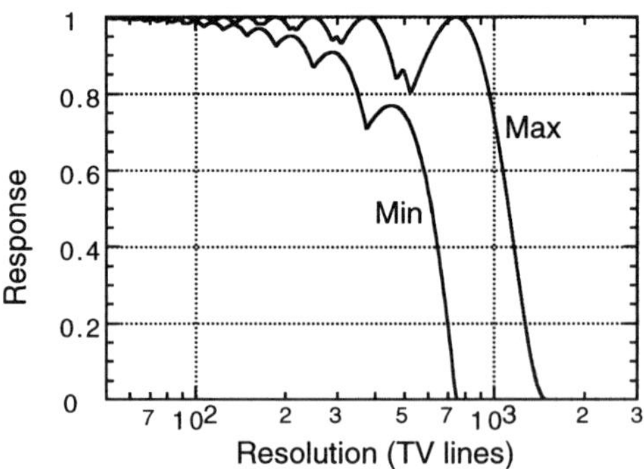

(b) Mask-aperture MTF

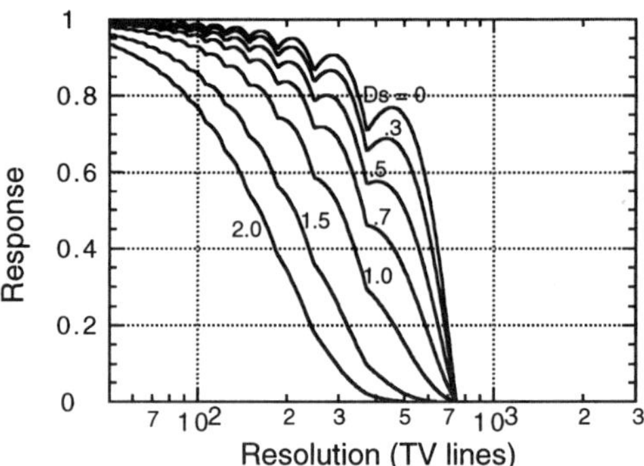

(c) Overall MTF

FIGURE 88. The MTF characteristics.

necessary to reduce the shadow-mask pitch in inverse proportion to the spatial frequency increase.

In the case of the color phosphor triads of a dot screen the finest pitch commercially available is 0.21 mm, with a pitch of even 0.15 mm having been used experimentally (Ando *et al.*, 1985). However, for practical use, the lower limit is currently 0.26 mm. Excessive reduction of the pitch may not allow a proportionate reduction in the width of the black matrix area because of manufacturing tolerances, thus resulting in a loss of brightness. As already mentioned, another obstacle to obtaining smaller pitch arises from the etching limitations of the shadow mask.

In data display tubes, regardless of their lower brightness and beam-current requirements, spherical aberration is the major factor limiting spot size. To reduce the aberration a gun with multistage focusing is generally used. However, with some degree of spot defocusing or deformation due to deflection occuring at the corners, especially in the case of a self-converging deflection system, it is necessary to employ dynamic focusing or dynamic astigmatism correction to minimize this. To obtain a sufficiently small spot, it may also be necessary to reduce the diameter of the control-grid aperture. In order to obtain the necessary beam current, however, this may result in a current density at the cathode surface that exceeds the permissible limit of a conventional oxide cathodes. In this case the use of a so-called dispenser (impregnated) cathode is required. (A more detailed discussion of gun problems is provided in the following chapter "Electron Gun Systems for Color Cathode Ray Tubes" by Suzuki.)

2. *Screen Size*

As mentioned earlier, a tube with a size of about 34 cm (13 V) is believed to be ideal for many data display applications. However, for the viewer to benefit from the full resolution of the input signal, tubes with a larger screen size such as 41 cm (16 V), 48 cm (19 V) or 51 cm (20 V) have come into widespread use. In the case of CAD and CAM applications, where even more information is displayed, 48–51 cm tubes appear essential; for applications such as air traffic control displays, still larger rectangular screens with 68 cm (27 V) along the diagonal as well as tubes with square-shaped screens (70 cm diagonal) have come into use.

3. *Brightness*

As mentioned in Section XIII.A, to prevent eye fatigue data display tubes are normally operated at brightness levels and contrast ratios considerably less than those of TV applications. In particular, for data display applications the brightness of the highlight areas is limited to about 100 cd/m^2 and the contrast limited to about 3–10 compared with a brightness of about 300 cd/m^2 and contrast of 30–100 in TV applications.

4. *Deflection Yoke Requirements*

The requirements on the deflection yoke are especially severe in the case of data display tubes. The beam convergence in such tubes is as important as the resolution, making it advisable to maintain the misconvergence less than the center-to-center spacing of adjacent mask apertures. As explained in Section VII, various technologies have been developed to accomplish this. Also, as explained in Section VII, the higher line-scanning frequencies used in high-resolution displays cause many technical difficulties that need to be resolved.

Because the adjustments required for obtaining good convergence and minimum raster distortion involve difficult procedures, this work is often done in the tube manufacturer's plant rather than by normal operators in a monitor assembly line. In this process, referred to as integrated tube and components (ITC), the tubes are assembled with their deflection yokes and necessary neck components, then properly adjusted, fixed, and shipped to their customers.

B. *High-Definition Television Tubes*

For high-definition television (HDTV) where a larger picture size and higher resolution are required new tubes have been developed. These have a screen with an aspect ratio of 16:9 and an increased number of scanning lines (1125 lines in Japan and 1050 lines in the United States).

Although the ideal diagonal screen size for this purpose is 1.5 m (60 in) or larger, this is currently achieveable with CRTs only by means of projection technologies. However, smaller direct view systems are also of interest and for this purpose tubes with 86, 76 and 66 cm diagonal size with a 16:9 aspect ratio have been standardized in Japan.

Aside from the new aspect ratio of the screen this requires redesign of the electron gun, shadow-mask and screen. As in the case of data display tubes a multistage focusing gun with dynamic focus and dynamic astigmatism correction are required. However, in the case of HDTV, as in normal TV, since the peak beam current for white areas reaches as high as 6 mA or more compared with 1 or 2 mA for data display tubes, the gun design has to be appropriatley modified.

The phosphor triad pitch also has to be reduced for HDTV pictures. In the case of an 86 cm diagonal tube, for instance, a dot structure with a 0.4 mm triad pitch is used for broadcast monitors and a stripe structure with a 0.74 mm triad pitch is used for normal HDTV receivers.

In some cases tubes with a conventional resolution and a wide aspect ratio (16:9) are of interest because they are capable of displaying movie pictures with a wide aspect ratio that are broadcast on existing TV channels with a letter box format. Tubes for this format are now being produced, especially in Japan.

XIII. Human Factors and Health Considerations

A. *Ergonomic Issues*

Visual display terminals (VDTs), differing from TV displays, are generally used in business offices and generally viewed at close range for long periods. Because of this, ergonomic issues regarding VDTs are often discussed, particularly for reducing eye fatigue and improving legibility. The International Organization for Standardization (ISO), for example, has accepted a standard in its document ISO 9241-3 (1992) that specifies that the height of characters should subtend at least approximately 20 to 22 minutes of angle at a normal viewing distance. This means the character height should be at least 2.6 mm for a viewing distance of 45 cm.

Excessive brightness and contrast can also increase eye fatigue. For data display applications the ideal brightness of white highlight areas is believed to be about 100 cd/m^2 and the contrast ratio about 3–10 (ISO 9241-3, p. 7, 1992). With the growing use of multimedia applications television-type images are often presented on data display monitors along with the normal data. For this reason and also because of the limited brightness requirement of data displays the transmittance of the glass (faceplate) panels currently used is reduced to about 50% (see Section X.A), almost the same as in the case of TV applications. In addition, a color-selective filter may be provided on the panel surface to absorb ambient light more effectively and increase the contrast (see Section X.B). Other ergonomic issues are mentioned in what follows.

1. *Surface Reflection*

Because the specular reflections on the CRT faceplate panel can be a major distraction for viewers reading information, an antireflection coating or a nonglare coating is provided on the panel surface (as discussed in Section X.B).

2. *Flicker*

If the interlaced scanning used in commercial television is employed for a data display, this causes characters on the screen to appear to jitter up and down by one line at the frame rate (either 30 or 25 Hz, for example). This effect is called dot flicker and can be avoided by using progressive scanning (whereby each successive line of a frame is scanned sequentially without interlacing). However, such scanning exacerbates large-area flicker if the frame rate is too low. The frame rate above which the human eye can no longer sense flicker (the critical flicker frequency) is believed to be about 70–80 Hz, but in some cases may be as high as 90 Hz for bright images. To some extent, however, the critical flicker frequency depends on the individual observer. The frame rate normaly employed in present displays is 70–80 Hz. Thanks to improvements in circuit technology the cost of doing this has become less of a problem. To some extent, but not entirely, the flicker can be reduced by employing long persistence phosphors such as those listed in Table I. Such phosphors, however, are no longer used due to the increase in frame rates.

3. *Phosphor Emission Color*

Although data display tubes generally use the same color phosphors as television tubes these may not always be appropriate when using a CRT solely for character display purposes. Although blue phosphors with good color purity are used in television in order to reproduce realistic and natural images, characters displayed in this color are extremely difficult to read because of the lower visual sensitivity of the human eye at these wavelengths and the resulting low brightness. This problem can be ameliorated by employing blue phosphors (sky blue color) with a longer wavelength and less saturated emission. However, such phosphors are no longer employed because they are not suitable for multimedia applications requiring natural-color images. In any case the desired sky blue color, if necessary, can be synthesized by generating an appropriate mixture of light by means of the three conventional phosphors of a color tube.

4. *Static Charge*

Because the CRT displays used for computer monitors are usually viewed at close range by the operator, the screen is often touched by their fingers. As explained in Section X.B the resulting discharge current may cause a minor discomfort. This can be avoided by making the surface of the faceplate panel slightly conductive as discussed in detail in Section X.B.

B. Suppression of Undesirable and Hazardous Radiation

1. *X-Ray Emission*

In order to control the X-ray emission from the CRT so that it is below the potential health hazard level, the allowable upper limits have been set by regulations in major countries. For instance, the United States Department of

Health and Human Services (HHS) specifies (H.E.W., 1976) a limit of 36 pA/kg (0.5 mR/h). Although other countries have adopted this limit as their standard, Germany has its own more stringent standard through the Röntgenverordnung (Röv, 1987), setting a limit of 1 mSv/h (approximately equivalent to 7.2 pA/kg). X-ray absorbing glasses are currently used in all CRT bulbs to satisfy these limits (see Section VIII.A). The X-ray absorption coefficient and thickness of the glass must thus be chosen for each bulb design. The upper limit of X-ray emission allowed for each bulb design is specified in the EIA (U.S.), EIAJ (Japan) or Pro Electron (Europe) standards. Figures 89 and 90, for example, indicate the standards established by the EIAJ for 34 cm tubes (EIAJ, 1992). Because no X-ray absorbing glass is available for the region near the anode button of the glass bulb, two lines are shown, one applicable to the anode button portion only and the other to the remaining portion. The limit lines for the entire bulb are thus only effective when the anode button portion is shielded by an appropriate external material.

Figure 89 shows the relationship between the anode voltage and the allowable anode current required to satisfy the X-ray emission limit at 36 pA/kg. Figure 90 indicates the maximum X-ray level as a function of the anode voltage for an anode current of 300 mA for both the entire tube and the anode button. In all cases, the allowable X-ray level is not permitted to exceed the given limit even in the failure mode of the tube. For example, 34 cm tubes are generally operated at an anode voltage of about 25 kV. Assuming, for

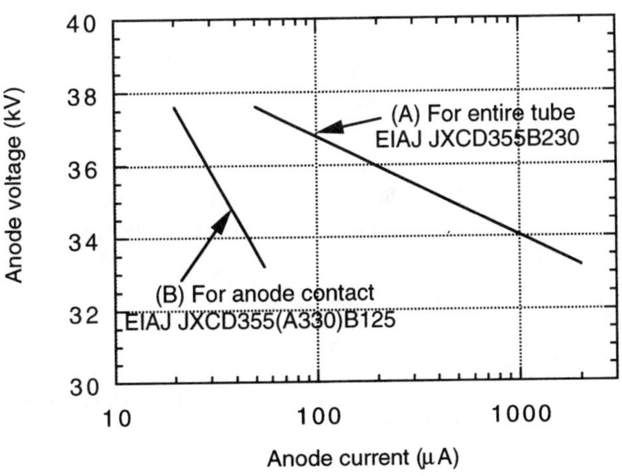

FIGURE 89. X-ray radiation limit curves for maximum allowable exposure of 36 pA/kg (0.5 mR/H) for 34 cm tube; (A) for entire tube; and (B) for anode contact button area. Note: All equipment may be operated below curve (A) provided an appropriate shield is used over the anode button.

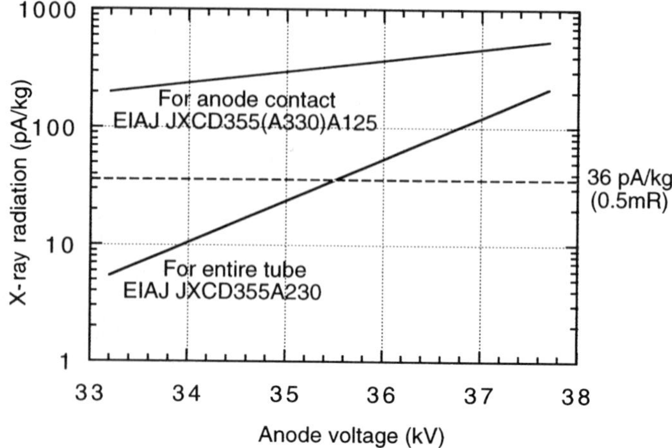

FIGURE 90. X-ray radiation limit curve for a constant anode current of 300 mA for 34 cm tube.

example, that the voltage rises to 33 kV due to failure of the high-voltage stabilizer the radiation must still be within the limit as shown by Fig. 90. It should be noted that the forementioned limits on X-ray radiation apply not only to color tubes but all other CRTs.

2. *Low-Power Emission of Magnetic and Electric Fields*

Some reports based on animal studies have suggested that prolonged exposure to electromagnetic waves, regardless of how weak in power or low in frequency, may result in undesirable effects in the human body such as abnormal pregnancies or birth deformities. Examples of this are the stray magnetic fields often produced by the deflection yoke and flyback transformer. Although numerous other studies have shown no causal relation between exposure to such fields and undesirable effects, the current trend is to preclude any possibility of ill effects by adopting strict standards. The first to introduce standards regulating the allowable magnetic and electric fields were the Scandinavian countries. Table V indicates the Swedish regulations

TABLE V
SWEDISH REGULATION (MPR II) ON ALLOWABLE STRAY FIELDS FROM VDTs

Frequency range		5–2 kHz	2–400 kHz
Electromagnetic field		250 nT max	25 nT max
Electrostatic field		25 V/m max	2.5 V/m max
Background field maximum	Electromagnetic	40 nT max	5 nT max
	Electrostatic	2 V/m max	0.2 V/m max

adopted in March 1987, on low-power emission of magnetic and electric fields from VDTs (MPR, 1990). Because the allowable magnetic field is approximately 1/1000 of the earth's field, measurement must be made in a well-shielded room. The methods used to reduce the magnetic field generated by the deflection yoke to meet this requirement, for example, are discussed in Section VII.

XIV. Concluding Remarks

During the past 20–25 years the performance of color cathode-ray tubes has improved greatly, largely in response to the requirements of computer and other multimedia applications. To satisfy these requirements almost every aspect of CRT performance was carefully scrutinized and numerous significant improvements were made during this period. In more recent years, in anticipation of the commercialization of high-definition television additional technical advances were also made.

Paralleling the advances in CRT technology during the past two decades an entirely new display technology has evolved, making use of liquid-crystal materials. This has led to the advent of the laptop computer, an application for which the conventional CRT is unsuited because of its inherent bulk and power consumption. The success of these small liquid-crystal displays has, in turn, stimulated intensive efforts to develop larger-sized liquid-crystal displays for use in desktop computer monitors, now almost entirely dominated by color CRTs.

Although liquid-crystal displays are attractive because of their panel-like form, desktop computer monitors with sufficiently large liquid crystal displays are presently estimated to cost 3–5 times more than monitors using color CRTs. In general, it is believed that at least a decade will be required before the cost of liquid-crystal monitors will become low enough to supplant CRT monitors.

In the case of directly viewed home TV, where factors such as long life, low-cost, high-brightness and wide viewing angle are paramount, it seems even less likely that liquid-crystal displays will dominate the market during the coming decade. Because the worldwide TV market is increasing at a rate of about 4–5% per year, production of color CRTs may well increase during the next two decades.

In view of the importance of electronic displays for an ever increasing number of new applications, there have been continuing investigations of other display technologies in the hope that they will lead to a more "ideal" thin display device. One of the oldest of these technologies, making use of plasma or gas-discharge phenomena, has been receiving attention for more than 40 years. In spite of hopes that this technology would find major use in computer monitors, its most promising application now appears to be for large wall-type TV displays with a diagonal size greater than about 1 m. Aside from their high cost, displays of this size will probably constitute only a limited portion of the broad TV market and are not likely to supplant CRT color tubes in the near future for most TV applications where the image is directly viewed.

In the long term, even with its many advantages, the bulkiness of the CRT remains a major disadvantage in a market where compactness is becoming of increasing importance. Recognizing this situation, a number of serious attempts were made in the past to produce a flat CRT. In 1957 a panel-type display about 30 cm × 30 cm in size was developed by Aiken (1957), in which an electron beam was injected from the side into the vacuum space between two close-spaced glass plates. By applying time-varying voltages to sets of electrodes below and on the plates the beam could be deflected and then sharply bent toward the phosphor screen on which a raster pattern was scanned. Although monochrome images of TV quality were demonstrated, this approach was discontinued because it was not suited for producing color images and required relatively thick glass plates to withstand the vacuum pressure.

More recently, Matsushita (Yamamoto *et al.*, 1994) demonstrated a prototype model (about 28 cm × 21 cm) of a flat CRT (called MDS) in which multiple beams of electrons from an array of heated cathodes were used to produce high-quality color images in a line-at-a-time fashion. However, because of its relatively complex structure and high cost this device was not commercial.

An attempt to develop a flat CRT was reported in 1997 by workers at Philips (van Gorkom *et al.*, 1997). In this approach the display structure was contained between two flat glass plates and electrons emitted from a horizontal wire cathode along the bottom edge were launched into a set of vertical channels. By means of a set of horizontal electrodes, electrons could then be extracted one row at a time from the channels and, with the aid of an adjacent set of vertical electrodes, the current flow to the corresponding row of phosphor elements on the front plate could be modulated. Of particular interest is the fact that the vertical channels and substrate for all the electrodes were made of glass, thus making possible a distributed internal support structure of insulating material and enabling thin outer plates to be used. Although experimental panels of this type with a 43 cm diagonal size and 1 cm overall thickness were developed that could display good quality color TV images, this approach is no longer being pursued, presumably because of economic considerations.

The most promising approach today for producing a panel display based on vacuum technology appears to be one in which a two-dimensional array of tiny field emitters is provided on the surface of a glass plate and a phosphor screen deposited on the surface of a second close-spaced glass plate maintained at a positive potential. In operation, sets of X and Y electrodes mounted behind and in front of the emitters control the current flow to the phosphor elements one row at a time. Although relatively high-quality displays have been demonstrated with small panels (Tanaka *et al.*, 1997), one of the problems is that of obtaining sufficient brightness. If thin plates are used and an array of insulating spacers are provided between them to support the vacuum, the voltage that can be applied to the phosphor screen is limited. On the other hand, if the spacing is increased to allow higher voltage (for example, 5 kV) a more complex spacer design is required to prevent breakdown.

In conclusion, although conventional CRTs are being subjected to increasing competition from new types of panel displays, it is possible that the vacuum technology already developed for CRTs will, in the end, also provide the means for creating thin panel displays that can supplant the CRT.

Acknowledgment

The author wishes to express his sincere gratitude to Dr. B. Kazan for his invaluable editorial guidance in preparing the manuscript and his help in improving the English. I would also like to express my sincere thanks to the many researchers and engineers who contributed helpful information for this chapter: especially Prof. H. Yamamoto of Tokyo Engineering University on phosphor materials; Mr. K. Kikuchi of Nippon Electric Glass on glass bulbs; Mr. H. Kawamura of Hitachi on glass surface treatment and many other engineers at Hitachi where I was employed when I started work on this article and at LG Electronics where I am currently employed. Finally, I would like to extend my thanks to Mr. M. Fukushima who first encouraged me to write this chapter when he was at Hitachi.

Appendix: Some Specifications of Representative Tube Types

Although more than several hundred tube types are being produced and marketed throughout the world, specifications of representative tubes of three sizes are given here. Abbreviated specifications and drawings are shown in following pages.

Color Picture Tube for TV Use Hitachi **A68KSA30X10** (27V)

General data	
ELECTRICAL	
Electron guns	Unitized (one piece) triple aperture electrodes: center beam (green), side beams (blue, red)
Heater current at 6.3 V	680 mA
Focusing method	Electrostatic
Focus lens	Bipotential Unipotential
Convergence method	Magnetic
Deflection method	Magnetic
Deflection angle (diagonal, approx.)	110°
OPTICAL	
Light transmission at center (approx.)	39.0%
Screen, on inner surface of faceplate	Aluminized, tricolor phosphor-stripe type, black matrix screen
Phosphor (three separate phosphors)	Medium short persistence red, green, blue (WTDS code X)
Arrangement	Vertical line trios
Spacing between centers of adjacent stripe trios at screen center (approx.)	0.79 mm
MECHANICAL	
Overall length	450.7 ± 6.5 mm
Useful display screen dimensions (projected):	
Diagonal	676.0 mm
Horizontal axis	540.8 mm
Vertical axis	405.6 mm
Area	2,193 cm^2
Implosion protection	Thermal clamping band type
Weight (approx.)	30 kg

Ratings and electrical data	
MAXIMUM AND MINIMUM RATINGS (design-maximum values)	
Anode voltage	32,000 V absolute max. 20,000 V min.
Total anode current	
Long-term average	1,500 max. μA
EXAMPLE OF USE OF DESIGN RANGES	
Anode voltage	27,500 V
Grid no. 3 (focusing electrode) voltage	7,150 to 8,250 V
Heater voltage	6.3 V

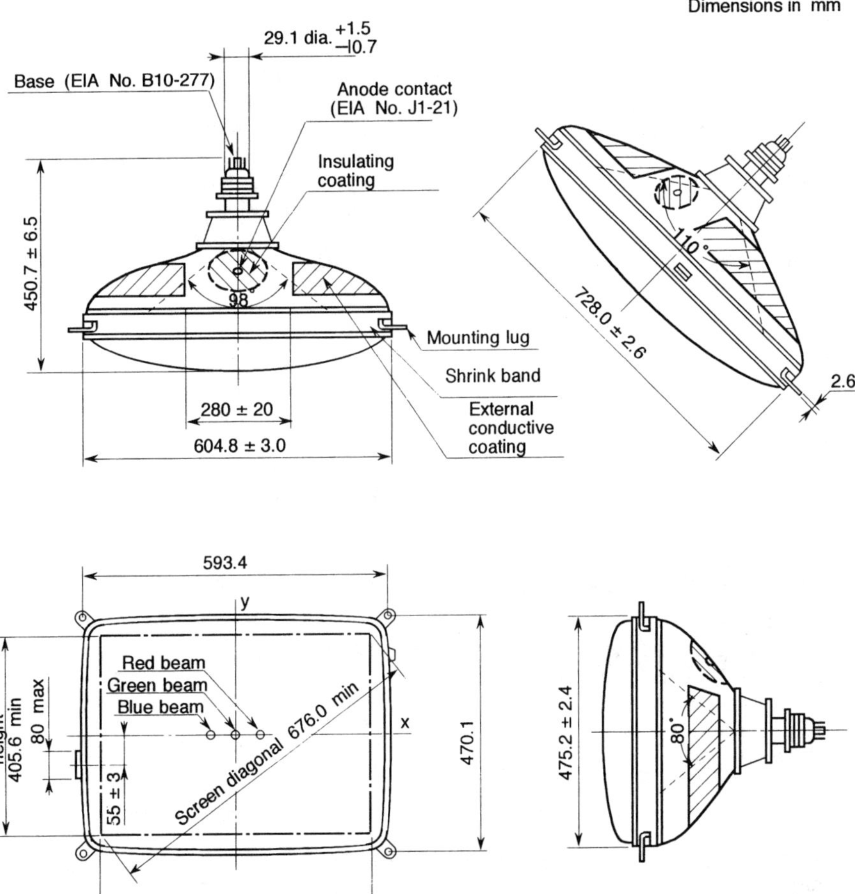
Dimensional Outline of
A68KSA30X10
Dimensions in mm
29.1 dia. +1.5 −0.7
Base (EIA No. B10-277)
Anode contact
(EIA No. J1-21)
Insulating
coating
450.7 ± 6.5
98°
110°
728.0 ± 2.6
Mounting lug
Shrink band
2.6
External
conductive
coating
280 ± 20
604.8 ± 3.0
593.4
y
Red beam
Green beam
Blue beam
Screen diagonal 676.0 min
x
470.1
475.2 ± 2.4
80°
Screen
height
405.6 min
80 max
55 ± 3
Screen width 540.8 min

FIGURE A.1

Color Data Display Tube for Monitor Use Hitachi **M51LCJ180X65(U)** (20V)

General data	
ELECTRICAL	
Electron guns	Unitized (one piece) triple aperture electrodes: center beam (green), side beams (blue, red)
Heater current at 6.3 V	320 mA
Focusing method	Electrostatic and dynamic focusing
Focus lens	Unipotential Bipotential
Convergence method	Magnetic
Deflection method	Magnetic
Deflection angle (diagonal, approx.)	90°
OPTICAL	
Light transmission at center (approx.)	51%
Screen, on inner surface of faceplate	Aluminized, tricolor phosphor-dot type, black matrix screen
Phosphor (three separate phosphors)	Medium short persistence red, green, blue (WTDS code X)
Arrangement	In-line trios
Spacing between centers of adjacent dot trios (approx.)	0.26 mm
Horizontal pitch	0.22 mm
MECHANICAL	
Overall length	447.5 ± 6.5 mm
Useful display screen dimensions (projected):	
Diagonal	508.0 mm
Horizontal axis	406.4 mm
Vertical axis	304.8 mm
Area	1,239 cm^2
Implosion protection	Thermal clamping band type
Weight (approx.)	18 kg
Ratings and electrical data	
MAXIMUM AND MINIMUM RATINGS (absolute-maximum values)	
Anode voltage	30,000 V absolute max. 22,000 V min.
EXAMPLE OF USE OF DESIGN RANGES	
Anode voltage	27,500 V
Grid no. 5-1 (static focusing electrode) voltage	6,500 to 7,300 V
Grid no. 5-2 (dynamic focusing electrode) voltage	Static + dynamic controlled 650 V
Heater voltage	6.3 V

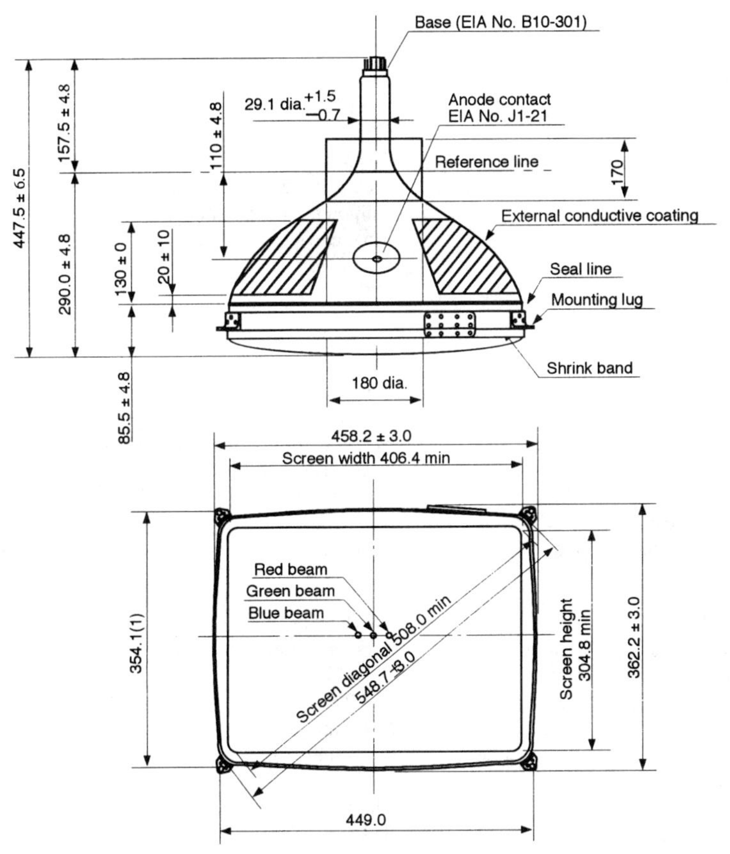
Dimensional Outline of
M51LCJ180X65(U)
Dimension in mm
Base (EIA No. B10-301)
29.1 dia. +1.5 −0.7
Anode contact
EIA No. J1-21
Reference line
170
External conductive coating
Seal line
Mounting lug
Shrink band
180 dia.
447.5 ± 6.5
157.5 ± 4.8
290.0 ± 4.8
110 ± 4.8
130 ± 0
20 ± 10
85.5 ± 4.8
458.2 ± 3.0
Screen width 406.4 min
Red beam
Green beam
Blue beam
Screen diagonal 508.0 min
548.7 ± 3.0
354.1(1)
Screen height 304.8 min
362.2 ± 3.0
449.0

FIGURE A.2

COLOR PICTURE TUBE FOR HDTV AND WIDE ASPECT RATIO (16:9) USE
HITACHI **W76LKS132X** (30W)

General data	
ELECTRICAL	
Electron guns	Unitized (one piece) triple aperture electrodes: center beam (green), side beams (blue, red)
Heater current at 6.3 V	320 mA
Focusing method	Electrostatic and dynamic focusing
Focus lens	Unipotential Bipotential
Convergence method	Magnetic
Deflection method	Magnetic
Deflection angle (diagonal, approx.)	106°
OPTICAL	
Light transmission at center (approx.)	35.5%
Screen, on inner surface of faceplate	Aluminized, tricolor phosphor-stripe type, black matrix screen
Phosphor (three separate phosphors)	Medium short persistence red, green, blue (WTDS code X)
Arrangement	Vertical line trios
Spacing between centers of adjacent stripe trios at screen center (approx.)	0.74 mm
MECHANICAL	
Overall length	517.4 ± 6.5 mm
Neck diameter	32.5 mm nom.
Useful display screen dimensions (projected):	
Diagonal	756.2 mm
Horizontal axis	651.9 mm
Vertical axis	370.7 mm
Area	2,443 cm^2
Implosion protection	Thermal clamping band type
Weight (approx.)	41 kg

Ratings and electrical data	
MAXIMUM AND MINIMUM RATINGS (design-maximum values)	
Anode voltage	35,000 V absolute max. 20,000 V min.
Total anode current	
Long-term average	2,000 max. μA
EXAMPLE OF USE OF DESIGN RANGES	
Anode voltage	32,000 V
Grid no. 3, 5-1, 5-3 (static focusing electrode) voltage	8,320 to 9,600 V
Grid no. 5-2, 5-4 (static focusing and dynamic focusing electrode) voltage	8,320 to 9,600 V
Heater voltage	6.3 V

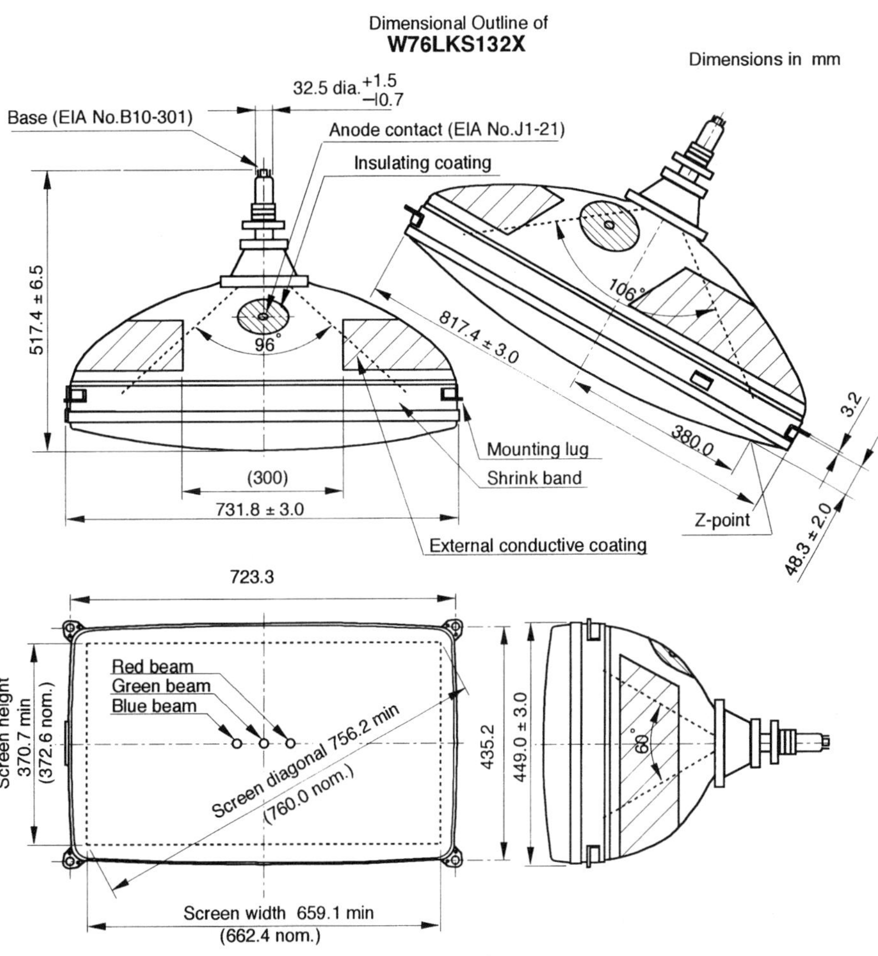
Dimensional Outline of
W76LKS132X
Dimensions in mm
32.5 dia. +1.5 −0.7
Base (EIA No.B10-301)
Anode contact (EIA No.J1-21)
Insulating coating
517.4 ± 6.5
96°
106°
817.4 ± 3.0
380.0
3.2
48.3 ± 2.0
Mounting lug
Shrink band
Z-point
(300)
731.8 ± 3.0
External conductive coating
723.3
Red beam
Green beam
Blue beam
Screen height 370.7 min (372.6 nom.)
Screen diagonal 756.2 min (760.0 nom.)
435.2
449.0 ± 3.0
60°
Screen width 659.1 min (662.4 nom.)

FIGURE A.3

References

Aiken, W. R. (1957). A thin cathode-ray tube, *Proc. IRE*, **45**, 1599–1604.

Alig, R. C., and D'Amato, R. J. (1989). Computer-aided design and analysis of magnetic shields for color picture tubes, *SID 89 Digest*, pp. 34–37.

Ando, K., Osawa, M., Shimizu, T., Maruyama, T. and Fukushima, M. (1985). A flicker-free 2448×2048 dots color display, *SID 85 Digest*, pp. 338–340.

Barbin, R. L., and Hughes, R. H. (1972). New color picture tube system for portable TV receivers, *IEEE, Broadcast Telev. Receivers*, **18**, 193–200.

Bauder, R. C., and Ragland, F. R. (1990). An improved shadow-mask support system for large-size CRTs, *SID 90 Digest*, pp. 426–429.

Born, M., and Wolf, E. (1997). *Principles of Optics*, Cambridge University Press, New York, p. 126.

Bril, A., and Klasens, H. A. (1955). Phosphors for tri-color television tubes, *Philips Res. Rep.*, **10**, 305.

Buchsbaum, W. H. (1966). GE 11-inch color TV: The new look in color receivers, *Electron World*, **75**(3), 39–41.

Daiku, N., and Okada, K. (1988). Panel for CRT, *Japanese Utility Model* (laid open), Sho-63-133049.

Dietch, L., Palac, K., and Chiodi, W. (1986). Performance of high-resolution flat tension mask color CRTs, *SID 86 Digest*, pp. 322–323.

Donofrio, R. L. (1995). Corner lock suspension, *Information Display*, Nov. 1995, pp. 12–16.

EIAJ (1992). JXCS 370–001.

Elst, J. H. R. M., and Wielenga, D. K. (1977). The finite-element method and the ASKA program, applied in stress calculations for television picture tubes, *Philips Tech. Rev.*, **37**, 56–71.

Enstrom, R. E., Stepleman, R. S., and Appert, J. R. (1978). Application of finite element method to the analysis of stresses in television picture tubes, *RCA Review*, **39**, Dec.

Fiore, J. P., and Kaplan, S. H. (1964). Cathode-ray tube with color dots spaced by light-absorbing areas, *U.S. Patent*, 3,146,368.

Fujimura, T. (1970). Exposure system for color picture tube, *Japanese Patent*, Sho 45-23656.

Fujimura, T. (1988). Mis-landing pattern eliminatable by a smooth surface lens system, *J. Television Engineers of Japan*, **42**(12), 1364–1371.

Ghosh, A., and Gulati, S. T. (1995). Mechanical realiability of CRTs, *SID 95 Digest*, pp. 36–38.

Gulati, S. T. (1993). Stress corrosion in silicate glasses and its impact on CRT panel design, *SID 93 Digest*, pp. 39–42.

Haantjes, J., and Lubben, G. J. (1959). Errors of magnetic deflection II, *Philips Research Reports*, **14**, 65–97.

Hamano, E., Koba H., Mori, H., Kawata, Y., and Nakagawa, K. (1984). A 15-in. full-square CRT, *SID 84 Digest*, pp. 332–335.

Hase, T., Kano, T., Nakazawa, E., and Yamamoto, H. (1990). Phosphor materials for cathode-ray tubes, *Adv. Electronics Electron Physics*, **79**, 271–373.

Hayama, H., Aoyama, T., Utsumi, T., Miura, Y., Suzuki, A., and Ishiai, K. (1994). Anti-reflection and anti-static coating for CRTs, *National Technical Report*, **40**, 1(2), 90–96.

Heijnemans, W. A. L., Niewwendijk, J. A. M., and Tink, N. G. (1980). *Philips Tech. Rev.*, **39**(6/7), 154–171.

H.E.W. [The U.S. Dept. of Health, Education and Welfare: presently Health and Human Services (H.H.S.)] (1976). Regulations for the Administration and Enforcement of the Radiation Control for Health and Safety Act of 1968.

Hondoh, M., Kumazawa, T., and Yoneyama, M. (1978). Mechanism of color CRT impact breakage, *Japan Machinery Soc.*, No. 44-386, pp. 3362–3370.

Iguchi, Y., Utsumi, I., Saito, T., Murakami, K., Inoue, T., and Nomura, K. (1997). A super-flat Trinitron color CRT, *SID 97 Digest*, pp. 355–358.

Ikeda, H., Okuda, S., Tani, Y., Tanabe, S., Hattori, A., and Takita, H. (1995). New chassis and auto-degauss coil for a CRT display monitor, *Asia Display '95*, pp. 511–514.

Infante, C. (1985). CRT systems, *SID Seminar Lecture Notes*, **1**, 1.

Ishii, T., Ohta, K., and Ueba, Y. (1983). New non-glare coating for color display tubes, *Japan Display '83*, pp. 120–123.

Ishiyama, M., Chiba, J., and Suzuki, Y. (1975). Browning of glass by electron bombardment, *Reports Res. Lab.*, Asahi Glass Co., Ltd., **25**, 1.

ISO 9241-3 (1992). Ergonomic Requirements for Office Work with Visual Display Terminals (VDTs)—Part 3: Visual display requirements, (ISO: the International Organization for Standardization).

Itoh, J., Yoshioka, S., Nakamura, K., and Umino, K. (1987). Structural evaluation for large size CRTs, *Material Conference* 87–16.

Itoh, J., Yoshioka, S., Hijikata, A., Ohsugi, H., Nakamura, K., and Umino, K. (1992). Dynamic behavior of CRT under impact loading, *J. Soc. Mat. Sci. Japan*, **41**(469).

Itoh, T., Matsuda, H., and Onodera, M. (1995). Microfilter® color CRT, *SID 95 Digest*, pp. 25–27.

Janes, R. B., Headrick, L. B., and Evans, J. (1956). Recent improvement in the 21AXP22 color kinescope, *RCA Rev.*, **XV**, 155–157, 161–163.

Kawamura, H., Nishizawa, M., and Kobara, K. (1994). Technical trend of surface treatment of color CRT, *Display and Imaging*, Sept.

Kazan, B. (1985). Luminescent materials for displays: Comments on evolution and present status, *Displays*, April '85.

Klein, C. A. (1968). Band-gap dependence and related features of radiation ionization energies in semiconductors, *J. Appl. Phys.*, **39**, 2029.

Kuhlman, B. E., and Kurman, E. (1994). Manufacturing and measurement of multilayer conductive anti-reflective coating vacuum deposited directly onto cathode ray tube (CaRT™), *SID 1994 Display Manufacturing Technology Conference*, pp. 53–55.

Law, H. B. (1951). A three-gun shadow-mask color kinescope, *Proc. I.R.E.*, **39**, 1186–1194.

Leverenz, H. W. (1950). *An Introduction to Luminescence of Solids.* John Wiley & Sons Inc., New York.

Levine, A. K., and Palilla, F. C. (1964). A new highly efficient red-emitting cathodoluminescent phosphor (VO_4:Eu) for color television, *Appl. Phys. Lett.*, **5**, 118.

Levy, P. W. (1960). The kinetics of gamma-ray induced coloring of glass, *J. Am. Ceramics Soc.*, **43**, 389.

McLellan, G. W., and Shand, E. B. (1984). *Glass Engineering Handbook*, McGraw-Hill Book Co., New York, 18-1–18-5.

Meyer, V. D. (1970). Absorption of electron kinetic energy by inorganic phosphors, *J. Appl. Phys.*, **41**, 4059.

Morrell, A. M., Law, H. B., Ramberg, E. G., and Herold, E. W. (1974). *Color Television Picture Tubes*, Academic Press, pp. 8, 42–44, 48–50, 50–62, 66–68, 68–79, 80–85, 86, 117–128, 130–133.

MPR (1990). MPR 1990, Swedac.

Nakamichi, H. (1979). Reinforcement technique for cathode ray tubes, *National Technical Report*, **25**(2), Apr. 1979, 277–284.

Nakanishi, T., Hasuno, H., Chiba, T., and Haga, A. (1990). A simulation of the influence of a color CRT under terrestrial magnetic field, (IDYG0-30), *ITE Tech. Report*, **14**(4), 19–24.

New Glass Handbook (1991). Maruzen, pp. 407–410.

Ogura, I. (1966). The correctable condition of exposure lens for color picture tube and a design of blue correction lens, Hitachi Internal Technical Memorandum.

Ohishi, I. (1970). Seminar on solid-state image transducers, *Inst. of TV Engineers of Japan*, Doc. No. 47, Feb. 1970.

Ohta, K. (1975). "CRT," *Japanese Utility Model, Japanese Patent Document*, Sho-50-26277.

Okada, Y., and Ikegaki, M. (1983). A structure analytical study on shadow mask thermal deformation of color CRT, *Japan Display* '83, pp. 20–22.

Ono, Y., and Ohtani, Y. (1992). A new antireflective and antistatic double-layered coating for CRTs, *SID 92 Digest*, pp. 511–513.

Onodera, M., and Matsuda, H. (1993). Color display tube with high-contrast and antireflection coating, *Toshiba Review*, **48**(8), 639–642.

Pfahnl, A. (1961). Aging of electronic phosphors in cathode ray tubes, *Adv. Electron Tube Techniques*, Pergamon Press, New York, pp. 204–208.

Raue, R., Vink, A. T., and Welker, T. (1989). Phosphor screen in cathode-ray tubes for projection television, *Philips Tech. Rev.*, **44**, 335–347.

Rock, F. C. (1969). Antireflection coating and assembly having synthesized layer of index of refraction, *U.S. Patent*, No. 3,432,225.

RöV (Röntgenverordnung) (1987). *Bundesgesetzblatt*, 1987, Teil 1.

Royce, M. R., and Smith, A. L. (1968). Rare-earth oxysulfides—A new family of phosphor hosts for rare-earth activators, *Extended Abstr. Electrochem. Soc.*, (Spring Meeting), **34**, 94.

Sakka, S. (1985). *Handbook of Glass*, Asakura Book Co., Japan, pp. 128–131.

Scobey, M. A., Seddon, R. I., Seeser, J. W., and Austin, R. R. (1989). Magnetron sputtering apparatus and process, *U.S. Patent*, No. 4,851,095.

Shimizu, Y., Kobayashi, H., Iwasaki, Y. Wada, M., and Daiku, N. (1981). A new color picture tube with improved color fidelity and contrast, *IEEE Chicago Spring Conference on Consumer Electronics.*

Shimoma, T., Okada, H., and Fukuda, K. (1978). *Tech. Group Electron. Inst. Electr. Eng. Jpn.*, No. 30.

Shirai, S. Fukushima, M., Miyamoto, S., and Miyazaki, M. (1987). Quadrupole lens for dynamic focus and astigmatism control in an elliptical aperture lens gun, *SID 87 Digest* XVIII, pp. 162–165.

Shiramatsu, N., and Inoue, A. (1994). Removing moiré patterns from shadow-mask CRTs, *Information Display*, 6/94, pp. 12–16.

Sluyterman, A. A. S. (1993). Resolution aspects of shadow-mask pitch for TV applications, *SID 93 Digest*, pp. 340–343.

Strauss, P., Prazak, C. J., and Lee, S. D. (1986). Construction of a flat tension mask color CRT, *SID 86 Digest*, pp. 324–326.

Suzuki, H., Mitsuda, K., Muranishi, H., Iwasaki, K., and Ashizaki, S. (1989). 27V-in flat-square high-resolution color CRT for graphic displays, *Japan Display*, '89, pp. 554–557.

Suzuki, H. (1999). Electron-gun systems for color cathode ray tubes, *Advances in Imaging and Electron Physics*, Academic Press, Boston, Vol. 105, 267–404.

Takano, Y. (1983). Recent Advances in Color Picture Tubes, *Advances in Image Pickup and Display*, Academic Press, New York, Vol. 6, pp. 1–15.

Taki, A., and Arimoto, N. (1996). Development of 17-in. 'pure flat' color monitor tube, *SID 96 Digest*, pp. 879–882.

Tamatani, M. (1980). Gold-activated zinc sulfide phosphors, *Gold Bulletin*, **3**, 98–104.

Tanaka, M., Takayama, K., Azeta, A., Yano, K., and Kishino, T. (1997). A new structure and driving system for full-color FEDs, *SID 97 Digest*, pp. 47–51.

Tawara, T. (1987). Toshiba new development of C3 CRT and picture enhancement technology, *TV Technol.*, Oct. 1987, pp. 43–45.

Tohda, T., Arimoto, N., Hayama, H., and Takahashi, T. (1992). Anti-glare, anti-reflection and anti-static (AGRAS) coating for CRTs, *Japan Display*, '92, pp. 289–292.

Tokita, K., Koshigoe, S., and Sone, T. (1984). Full square color picture tube, *Toshiba Rev.*, **39**, 1.

Toshiba Product Catalog (1989). Bazooka-V2/V1, Oct. 1989.

Trond, S. S. (1979). Filter phosphor, *RCA Engineer*, **25**(2), 19–24.

van Gorkom, G. G. P., Baller, T. S., Dessens, P. A., Hendriks, B. H. W., Lambert, N., Ligthart, H. J., Montie, E. A., Thomas, G. E., Trompenaars, P. H. F., and de Zwart, S. T. (1997). A new thin CRT, *SID 97 Digest*, pp. 235–238.

Wittke, J. P. (1987). Moiré consideration in shadow-mask picture tubes, *SID 87 Digest*.

Yamamoto, H., Matsukiyo, H., Morita, Y., and Uehara, Y. (1987). Factors inducing degradation of $InBO_3:Tb^{3+}$ by electron irradiation, *Extended Abstr. Electrochem. Soc.* (172nd Meeting), No. 1240.

Yamamoto, H., and Matsukiyo, H. (1991). Problems and progress in cathode-ray phosphors for high-definition displays, *J. Lumin.*, **48 & 49**, 43–48.

Yamamoto, K., Yokomakura, M., Inohara, S., and Nonomura, K. (1994). A 14-in. color flat-panel display using filament cathodes, *SID 94 Digest*, pp. 381–384.

Yamazaki, E., Maruyama, K., and Ogura, I. (1971). Correction lens, *U.S. Patent*, No. 3,628,850.

Yamazaki, E., Maruyama, K., Udeda, T., and Ogura, I. (1973). A segmented lens for improving color television dot pattern, *SMPTE*, **82**(3), 149–150.

Yamazaki, E. (1993). CRT display, *SID Seminar Lecture Notes 1993*, **II**, p. F-5/7.

Yonai, F., Yamazaki, E., Yamaguchi, A., and Hirai, R. (1986). High-performance square picture tube, *Japan Display '86*, late-news paper (poster).

ADVANCES IN IMAGING AND ELECTRON PHYSICS, VOL. 105

Electron Gun Systems for Color Cathode Ray Tubes

HIROSHI SUZUKI

Matsushita Electric Industrial Co., Ltd., AVC Products Development Laboratory, AVC Company, 1-1 Matsushita-cho, Ibaraki, Osaka 567-0026, Japan

I. Historical Introduction

During the period from about 1930 to 1950 relatively advanced gun designs were developed for black-and-white television cathode ray tubes (CRTs). However, these were not entirely satisfactory for use in the new shadow-mask color tubes introduced around 1950. Because of the need to use guns with a

ISBN 0-12-014747-5
ISSN 1076-5670/99 $30.00

limited diameter both the brightness and the resolution of these tubes were noticeably inferior to that of black-and-white tubes. It was not until almost two decades later that this problem was solved with the introduction of computer-aided design methods. This enabled small-diameter guns to be made that could perform as well as the larger-diameter older types of guns, thus significantly increasing the brightness and resolution obtainable from color tubes.

It is of historic interest to note that the means for generating an electron beam have gone through several major stages. In the earliest CRTs, such as were first demonstrated by Braun (1897) and explored by others during the following two decades, the concept of an electron gun hardly existed. In these tubes a high voltage (e.g., 20–40 kV) was applied between a cold cathode electrode and an anode wire inserted into the tube neck about 10 cm away. Because of the residual air left in the tube (approximately 5×10^{-3} torr) a gas discharge was created, generating high-energy electrons, which were drawn toward the anode. By inserting a disc with a small aperture beyond the anode, most of the electrons continuing past the anode were blocked except for a narrow pencil of electrons, which constituted the electron beam.

Between 1903 and 1904 Wehnelt (1904) found that the electron emission from a heated platinum strip could be greatly enhanced if it were coated with an oxide of one of the alkaline-earth elements Ba, Ca, or Sr. Unfortunately, in the poorly evacuated diodes used by Wehnelt the cathodes deteriorated rapidly as a result of positive ion bombardment, making such cathodes of limited interest. In separate experiments a few years later Wehnelt also showed that a negatively biased cylindrical electrode placed in front of the cathode could modulate the beam current reaching the anode. In the years following, such electrodes (now referred to as control grids) were frequently referred to as Wehnelt cylinders.

About two decades later, in 1922, commercial oscilloscope tubes based on Wehnelt's hot cathode were developed by Johnson (1992) at the Western Electric Co. These tubes, first evacuated and then filled with low-pressure argon (approximately 10^{-2} torr), relied on the electron emission from the cathode rather than a cold discharge to produce the electron beam. To minimize the number of positive ions generated in the space between the cathode and anode (which was in the form of a tiny metal tube) and to prevent a gas discharge from building up, the spacing between the electrodes was made very small (about 1 mm) and a potential of only 350 V was applied between cathode and anode. However, in its relatively long trajectory after emerging from the anode, the electron beam generated positive ions, which accumulated along its path and neutralized the mutual repulsion of the electrons. The spot size of the beam was thus reduced to approximately one-tenth the size it would have had in a fully evacuated tube.

During the following decade extensive research was carried out on gas-focused tubes, especially in Germany, largely with TV displays in mind. With a Wehnelt cylinder added for current modulation and operating at an anode voltage of a few kilovolts, such tubes were able to display 400-line TV images with a good gray scale. However, serious practical problems arose because of their limited brightness, the dependence of the focusing action on the beam current and the limit on scanning speed resulting from the time required (about 1 μs) for ion buildup along each new path of the deflected beam.

During the period from about 1925 to 1930 several important developments occurred. In 1926 Bush (1926) showed that a short axially symmetric magnetic field acted on electron rays in a manner similar to that of an optical lens on light rays, thus laying the groundwork for the field of electron optics. Several groups began separately to explore methods of producing narrow electron beams in high-vacuum tubes using heated oxide cathodes as the electron source. In some of these tubes it was shown that a rudimentary form of beam focusing could be achieved by a potential difference applied to adjacent cylinders or aperture electrodes as well as by a magnetic lens.

Of the various competing approaches to the high-vacuum CRT the most successful one was developed in 1929 by Zworykin (1929). This contained an electron gun consisting of an indirectly heated oxide cathode, a negative control grid, and an adjacent low-voltage anode maintained at several hundred volts. To converge the beam emerging from the gun a second anode was added in the form of a conductive coating applied to the inner surface of the tube wall, extending from the phosphor screen to the low-voltage anode. By applying about 5 kV to the second anode the beam was not only accelerated but also focused to a small spot by the lens action created in the region between the two anodes. The superior performance of this tube, with its high brightness and small spot size as well as its relative simplicity, was rapidly recognized and served to accelerate the introduction of black-and-white television during the next decade.

As a result of an extensive research program at RCA, by 1950, the three-gun shadow-mask tube was developed, opening the way for commercial color TV. Initially the guns used for such tubes were basically similar to those employed in black-and-white tubes, consisting of an assembly of cylindrical electrodes. However, in order to fit three such guns into a tube neck of acceptable size, it was necessary to reduce their diameters. This resulted in an increase in spherical aberration of the gun lenses, degrading the spot size of the beams and limiting the useful amount of current that could be obtained.

To maximize the diameter of the guns that could fit into a neck of a given size they were mounted in a delta or triangular arrangement. However, when deflected by a common magnetic yoke the convergence of the beams at the screen was seriously disturbed. To minimize this problem special auxiliary magnetic components were added together with dynamic correction circuits, which required relatively complex adjustment procedures. These problems were further exacerbated by the fact that the characteristics of the individual guns differed from each other due to the limited accuracy of the finished guns.

In 1965 an in-line gun arrangement was introduced by General Electric Co. (Buchsbaum, 1966) where the three beams were arranged in a side-by-side manner in a horizontal plane using individual guns similar to those in the delta gun arrangement. Although dynamic convergence was still needed, the convergence circuits and their adjustments were considerably simplified compared to the delta-gun tubes. A completely new gun design was introduced in 1968 by Sony for their Trinitron system (Yoshida *et al.*, 1968) in which three horizontal in-line beams were directed toward the center of a single large-diameter main focusing lens. These beams were then reconverged at the screen by electrostatic deflection plates positioned beyond this lens, to which simplified dynamic convergence voltages were applied. Because of the large size of the common focusing lens the spherical aberration was reduced and an improvement in spot size was obtained.

In 1972 a major improvement in gun design was introduced by RCA (Barbin and Hughes, 1972) in the form of a unitized in-line gun structure. This consisted of an assembly of metal plates, each provided with three apertures in registry with those of the other plates. Using separate cathodes, three electron beams were created, each traversing a set of successive apertures. By applying appropriate potentials to the aperture plates electron-optical lens action was obtained similar to that obtained from the cylindrical electrodes of previous guns. Because the size and position of the apertures of the plates could be precisely controlled during fabrication, variations of the individual beams with respect to each other were considerably reduced. With this unitized structure the beams could also be positioned closer to each other, allowing a reduction in neck size and improving their convergence characteristics.

The in-line beam arrangement resulted in another important improvement in color tubes, allowing a new type of self-converging deflection yoke to be used. With its specially shaped magnetic field this yoke was now able to maintain the three beams converged at the screen without the need for the additional magnetic components and associated dynamic correction circuitry required with the delta gun arrangement. Although the basic principle of self-convergence with in-line beams was known (Haantjes and Lubben, 1959), its practical realization was first accomplished by Barbin and Hughes using their unitized close-spaced in-line gun and a neck diameter reduced from 36 to 29 mm. Also they used a new precision static toroid (PST) yoke whose precisely wound toroidal coils enabled an exact control of the magnetic field shape. However, the use of the narrower neck and self-converging yoke challenged gun designers to achieve improved focus with smaller diameter gun lenses and nonuniform magnetic fields.

During the 1970s, in order to reduce the spot size, particularly at increased current levels, reduction in lens aberration and space-charge repulsion effects were major objectives in gun design. For this purpose additional electrodes were frequently inserted with appropriate potentials applied to control the diameter of the beam in the focusing lens region of the gun. In designing guns experimental techniques were frequently used (Zworykin and Morton, 1954). In one approach the potential distribution between electrodes was determined using simulated electrodes in an electrolytic tank or a resistance network board (Francken, 1959/60). From this information and the use of graphical or mathematical procedures the approximate electron trajectories could be determined for a particular electrode configuration. However, aside from the limited accuracy and tediousness of such approaches they were generally limited to axially symmetric or 2D structures. In parallel with these methods, the use of computer programs became more and more popular to simulate beam trajectories in practical gun designs. However, their application was largely limited to 2D electrode models.

In 1980 a radically new gun design was introduced in which the lenses were no longer axially symmetric (Hosokoshi *et al.*, 1980). In this gun the apertures of the main focusing lens were enlarged to the point where they overlapped each other, cutting through part of the wall that separated them. To compensate for the distortion of the lens field that this caused, small auxiliary electrodes were placed behind both the anode and focusing electrode plates whose apertures overlapped. This resulted in lenses whose effective diameter was enlarged and that had reduced spherical aberration compared to the previous axially symmetric lenses. Also, as the spacing between beams was not increased the desired small beam convergence angle was retained. The superior spot performance obtained with this gun design soon stimulated the design of various other types of guns with asymmetric lenses. In one form or another such guns are employed in essentially all present-day color tubes.

A more recent improvement in the electron gun design has been the addition of quadrupole lenses, which also use noncircular apertures. These lenses serve to compensate for the vertical over-focusing of the spot and some of the distortions in spot shape introduced by the magnetic field of the self-converging deflection yoke. In one form the electrodes of the quadrupole lenses are maintained at a fixed potential (Barten and Kaashoek, 1978) while in another form the potentials applied to the electrodes are modulated synchronously with the deflection field (Ashizaki *et al.*, 1986).

During the 1980s computer programs became a powerful tool for the design of new electron guns. These programs not only provide information more rapidly and accurately on the beam trajectories for axially symmetric lenses but also provide 3D information for axially nonsymmetric gun structures (Bechis *et al.*, 1989; Aalders *et al.*, 1989). Such programs are particularly important in developing guns of the latter type, which would otherwise be almost impossible to design. Of additional importance is the fact that these programs can also take into account space charge and thermal emission velocities as well as the distorting effects of the magnetic deflection field on the spot shape.

Since the 1920s thermionic cathodes coated with alkaline-earth oxides have been used in essentially all CRTs. However, with the more recent demand for CRTs with larger size and higher resolution, particularly for HDTV applications, cathodes capable of supplying a higher current density have been found necessary. This has revived interest in a form of dispenser cathode developed in 1955 in which a mixture of barium oxide and alumina is impregnated into a porous tungsten plug (Levi, 1957/58). Such cathodes are capable of producing an emission current density several times greater than that of the conventional oxide cathode. Another cathode of interest is an oxide type of cathode to which a material such as scandium oxide powder has been added (Saito *et al.*, 1990). Although this cathode is capable of producing a current density only about 50% higher than the conventional oxide cathode, its structure is relatively simple and, unlike the more costly dispenser cathode, does not require an increase in operating temperature.

Since the high-tension voltage required on the anode and phosphor screen can cause destructive high-voltage discharges within the tube as well as damage to the external circuits, improvements have been made in tube design to prevent such damages from occurring (Hernqvist, 1981) and to properly dissipate their energy if they should occur (Gerritsen, 1978). Such improvements are of particular importance in the case of the newer large-screen and high-resolution tubes in which anode-voltages up to 32 kV are used.

II. Electron Beam Formation and Factors Limiting Current Density of Spot

A. *Introduction*

To produce a CRT image with high-resolution and high-brightness the electron beam must carry a large amount of current and be concentrated into a small area at the phosphor screen, in effect requiring a high current-density spot. As discussed in the sections that follow, there are several factors that limit the current density of the electron beam spot at the phosphor screen. These factors are the thermal velocity spread of electrons emitted from the cathode, mutual repulsion of the negatively charged electrons in the beam, and electron-optical aberrations of the lens system of the gun. A further limitation occurs due to distortion of the beam as a result of the deflection process. These limitations severely restrict the current density of the electron beam spot in practical tubes.

In Section II some fundamental aspects of electron beam and spot formation will be discussed with emphasis on color CRTs. The topics covered are: beam formation in the electron gun (Section II.B); modulation of the beam current (Section II.C); factors limiting the current density of the spot (Section II.D); spot growth due to magnetic deflection (Section II.E); and examples of limits achieved in present tubes (Section II.F).

B. Beam Formation in the Electron Gun

The electron-optical arrangement of a typical electron gun system (bipotential type, see Section III.B) is schematically shown in Fig. 1. This consists of the cathode K, control grid G_1 and electrodes G_2, G_3, and G_4. The potentials of these electrodes will be denoted by V_c, V_{g1}, V_{g2}, V_{g3}, and V_{g4}, respectively. (Unless otherwise indicated V_c is maintained at ground potential.) In some cases, V_{g3} is written as V_{foc} since it is used to adjust the focus and V_{g4} written as V_a the final anode voltage.

In all present-day color CRTs the source of electrons is an indirectly heated thermionic cathode having a flat surface. (See Section IV for a detailed discussion of cathodes.) Because of the shape of the equipotential surfaces created in front of the cathode by the potentials applied to the neighboring electrodes G_1 and G_2, the electrons emitted from the cathode are converged to a small diameter, referred to as the crossover, at a point O a short distance in

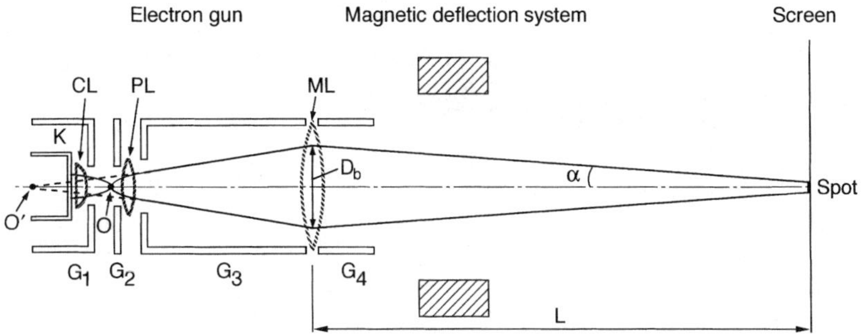

FIGURE 1. Electron-optical arrangement of a typical electron gun and its magnetic deflection system. The main lens, prefocus lens, and cathode lens are represented by ML, PL, and CL, respectively.

front of the cathode. The converging field in front of the cathode is sometimes called the cathode lens denoted by CL. The group of electrodes, K, G_1, and G_2 are referred to as the triode section.

After leaving the crossover the diverging beam of electrons then passes through the prefocus lens (PL), which is created in the region near the exit of the G_2 aperture as a result of the potential difference between V_{g2} and V_{g3}. Because of its converging action this lens reduces the divergence angle of the beam. At the main focusing lens or main lens (ML), created by the potential difference between V_{g3} and V_{g4}, the diverging electron beam is converged, arriving at the phosphor screen as a small spot. As shown, the diameter of the beam at the ML is denoted by D_b, the distance between the ML and the screen by L and the semiangle of the converging beam by α. (The shadow-mask employed in an actual color tube is omitted here.)

As shown in Fig. 1, the converging action of the PL causes the crossover, as seen from the ML, to appear as if it originated at point O′. This point is referred to as the virtual crossover. In the terminology of electron optics the virtual crossover is the object that is imaged by the main lens as a spot on the screen. In broad terms, the electron gun of a CRT can be considered to consist of two main parts, namely the ML and the beam-forming region which includes the triode section (K, G_1 and G_2) together with the PL.

The beam-forming region is shown in greater detail in Fig. 2 somewhat exaggerated for the sake of clarity. As a result of the penetration of the electrostatic field into the G_1 aperture, the equipotential surfaces near the

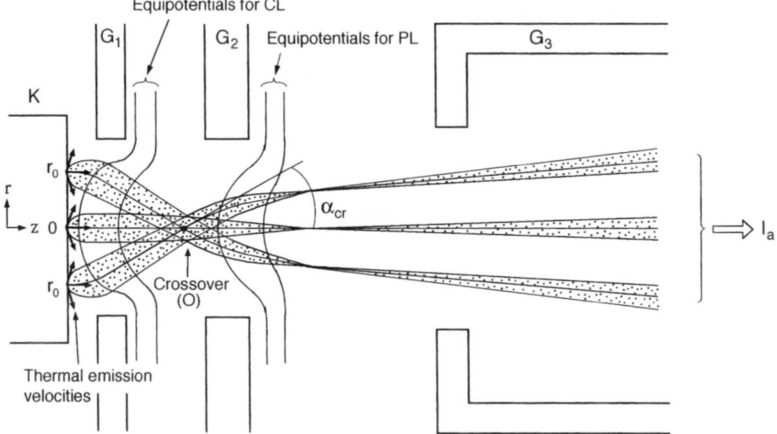

FIGURE 2. Beam-forming region of gun showing electron trajectories originating at the center and two off-axis points of the cathode. Thermal velocity effects are indicated by the small arrows at these points.

cathode curve downward toward the cathode (as indicated by the thin-line curves) producing the cathode lens (CL) shown in Fig. 1. In accordance with this, the field strength at the cathode is not uniform over the surface decreasing from its center. (More detailed discussions about the field strength distribution at the cathode are given in Section II.C.)

The electrons emitted from each point on the cathode have a spectrum of thermal velocities with a Maxwellian distribution (see, e.g., Moss, 1968a; Klemperer and Barnett, 1971a). At a cathode temperature of approximately 1000°K (typical of normal operation) these energies are of the order of 0.1 eV on average. As a result, narrow bundles of rays or beamlets are emitted from each point on the cathode rather than single rays. This is indicated in Fig. 2 by the small horizontally oriented arrows at the surface of the cathode where the three beamlets are shown.

The central ray of each beamlet, referred to as the principal ray, leaves the cathode in the normal direction. Assuming no lens aberrations, the principal ray crosses the axis at a common crossover point O. However, as a result of the thermal spread of velocities within each beamlet, the beam at the crossover has a finite width with no sharply defined edge. In addition, although not shown in the figure, the mutual repulsive forces of the electrons and aberrations of the lens system of the triode section further increase the size of the crossover. The angle of divergence of the beam at the crossover is α_{cr}, as shown, the half angle subtended by the principal rays of the outermost beamlets. The symbol r_0 in the figure represents the radius of the emitting area of the cathode as discussed later in the text. Beyond the crossover the beam angle is reduced by the curved equipotential surfaces near the exit of the G_2 aperture (indicated by the thin-line curves), which produce the PL shown in Fig. 1. All the beamlets from the entire emitting area of the cathode constitute the total cathode current, which is essentially the same as the beam (or anode) current I_a (of a single beam) since in the normal operation of shadow-mask CRTs no beam current is intercepted by the electrodes.

C. Modulation of the Beam Current

For modulating the beam current I_a a variable negative voltage $-V_{g1}$ can be applied to G_1 while holding the cathode at zero potential (grid drive). Alternatively, a variable positive voltage V_c can be applied to the cathode while holding G_1 at zero potential (cathode drive). The resulting beam current characteristic (or modulation characteristic) as a function of $-V_{g1}$ or V_c is schematically shown in Fig. 3. (It should be noted that positive voltages V_c applied to the cathode as well as negative voltages $-V_{g1}$ applied to G_1 are both measured to the left.)

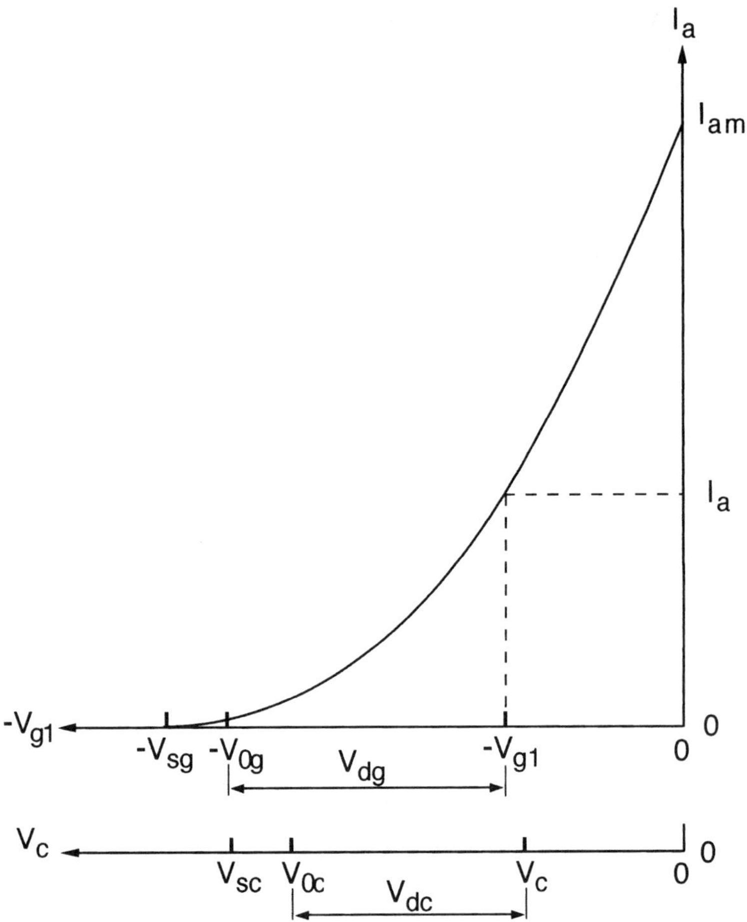

FIGURE 3. Beam current I_a (of a single beam) as a function of G_1 voltage $-V_{g1}$ or cathode V_c voltage. Also shown are Laplace cutoff voltages $-V_{0g}$ and V_{0c}, visual cutoff voltages $-V_{sg}$ and V_{sc}, and drive voltages V_{dg} and V_{dc} for grid drive and cathode drive, respectively.

The cutoff voltage is defined as the value of V_c or V_{g1} at which the electrostatic (Laplace) field strength falls to zero at the center of the cathode. This cutoff voltage is sometimes called the Laplace cutoff voltage, V_0, and plays an important role in deriving the various quantities relating to the beam current characteristic as discussed in the following. As will be explained later, the value of the cutoff voltage V_0 is found to be different for cathode drive (cathode cutoff voltage) and grid drive (grid cutoff voltage). These values are denoted by V_{0c} and V_{0g}, respectively, as shown in Fig. 3. (It should be noted that due to the thermal velocities of the emitted electrons the beam current is not zero at the Laplace cutoff voltages V_{0c} and V_{0g}.)

The drive voltage is defined as the difference between the cutoff voltage V_{0c} or V_{0g} and the voltage V_c or V_{g1}, respectively. As shown in Fig. 3, for grid drive the drive voltage is denoted by $V_{dg}(= V_{0g} - V_{g1})$ and for cathode drive by $V_{dc}(= V_{0c} - V_c)$. The beam current I_a of a CRT is generally found to be an exponential function of the drive voltage V_d of the form $I_a = CV_d^{\gamma}$, where the exponent γ is referred to as the gamma of the picture tube and C is a constant of proportionality. As discussed in what follows, the value of γ has been shown to be approximately 2.5 for a small drive voltage (small currents) and increases with increasing values of V_d.

The maximum beam current I_{am} (of a single beam) occurs at zero bias (maximum drive), being the same for grid drive and cathode drive. The ratio $I_{am}/V_0^{3/2}$ is called the drive factor K and sometimes referred to as the modulation constant or effective perveance. For grid drive K is denoted by $K_g(= I_{am}/V_{0g}^{3/2})$ and for cathode drive by $K_c(= I_{am}/V_{0c}^{3/2})$. Because a large value of K requires a small value of drive voltage for a given beam current, the gun geometry and drive conditions are usually chosen so that K is as large as possible. The cathode drive factor K_c is always greater than the grid drive factor K_g (to be discussed later in the text).

The spot cutoff voltage V_s is determined visually from experiments as follows: A static (undeflected) beam spot is focused on the screen in a dark room at a sufficiently small beam current (usually less than 1 μA). The beam current is then decreased slowly by increasing V_c or V_{g1} up to the value V_{sc} or V_{sg}, respectively, as shown in Fig. 3, at which the light output from the spot becomes just invisible. This corresponds to a very small beam current of the order of 10^{-5} μA (Hasker, 1966). As a result of their thermal velocities, the emitted electrons can overcome some amount of retarding-field in front of the cathode. The visual cutoff voltage V_s is thus always larger than the Laplace cutoff voltage V_0. The difference between the Laplace cutoff voltage and visual cutoff voltage has been shown to be proportional to $V_{0c}^{1/3}$ or $V_{og}^{1/3}$ (Hasker, 1966) and is usually of the order of 8 or 10 V.

1. *Field Strength at the Cathode and Cutoff Voltages*

The Laplace cutoff voltages V_{0g} and V_{0c} can be expressed in terms of electrode potentials and geometry-dependent constants of a gun, assuming an axially symmetric electrode system consisting of the cathode K, control grid G_1, and electrodes G_2 and G_3 (G_4 is not considered because its effect on the field at the cathode surface is negligible), in which there is no current flow. Using cylindrical coordinates with the radial distance denoted by r and axial distance by z, the cathode is assumed to be at the position

$z = 0$. The potential $V(r, z)$ at any point can then be expressed by a series expansion in r and the axial potential $\phi(z)$ and its derivatives derived from Laplace's equation (see, *e.g.*, Spangenberg, 1948c; Klemperer and Barnett, 1971b).

The series expansion is written as follows:

$$V(r, z) = \phi(z) - \frac{\phi''(z)}{4} r^2 + \frac{\phi^{(4)}(z)}{64} r^4 - \frac{\phi^{(6)}(z)}{2304} r^6 + R(r, z) \quad (1)$$

where the superscripts in parentheses indicate the order of the derivative and $R(r, z)$ represents terms higher than the sixth-order. Differentiating Eq. (1) with respect to z and putting $z = 0$, we obtain

$$\left.\frac{\partial V(r, z)}{\partial z}\right|_{z=0} = E(r) = \phi'(0) - \frac{\phi'''(0)}{4} r^2 + \frac{\phi^{(5)}(0)}{64} r^4 - \frac{\phi^{(7)}(0)}{2304} r^6 + R'(r) \quad (2)$$

where $E(r)$ is the field strength at the cathode.

In Eq. (2) it is noted that as the order of the terms increases their magnitude diminishes rapidly for small values of r. For the current purpose, only the first two terms will be retained. Consequently, the approximate field strength at the cathode can be expressed as a parabolic function of r as follows:

$$E(r) = a + br^2 \quad (3)$$

where a and b are constants given by

$$a = \phi'(0), \quad b = -\phi''(0)/4 \quad (4)$$

In a simulation with the aid of an analogue resistance network (Hasker and Groendijk, 1962), a parabolic distribution of the field strength at the cathode has been confirmed over a wide range of drive voltage except for the region close to zero-bias.

Since the axial potential $\phi(z)$ can be expressed as a linear combination of electrode potentials V_{g1}, V_{g2}, and V_{g3} and functional coefficients, $\varphi_1(z)$, $\varphi_2(z)$, and $\varphi_3(z)$ associated with these electrodes, we can write $\phi(z)$ for the case of grid drive ($V_c = 0$) as

$$\phi(z) = \varphi_1(z)(-V_{g1}) + \varphi_2(z)V_{g2} + \varphi_3(z)V_{g3} \quad (5)$$

The coefficients $\varphi_1(z)$, $\varphi_2(z)$, and $\varphi_3(z)$ are sometimes called the axial potential functions; $\varphi_1(z)$ represents the axial potential distribution when $-V_{g1} = 1$ V and $V_{g2} = V_{g3} = 0$; $\varphi_2(z)$ represents the similar distribution when $V_{g2} = 1$ V and $-V_{g1} = V_{g3} = 0$; $\varphi_3(z)$ represents the similar distribution when $V_{g3} = 1$ V and $-V_{g1} = V_{g2} = 0$. An example of these potential functions is given in Fig. 4.

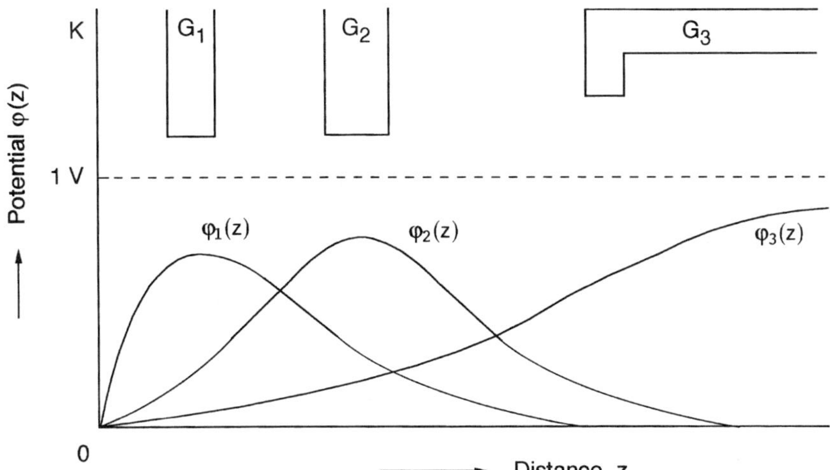

FIGURE 4. Axial potential distributions given by functions $\varphi_1(z)$, $\varphi_2(z)$, and $\varphi_3(z)$ for G_1, G_2, and G_3, respectively, when 1 volt is applied to any one of these electrodes while holding the others at zero.

By introducing Eq. (5) into Eq. (2) and retaining the first two terms, we obtain Eq. (6), which is similar to Eq. (3) but includes the electrode potentials, V_{g1}, V_{g2}, and V_{g3} as follows:

$$E(r) = -(a_1 + b_1 r^2)V_{g1} + (a_2 + b_2 r^2)V_{g2} + (a_3 + b_3 r^2)V_{g3} \quad (6)$$

where a_1, a_2, a_3, b_1, b_2, and b_3 are constants whose values are given by

$$\left.\begin{array}{ll} a_1 = \varphi_1'(0), & b_1 = -\varphi_1''(0)/4 \\ a_2 = \varphi_2'(0), & b_2 = -\varphi_2''(0)/4 \\ a_2 = \varphi_3'(0), & b_3 = -\varphi_3''(0)/4 \end{array}\right\} \quad (7)$$

These values depend upon gun geometry and can be determined for a particular gun geometry by simulation with the aid of an analogue resistance network (Francken, 1959/60) or a digital computer (Weber, 1967a,b). [Another method is possible for determining the field strength at the cathode where the field strength is approximated by a series expansion up to the fourth order of the radial distance r in which its coefficients are obtained analytically (van den Broek, 1986b).]

From Eq. (6) the grid cutoff voltage V_{0g} can be obtained as the value of V_{g1} for which the field strength at the center of the cathode $E(0)$ becomes zero for fixed values of V_{g2} and V_{g3}. Thus we obtain

$$V_{0g} = (a_2/a_1)V_{g2} + (a_3/a_1)V_{g3} = D_{1g}V_{g2} + D_{2g}V_{g3} \quad (8)$$

where quantities D_{1g} and D_{2g} are constants known as the "durchgriffs" or penetration factors for grid drive, which depend on gun geometry. In modern color CRT guns typical values of D_{1g} and D_{2g} are found to be of the order of 0.25 and 5×10^{-4}, respectively.

In the case of cathode drive the cathode is at a positive potential V_c instead of zero while V_{g1} is held at zero. As the potential difference between the cathode and G_2 changes with a change in V_c, the field strength produced at the cathode is different from that produced in grid drive where V_c is held at zero. Thus, for cathode drive Eq. (6) must be altered as follows:

$$E(r) = (a_1 + b_1 r^2)(-V_c) + (a_2 + b_2 r^2)(V_{g2} - V_c) + (a_3 + b_3 r^2)(V_{g3} - V_c) \tag{9}$$

which can be arranged as

$$E(r) = -\{a_1 + a_2 + a_3 + (b_1 + b_2 + b_3)r^2\}V_c + (a_2 + b_2 r^2)V_{g2} + (a_3 + b_3 r^2)V_{g3} \tag{10}$$

The cathode cutoff voltage V_{0c} is found as the value of V_c for which the central field strength $E(0)$ becomes zero for fixed values of V_{g2} and V_{g3}. Thus we obtain

$$V_{0c} = \frac{a_2 V_{g2} + a_3 V_{g3}}{a_1 + a_2 + a_3} = D_{1c} V_{g2} + D_{2c} V_{g3} \tag{11}$$

where the quantities D_{1c} and D_{2c} are the "durchgriffs" or penetration factors for cathode drive which depend on gun geometry. By virtue of Eq. (8), we can change Eq. (11) into the following form:

$$V_{0c} = \frac{(a_2/a_1)V_{g2} + (a_3/a_1)V_{g3}}{1 + a_2/a_1 + a_3/a_1} = \frac{D_{1g} V_{g2} + D_{2g} V_{g3}}{1 + D_{1g} + D_{2g}} \tag{12}$$

This equation can be written as

$$V_{0c} = \frac{V_{0g}}{1 + D_{1g} + D_{2g}} \tag{13}$$

From this it follows that the cathode cutoff voltage V_{0c} is always smaller than the grid cutoff voltage V_{0g} for fixed values of V_{g2} and V_{g3}. If we use the values of 0.25 and 5×10^{-4} for the penetration factors D_{1g} and D_{2g} for grid drive, respectively, as already mentioned here, the cathode cutoff voltage is found to be nearly 80% of the grid cutoff voltage. From a comparison of Eq. (11) with Eq. (12) the penetration factors D_{1g} and D_{2g} are related to those for cathode drive D_{1c} and D_{2c} as follows:

$$\left.\begin{aligned} D_{1c} &= \frac{D_{1g}}{1 + D_{1g} + D_{2g}} \\ D_{2c} &= \frac{D_{2g}}{1 + D_{1g} + D_{2g}} \end{aligned}\right\} \tag{14}$$

By making use of Eq. (13), we obtain a relationship between the cathode drive factor $K_c(= I_{am}/V_{0c}^{3/2})$ and the grid drive factor $K_g(= I_{am}/V_{0g}^{3/2})$ as

$$K_c = K_g(1 + D_{1g} + D_{2g})^{3/2} \tag{15}$$

The value of K_g is normally $3 \sim 4 \times 10^{-6}$A/V$^{3/2}$ depending on gun geometry (see, e.g., Moss, 1968b). If we use the values of 0.25 and 5×10^{-4} for D_{1g} and D_{2g}, respectively, the cathode drive factor K_c is found to be about 40% larger than the grid drive factor K_g. This is a major reason why cathode drive is almost exclusively used in modern color CRTs (see also Niklas *et al.*, 1957). Modern color CRT gun structures mostly employ a single common electrode as the control grid G_1 for all three beams, as discussed in Section III.D, with the drive voltages applied to the separate cathodes of the three beams. In the following discussions of this section we will deal with only cathode drive unless otherwise stated, although similar results would be found in the case of grid drive.

We now turn our attention to the value of $r = r_0$ on the cathode where the field strength falls to zero. From within this radius of the emitting area of the cathode directly beneath the G_1 aperture the beam current is drawn into the gun space as shown in Fig. 2. The radius r_0 can be obtained from Eq. (10) by putting $E(r_0) = 0$. After some rearrangement we can obtain the following equation:

$$r_0^2 = \frac{V_{dc}}{a_c + b_c V_c} \tag{16}$$

where a_c and b_c represent $-(b_2 V_{g2} + b_3 V_{g3})/(a_1 + a_2 + a_3)$ and $(b_1 + b_2 + b_3)/(a_1 + a_2 + a_3)$, respectively. (It should be remarked that Eq. (16) has been derived based on purely electrostatic considerations despite the fact that emitted electrons have a negative space charge that must be taken into account in a rigorous calculation.) The variation of r_0 as a function of V_{dc} depends on gun geometry. Its dependence on some of parameters of the triode section has been investigated by van den Broek (1986b).

By making use of Eq. (16) together with Eq. (10) we can derive the following equation for the field strength $E(r)$ at the cathode at a certain value of drive voltage:

$$E(r) = E(0)(1 - r^2/r_0^2) \tag{17}$$

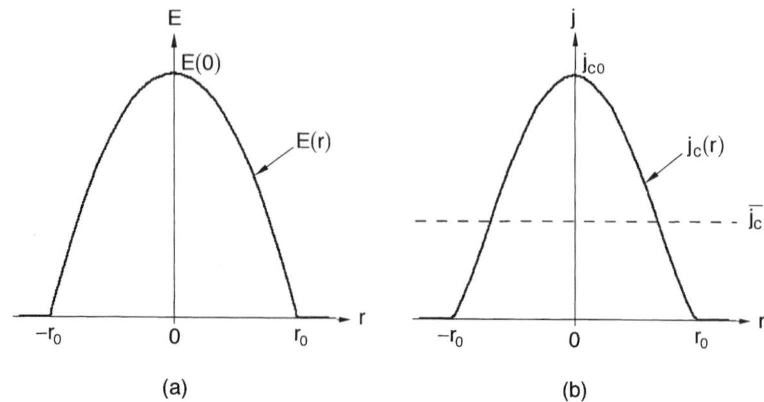

FIGURE 5. (a) Electric field strength distribution $E(r)$ at the cathode for $|r| \leq r_0$; (b) current density distribution $j_c(r)$ at the cathode where $\overline{j_c}$ is the average current density.

where $E(0)$ is the field strength at the cathode center and is given by

$$E(0) = (a_1 + a_2 + a_3)V_{dc} = (E_0/V_{0c})V_{dc} \tag{18}$$

in which E_0 equals $(a_1 + a_2 + a_3)V_{0c}$ and represents the central field strength for zero bias, that is, $V_c = 0$ for which $V_{dc} = V_{0c}$. As indicated, the field strength $E(r)$ decreases quadratically with the radial distance r from the center of the cathode and becomes zero at $|r| = r_0$. Beyond this distance $(|r| > r_0)$ the field strength is negative. The field strength distribution for $|r| \leq r_0$ is shown in Fig. 5a.

2. *Approximate Calculation of Beam Current*

Electron guns for CRTs are designed to operate so that the beam current is space-charge limited. Under this condition, the potential in front of the cathode is determined as follows: The negative space charge of the emitted electrons reduces the potential in front of the cathode so that the potential gradient becomes negative at the cathode surface. This will reduce the emission of most of the electrons and in turn will reduce the space charge, causing the potential gradient to increase again until equilibrium is reached. At this point only a fraction of the electrons, which are capable of being emitted from the cathode at the given temperature (temperature-limited or saturated emission current, see Section IV) is being used. Since the space-charge limited beam current depends on the (Laplace) field strength at the cathode as will be discussed, it can be controlled by changing the potential at either the cathode or G_1 as shown in Fig. 3.

In principle, calculation of the space-charge-limited beam current requires the solution of Poisson's equation. This solution can be obtained numerically for specific gun geometry by simulation with the aid of digital computers yielding the exact values of beam current characteristic (see, e.g., Weber, 1967a,b; Ogusu, 1969; Ninomiya *et al.*, 1971; van den Broek, 1986a; MacGregor, 1986; Bechis *et al.*, 1989; Aalders *et al.*, 1989). However, this generally involves complicated procedures. To obtain a more general insight into the beam current characteristic, for instance, its dependence on various factors such as applied voltages and gun geometry, it is preferable to use some simple approximate formulas. Approximate methods for calculating the space-charge-limited beam current for CRT electron guns have been investigated by various authors (Moss, 1946, 1968b; Ploke, 1951, 1952; Gold and Schwartz, 1958; Francken, 1959/60; Bessho, 1963; Hasker, 1966, 1972; van den Broek, 1986b).

a. Current Density at the Cathode. To calculate the current density at the cathode of a CRT gun, Ploke (1952) used an analogy with a simple parallel-plane diode as will be discussed here. In a planar diode, one electrode of which is an electron-emitting cathode and the other is the anode, the space-charge-limited current density j depends on the 3/2 power of the anode voltage, given by the following well-known equation referred to as Child's law:

$$j = 2.335 \times 10^{-6} \frac{V^{3/2}}{d^2} \tag{19}$$

Here j is the current density in A/cm^2, V is the anode potential in volts with respect to the cathode and d is the cathode-to-anode distance in centimeters. In this equation the thermal velocities of the emitted electrons and the finite value of the saturation current density at the cathode are neglected (Child, 1911; see also Spangenberg, 1948b; Herrmann and Wagener, 1951c).

In the absence of space charge, the uniform electrostatic (Laplace) field strength is given by

$$E = \frac{V}{d} \tag{20}$$

Equation (19) can thus be written as

$$j = 2.335 \times 10^{-6} \frac{E^{3/2}}{d^{1/2}} \tag{21}$$

This equation states that the space-charge-limited current density j is proportional to the 3/2 power of the (Laplace) field strength E at the cathode and inversely proportional to the 1/2 power of the cathode-to-anode distance d.

Equation (21) in turn can be used to calculate the beam current density at the cathode of an electron gun where the field strength varies over the surface of the cathode. Since there is no clear distance corresponding to the cathode-to-anode distance d of the planar diode, Ploke introduced the concept of the equivalent-diode distance explained as follows: With a slight change in Eq. (18), the central field strength $E(0)$ for the drive voltage V_{dc} can be expressed as

$$E(0) = \frac{V_{dc}}{1/(a_1 + a_2 + a_3)} = \frac{V_{dc}}{V_{0c}/E_0} \tag{22}$$

where a_1, a_2 and a_3 are constants given by Eq. (7), V_{0c} is the Laplace cutoff voltage and E_0 is the central field strength for zero bias. By analogy with Eq. (20), the drive voltage V_{dc} corresponds to the anode voltage V of the planar diode and the quantity $1/(a_1 + a_2 + a_3)(= V_{0c}/E_0)$ corresponds to the cathode-to-anode distance d of the planar diode. Using this analogy this quantity is defined as the equivalent-diode distance and expressed by the symbol d_e as follows:

$$d_e = \frac{1}{a_1 + a_2 + a_3} = \frac{V_{0c}}{E_0} \tag{23}$$

In other words, if a potential difference of V_{dc} exists between the cathode and anode in a planar diode whose anode-to-cathode distance is equal to d_e, the field strength is given by $E(0)$ in accordance with Eq. (22).

With $E_0 = V_{0c}/d_e$ from Eq. (23) substituted into Eq. (18), we obtain the following equation from Eq. (17) for the field strength $E(r)$ at the cathode:

$$E(r) = (V_{dc}/d_e)(1 - r^2/r_0^2) \tag{24}$$

Introducing Eq. (24) into Eq. (21) we obtain the following equation for the current density $j_c(r)$ at the cathode:

$$j_c(r) = j_{c0}(1 - r^2/r_0^2)^{3/2} \tag{25}$$

where j_{c0} is the current density at the cathode center given by

$$j_{c0} = 2.335 \times 10^{-6} \frac{V_{dc}^{3/2}}{d_e^2} \tag{26}$$

The shape of the current density distribution of Eq. (25), similar to the field strength distribution of Fig. 5a, is shown in Fig. 5b.

b. Beam Current Characteristic and Gamma γ. An integration of Eq. (25) over the emitting area of the cathode gives the total cathode current (of a single beam). Because the cathode current of shadow-mask CRTs is essentially the same as the beam (or anode) current I_a, we obtain the following relation:

$$I_a = \int_0^{r_0} j_c(r)2\pi r\,dr = \int_0^{r_0} j_{c0}(1 - r^2/r_0^2)^{3/2}\,2\pi r\,dr = \frac{2}{5}\pi r_0^2 j_{c0} \tag{27}$$

Using Eq. (26) for j_{c0}, we can obtain the following expression for the anode current:

$$I_a = 2.934 \times 10^{-6}\frac{V_{dc}^{3/2} r_0^2}{d_e^2} = C' V_{dc}^{3/2} r_0^2 \tag{28}$$

where C' is a constant for a given gun geometry and equal to $2.934 \times 10^{-6}/d_e^2$.

As mentioned, the beam current characteristic of CRTs is generally given by an exponential function of the drive voltage V_d as expressed by $I_a = C V_d^{\gamma}$. The exponent γ (gamma) is obtained as the slope of the double logarithmic plot of the beam current as a function of the drive voltage V_d as follows:

$$\gamma = \frac{\partial \log I_a}{\partial \log V_d} = \frac{V_d}{I_a}\frac{\partial I_a}{\partial V_d} \tag{29}$$

By making use of Eq. (28) together with Eq. (16) and inserting $V_d = V_{dc}$ (cathode drive) in Eq. (29), we thus obtain the following relation:

$$\gamma = 2.5 + b_c r_0^2 \tag{30}$$

This shows that γ is equal to 2.5 for $r_0 = 0$, namely, for $V_{dc} = 0$, and increases with increasing values of V_{dc} [see also Eq. (16)].

Generally, in either cathode drive or grid drive, γ has a value nearly equal to 2.5 for small values of V_d, that is, for small beam currents, and increases with increasing V_d. The particular dependence of γ on V_d is determined by the gun geometry and operating conditions (Gold and Schwartz, 1958). Moss (1968b) derived a semiempirical formula for the beam current characteristic from numerous experiments, showing that the beam current is proportional to $V_d^{7/2}$, thus giving the value of γ of 3.5 over the whole current characteristic. It should be noted that in television displays the raster cutoff voltage V_r (not shown in Fig. 3) for visible light output of the scanned raster (which corresponds to a beam current of a few microamperes) should be used in the determination of the drive voltage V_d instead of the Laplace cutoff voltage V_0. However, since under normal ambient light conditions of television display V_r is close to $V_0(V_0 \leq V_r < V_s$ generally holds), the use of V_0 for this purpose is still appropriate.

c. Definition of Cathode Loading. By virtue of Eq. (27) the peak current density j_{c0} at the cathode center is expressed as follows:

$$j_{c0} = 2.5\frac{I_a}{\pi r_0^2} = 2.5\overline{j_c} \tag{31}$$

where $\overline{j_c}$ signifies the average current density over the entire emitting area shown in Fig. 5b. The values of the peak and average current densities j_{c0} and $\overline{j_c}$, respectively, are frequently used to represent the loading level of a cathode. These values are usually expressed in units of A/cm^2. With the beam current I_a (of a single beam) expressed in milliamperes and the radius of the cathode emitting area r_0 in mm, Eq. (31) can be written as follows:

$$j_{c0} = 0.25\frac{I_a}{\pi r_0^2} = 0.25\overline{j_c} \quad [\mathrm{A/cm^2}] \tag{32}$$

In the discussions to follow, the central cathode loading j_{c0} will be used to indicate the loading level of the cathode. As will be discussed in Section IV, the cathode loading is restricted by the allowable emission current density of the cathode. For example, for a conventional oxide-coated cathode, a central cathode loading j_{c0} of 1.4 A/cm^2 is considered to be an acceptable maximum for the long-term average current to obtain good operating life (Barten, 1989), although operation at somewhat higher current densities is allowable with special cathodes as discussed in Section IV. In the case of an impregnated or I-cathode, for example, good operating life can be obtained with a central cathode loading j_{c0} of approximately 4 A/cm^2 as the long-term average current (Endo *et al.*, 1990).

d. Effects of Thermal Velocities of Electrons. As already mentioned here, the space-charge-limited current density j in a planar diode given by Eq. (19) (Child's law) has been derived without taking into account the thermal velocities of the electrons and the finite value of the saturation current density of the cathode. An exact analysis for the space-charge-limited current in a planar diode by Langmuir (1923), taking into account the initial velocities of the electrons and the saturation current density of the cathode, shows that a potential minimum is formed in front of the cathode, which limits the current reaching the anode. This occurs because only electrons emitted with axial initial velocities larger than that corresponding to the depth of the potential minimum can pass the minimum point, the other electrons being reflected back to the cathode.

The resulting equations for the space-charge-limited current density in a planar diode as a function of the anode potential V with respect to the cathode and the cathode-to-anode distance d are quite complicated. However, the current density j can be developed in a series expansion and, truncated after the second term, yields an approximate expression as follows:

$$j = 2.335 \times 10^{-6} \frac{(V - V_m)^{3/2}}{(d - d_m)^2} \left\{ 1 + \frac{0.0247 T^{1/2}}{(V - V_m)^{1/2}} \right\} \tag{33}$$

where j is in A/cm^2, V_m is the potential of the minimum in volts ($V_m < 0$), d_m is the distance between the cathode and the potential minimum in centimeters, and T is the cathode temperature in °K.

Both V_m and d_m are given by complicated functions of the current density j, the cathode temperature T, and the saturation current density j_s given by Eq. (80) in Section IV. The value of V_m is of the order of -0.1 to -0.5 V and the value of d_m is of the order of several microns in normal operation of CRTs. Equation (33) mentioned here seems not to be significantly different from Eq. (19) because V_m and d_m are small. However, V_m, although small, is significant because it determines the fraction of the emitted electrons that pass the potential minimum and reach the anode (see also, e.g., Hasker and van Dorst, 1989; Amboss, 1993).

It has been shown that the calculated beam current with the aid of Eq. (33) is appreciably larger for low beam currents (less than 0.5 mA) than those calculated based on Eq. (19) due to the influence of the initial thermal velocities (Loty, 1984; van den Broek, 1986b). The effect of the initial thermal velocities on the current density distribution at the cathode, cathode loading and beam current has been investigated by Hasker (1972), applying Langmuir's theory for the planar diode, for an electron gun with the aid of the Ploke's model already discussed here. Results of the investigation reveal that due to the initial thermal velocities the calculated current density distribution at the cathode deviates appreciably from that given by Eq. (25), which was derived without taking into account initial velocities. It has been shown that the peak-to-average ratio $j_{c0}/\overline{j_c}$ of the current density distribution at the cathode, which has the value of 2.5 in Eq. (31) already discussed here, ranges from 1.9–2.6 depending on the gun geometry and operating conditions.

D. Factors Limiting Current Density of Spot

Of major interest in a practical CRT is obtaining a spot with a high current density, especially in the case of tubes designed for high-brightness and high-resolution applications. The current density is limited by various factors, of which the most important are the thermal velocity spread of the electrons emitted from the cathode (known as the Langmuir limit), the mutual repulsion of electrons in the beam, and electron-optical aberrations of the lens system of the gun. These factors will be discussed in the following Sections D.1, D.2, and D.3, respectively. A further limitation, discussed in Section II.E, occurs due to distortion of the beam by the magnetic deflection field.

The process by which the current density is determined by the combined effect of the forementioned factors, however, is too complicated to allow exact calculation by a simple analytic approach. However, for practical purposes of gun design a simple approximate method for calculating the spot size is desirable. Such a method, which makes use of a linear combination of the individual factors limiting the spot size, will be discussed in Section D.4.

1. *Thermal Velocity Spread of Electrons*

For a Maxwellian velocity distribution of the emitted electrons it has been shown that the current density distribution of the beam at the crossover has the form of a Gaussian function (Law, 1937; 1942) and can be expressed as follows:

$$j_{cr}(r) = j_{cr0} \exp\left(-r^2/r_{cr}^2\right) \tag{34}$$

where r is the radial distance from the center of the crossover and r_{cr} is the value of r where the current density is reduced to 1/e of the peak current density j_{cr0} on the axis. Ideally, the current density distribution at the crossover should be unaffected by the spatial variation of emission density over the surface of the cathode. It should be noted that the crossover is not the image of the cathode surface but represents the exit pupil of the cathode lens. As in geometrical optics the crossover is the point of minimum beam diameter as discussed in Section II.B.

The spread of the initial thermal velocities of the emitted electrons places a limit on the current density at the crossover whose maximum value j_{cr0} can be expressed by the following relation:

$$j_{cr0} = j_c\left(1 + \frac{eV_{cr}}{kT}\right)\sin^2\alpha_{cr} \tag{35}$$

where j_c is the current density at the cathode (assumed to be uniform); V_{cr} is the potential at the crossover in volts; T is the temperature of the cathode in °K; α_{cr} is the half-angle of divergence of the beam at the crossover (see Fig. 2), e is the electronic charge 1.602×10^{-19} C; and k is Boltzmann's constant 1.380×10^{-23} J/°K.

Equation (35) was first derived by Langmuir (1937) and further developed by Pierce (1939). Langmuir's theory assumed a uniform emission density from the cathode, no aberrations in the lens system, and no space-charge repulsion within the beam. Although these three assumptions are far from satisfied in practice, this theory is still very important for many purposes in the design and study of electron guns.

The validity of Langmuir's theory in practical cases has been investigated in some detail (Moss, 1961, 1968c; Hasker, 1965). If, instead of a uniform current density j_c at the cathode, a nonuniform current density is assumed as in practical cases, an average current density $\overline{j_c}$ defined by Eq. (31) can be used as shown by Moss (1961). In addition, the second term within the parentheses of Eq. (35) eV_{cr}/kT can be written as 11600 V_{cr}/T, which is almost always much greater than 1. Also, as α_{cr} is usually very small we can use the approximation $\sin\alpha_{cr} \approx \alpha_{cr}$. Consequently, the expression of Eq. (35) for the maximum attainable current density at the crossover reduces to

$$j_{cr0} = \overline{j_c}(11600\, V_{cr}/T)\alpha_{cr}^2 \tag{36}$$

As indicated in Fig. 1, the crossover is imaged as a spot on the phosphor screen by the combined action of the prefocus lens PL and the main lens ML. Disregarding space-charge repulsion and lens aberrations, there is thus a one-to-one correspondence in current density distribution between the crossover and the beam spot. In view of this the spot also has a Gaussian form of the current density distribution expressed by

$$j_{sp}(r) = j_{sp0} \exp(-r^2/r_{sp}^2) \tag{37}$$

where r is the radial distance from the center of the spot at the screen and r_{sp} is the value of r where the current density of the spot is reduced to 1/e of the peak current density j_{sp0}.

The magnification M of the lens system, obtained from Lagrange's law (see, e.g., Klemperer and Barnett, 1971c), is as follows:

$$M = \alpha_{cr}\sqrt{V_{cr}}/\alpha\sqrt{V_a} \tag{38}$$

where α is the half-angle subtended by the beam spot at the screen (see Fig. 1), and V_a is the potential of the screen or anode voltage. The current density j_{sp0} of the spot can be obtained by the current density j_{cr0} of the crossover divided by the square of the magnification M. Thus, we obtain the following relation for the maximum attainable current density at the spot:

$$j_{sp0} = j_{cr0}/M^2 = \overline{j_c}(11600\, V_a/T)\alpha^2 \tag{39}$$

It should be remarked that the right-hand term of this equation includes no explicit form of lens magnification. This is due to the fact that the quantities V_{cr} and α_{cr} associated with the crossover in Eq. (36) have been canceled out by the lens magnification formula of Eq. (38), leaving only the quantities V_a and α involved with the spot at the screen.

An integration of Eq. (37) over the spot area gives the beam (or anode) current I_a (of a single beam) as follows:

$$I_a = j_{sp0} \int_0^\infty \exp(-r^2/r_{sp}^2) 2\pi r dr = \pi r_{sp}^2 j_{sp0} \tag{40}$$

As mentioned in Section II.B, in the normal operation of shadow-mask CRTs the beam current I_a is essentially the same as the cathode current, thus resulting in the following relation:

$$\pi r_{sp}^2 j_{sp0} = \pi r_0^2 \overline{j_c} \tag{41}$$

where r_0 is the radius of the emitting area of the cathode. By making use of Eq. (41) together with Eq. (39), we can obtain the radius r_{sp} for the minimum obtainable spot as follows:

$$r_{sp} = \frac{r_0}{\alpha} \left(\frac{T}{11600 V_a} \right)^{1/2} \tag{42}$$

The half-angle α of the converging beam can be written as $\alpha \approx D_b/2L$, where D_b is the diameter of the beam at the main lens and L is the distance from the main lens to the screen as shown in Fig. 1. By virtue of Eq. (31), the radius of the cathode emitting area r_0 can be expressed in terms of the central cathode loading j_{c0} (peak current density at the cathode) and the beam current I_a as

$$r_0 = \left(\frac{2.5 I_a}{\pi j_{c0}} \right)^{1/2} \tag{43}$$

Consequently, the minimum obtainable spot diameter $2r_{sp}$ (1/e value) determined by the thermal velocity spread of the emitted electrons can be expressed as follows:

$$2 r_{sp} = 3.3 \times 10^{-2} \frac{L}{D_b} \left(\frac{T I_a}{j_{c0} V_a} \right)^{1/2} \tag{44}$$

The spot diameter $2r_{sp}$ is often called the thermal spot size or the magnified crossover because it equals the diameter of the crossover ($2r_{cr}$) multiplied by the lens magnification M.

Although other definitions are also sometimes used, in practical applications the spot size is often defined as its width at the 5% value of the peak intensity (Barten, 1984). This definition will be used for the thermal spot size in the following discussions unless otherwise stated. Because the 5% value equals 1.73 times the 1/e value as indicated for a Gaussian distribution of the thermal spot shown in Fig. 6, the diameter of the thermal spot d_{th} at

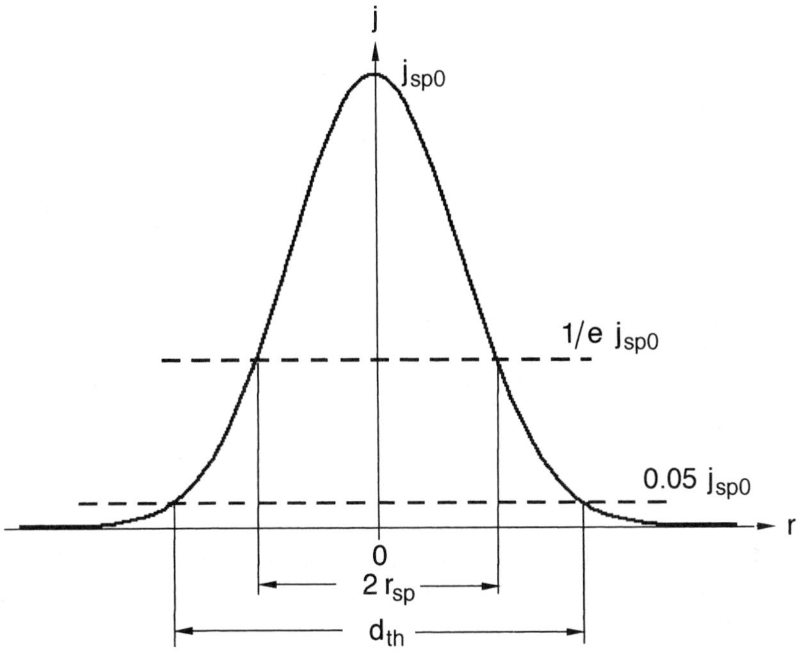

FIGURE 6. Gaussian distribution of current density of spot resulting from thermal energies indicating 1/e width $2r_{sp}$ and 5% width d_{th}.

the 5% value is written as

$$d_{th} = 5.7 \times 10^{-2} \frac{L}{D_b} \left(\frac{TI_a}{j_{c0} V_a} \right)^{1/2} \equiv C_{th}/D_b \tag{45}$$

As indicated, the spot size d_{th} is inversely proportional to the beam diameter D_b at the main lens so that the expression C_{th}/D_b is used for the sake of convenience in later discussions. At a given beam diameter D_b the thermal spot size varies as the inverse square-root of the cathode loading j_{c0}, resulting in a decrease in the thermal spot size with an increase in the cathode loading. This takes place with a decrease in the radius of the emitting area r_0 at a constant current in accordance with Eq. (43). Actually, a smaller G_1 aperture is used in gun design or the gun is operated with an increased cutoff voltage together with an increase in the drive voltage. However, as mentioned, this depends on the emission capability of the cathode.

2. *Space-Charge Repulsion*

Aside from the current density limit of the spot imposed by thermal emission velocities the current density is further limited by the mutual repulsion

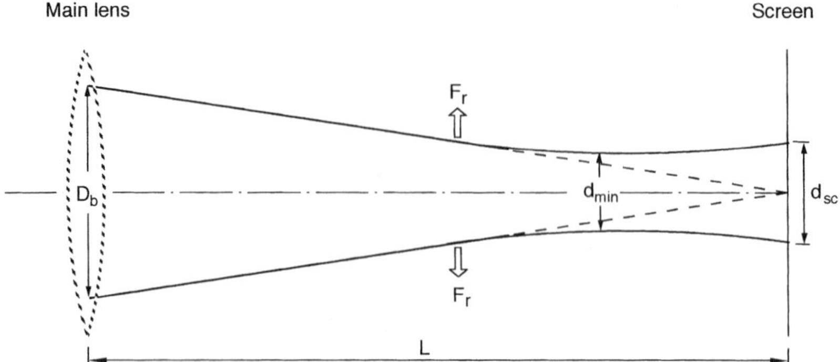

FIGURE 7. Space-charge-limited spot size d_{sc} at the screen due to space-charge repulsion in drift space between main lens and screen. Minimum beam diameter $d_{\min}$ occurs before the screen.

between electrons during their transit time from the cathode to the phosphor screen. As expected, the effect of this repulsion becomes increasingly significant at relatively high current levels of the beam and in regions where the transit time of the electrons is long. In CRT guns this effect may be pronounced in the relatively long field-free drift region between the main lens and the phosphor screen (Thompson and Headrick, 1940) and in the beam-forming-region near the cathode where the electron velocities are low (Aalders *et al.*, 1989).

Figure 7 illustrates the longitudinal section of the electron beam carrying a current I_a in the field-free drift space between the main lens of the gun and the phosphor screen. Assuming the lens to be ideal, that is, free from aberrations, it will impart an inward radial momentum to the electrons proportional to their radial position as they pass through the lens region where the beam diameter is denoted by D_b. Neglecting space-charge effects in the drift region and the thermal spread of velocities, the electron beam would converge to a point at the screen, following the dashed lines shown in the figure.

If space charge is considered, an outward radial force F_r will be added to the converging electrons because of their mutual repulsion. This force increases as the beam converges so that the inward radial velocities imparted to the electrons by the lens falls to zero at a point where the beam has a minimum diameter $d_{\min}$. As shown by the solid line, beyond this point the beam starts to diverge again, resulting in a spot size d_{sc} at the screen, which is the minimum attainable spot diameter limited by space-charge repulsion.

To derive equations for the trajectories of the beam subjected to space-charge repulsion the following assumptions are made:

1. The electrons are uniformly distributed throughout the cross section of the beam and the trajectory of each electron does not cross that of the others during its transit.
2. The axial velocity of the electrons is constant. As the thermal velocities of the electrons are a very small fraction of their axial velocity, they can be ignored.
3. The difference in potential between the center and the boundary of the beam resulting from space charge is also very small compared to the beam voltage.

By applying Gauss's law to the beam at any axial position z, the radial electric field E_r on the outer surface of the beam at a radial distance r can be expressed as follows:

$$E_r = \frac{I_a}{2\pi\varepsilon_0 r\sqrt{2(e/m)V_a}} \tag{46}$$

where ε_0 is the dielectric constant of free space 8.85×10^{-12} F/m; e is the electronic charge 1.602×10^{-19} C; m is mass of electron 9.107×10^{-37} kg; I_a is the beam current (of a single beam) in amperes; and V_a is the potential in the drift space in volts. The radial field E_r causes the outermost electrons of the beam to be accelerated in the radial direction in accordance with the relation

$$\frac{d^2r}{dt^2} = (e/m)E_r \tag{47}$$

The radial velocity dr/dt can be related to the slope dr/dz by using the axial velocity $\sqrt{2(e/m)V_a}$ as follows:

$$\frac{dr}{dt} = \sqrt{2(e/m)V_a}\,\frac{dr}{dz} \tag{48}$$

and the radial acceleration can be expressed as

$$\frac{d^2r}{dt^2} = 2(e/m)V_a\frac{d^2r}{dz^2} \tag{49}$$

Consequently, by making use of Eqs. (46), (47), and (49), the equation of the trajectory of electrons at the outermost edge of the beam is given by

$$\frac{d^2r}{dz^2} = \frac{I_a}{4\pi\varepsilon_0\sqrt{2(e/m)}V_a^{3/2}r} = 1.52 \times 10^4\frac{I_a}{V_a^{3/2}r} \tag{50}$$

As an example of the (numerical) solution of Eq. (50), Fig. 8 shows electron trajectories (9-pairs of solid-line curves) at the outermost edge of the beam with various angles at the main lens. We assume that the beam diameter D_b at

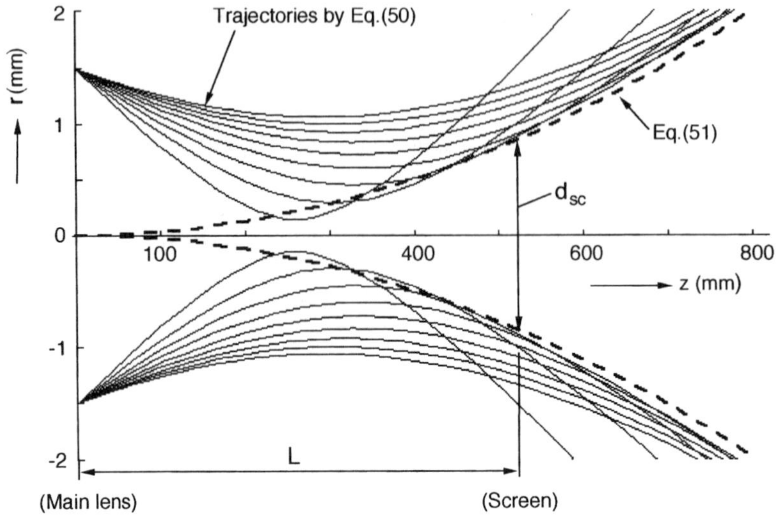

FIGURE 8. Outermost rays (solid lines) of the beam with various slopes at the main lens calculated by Eq. (50) whose envelopes (not shown) give the space-charge-limited spot size d_{sc} at the screen at a distance L from the main lens. Dashed-line curves show approximation given by Eq. (51).

the main lens is 3.0 mm, the beam current I_a is 4 mA, and the anode voltage V_a is 30 kV. In the figure each pair of trajectories corresponds to a different focusing condition of the main lens, thus covering a focusing of the beam from underfocusing to overfocusing. It follows from this that the envelopes formed by these trajectories determine the space-charge-limited spot size d_{sc}, as indicated, for any value of the main lens to screen distance L.

The heavy dashed-line curves in the figure represent the value of d_{sc} obtained by the following approximation formula (Alig, 1980):

$$d_{sc} = \frac{0.8\, I_a L^2}{2\pi\varepsilon_0\sqrt{2(e/m)}V_a^{3/2}D_b} = 2.42\times 10^4\,\frac{I_a L^2}{V_a^{3/2}D_b} \equiv C_{sc}/D_b \tag{51}$$

As seen in the figure this formula closely approximates the space-charge-limited spot size d_{sc} under some conditions. The approximation is good when the spot size d_{sc} at the screen does not differ too much from the beam diameter D_b at the main lens in a current range from a few tenths of a milliampere to a few milliamperes, a range typical of the currents used in color CRTs. Since d_{sc} in Eq. (51) has the same dependence on D_b as the spot size due to the thermal spread d_{th} (already mentioned) this approximation greatly simplifies the overall treatment of spot-size enlargement and is represented by C_{sc}/D_b for the sake of convenience in later discussions.

An analytical method for obtaining the minimum spot size d_{sc} at the screen due to the space-charge repulsion based on the same assumptions as already mentioned here was developed by Schwartz (1957). The resulting equation, shown in the following, gives the normalized value for d_{sc}:

$$\frac{I_a^{1/2}}{V_a^{3/4}} \frac{2L}{D_b} = F\left(\frac{d_{sc}}{D_b}\right) \tag{52}$$

In this equation, $F(d_{sc}/D_b)$ represents a universal function of the ratio of the minimum attainable spot diameter d_{sc} to the beam diameter D_b at the main lens. This function, shown in the form of a graph in Schwartz (1957) and also by Moss (1968a), permits an immediate calculation of d_{sc} as a function of the beam current and other parameters given in Eq. (52).

3. *Spherical Aberration of Lenses*

Although any axially symmetric electrostatic field acts as an electron lens, such lenses generally have considerable aberrations, which cause a further enlargement of the spot. Since such aberrations greatly increase with increasing width of the beam in the lens region it is desirable that electron rays directed toward the main lens, for example, should leave the beam-forming region of the gun with only a small angle. To obtain such an electron beam it is desirable to limit the emission of electrons to the central area of the cathode, although this results in an increase in the current density at the cathode or cathode loading. (Other methods for reducing aberrations by means of special prefocus lens designs will be discussed in Section III.C.)

The general properties of image formation and the geometric aberrations associated with it can be derived making use of the axial symmetry of the lens field (see, e.g., Zworykin *et al.*, 1945a; Grivet, 1972a). Assuming the electrostatic field to be continuous throughout the lens space, the coordinates of each point at the image plane can be expressed by a power series of the coordinates of the rays at the object plane and the aperture plane (equivalent-lens plane) as discussed in what follows. As shown in Fig. 9, the coordinate of an electron ray in the image plane is represented by $r_i(r_i^2 = x_i^2 + y_i^2)$, and those in the object plane and aperture plane by $r_o(r_o^2 = x_o^2 + y_o^2)$ and $r_a(r_a^2 = x_a^2 + y_a^2)$, respectively. Since r_i is a function of r_o and r_a, it can be expressed as a power series in a linear combination of the following terms:

$$r_o,\ r_a,\ r_o^2,\ r_a^2,\ r_o r_a,\ r_o^3,\ r_o^2 r_a,\ r_o r_a^2,\ r_a^3,\ \text{Res.}$$

where Res. represents residual terms higher than the third-order. To a first approximation, taking only first-order terms into account, that is, r_o and r_a, and neglecting higher-order terms of the power series, a sharp, geometrically similar image of the object is produced. This image is usually referred to as

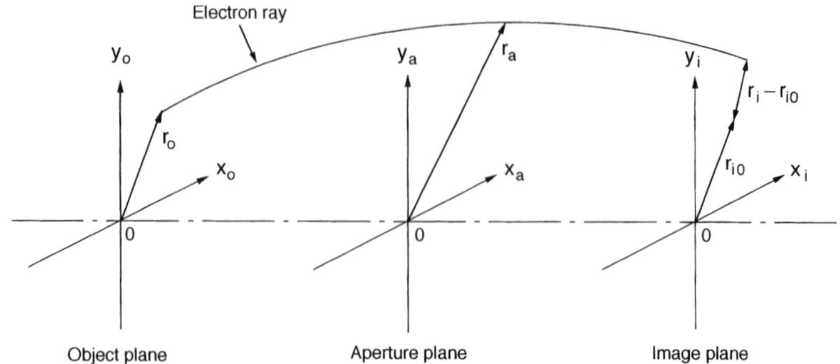

FIGURE 9. Radial coordinates of an electron ray; r_0 in the object, r_a in the aperture (equivalent lens), and r_i in the image plane. The Gaussian image is at r_{i0}, with the difference $\Delta r (= r_i - r_{i0})$ giving the aberration.

the first-order or Gaussian image but may also be referred to as the paraxial image. If we represent the coordinate of the Gaussian image by r_{i0}, then the difference $\Delta r (= r_i - r_{i0})$ will give the (radial or transverse) aberration.

Taking into account the fact that the electrostatic lens field is axially symmetric, the higher-order terms for the radial coordinate r_i of an electron ray in the image plane can only be odd terms, namely, third-order, fifth-order, and so on, as will be discussed. If we neglect the fifth- and higher-order terms, the aberrations given by the difference $\Delta r (= r_i - r_{i0})$ are represented by a linear combination of the third-order terms of r_o^3, $r_o^2 r_a$, $r_o r_a^2$, and r_a^3. It is well known that for electrostatic lenses there are five types of aberrations associated with these third-order terms, namely, distortion, astigmatism, field curvature, coma, and spherical aberration just as in the case of glass lenses (see Zworykin *et al.*, 1945a; Grivet, 1972a). Distortion is associated with r_o^3, astigmatism and field curvature are associated with $r_o^2 r_a$, coma is associated with $r_o r_a^2$, and spherical aberration is associated with r_a^3.

The spherical aberration is the only type of aberration that exists for a point object on the axis because it does not depend on r_o, but only on r_a. For this reason spherical aberration is also called the aperture defect and is determined mainly by the aperture size at the lens. In a CRT electron gun r_o corresponds to the radius of the crossover and r_a corresponds to the radius of the beam at the main lens $(= D_b/2)$. As normally $D_b/2 \gg r_o$, spherical aberration is the dominant aberration in the beam spot compared to the other four aberrations already mentioned here.

Spherical aberration is present in each lens of the gun system, namely, in the cathode lens, the prefocus lens, and the main lens as already mentioned here. In general, the spherical aberration of each of these lenses increases in an additive manner the beam spot size at the screen. Since the spherical

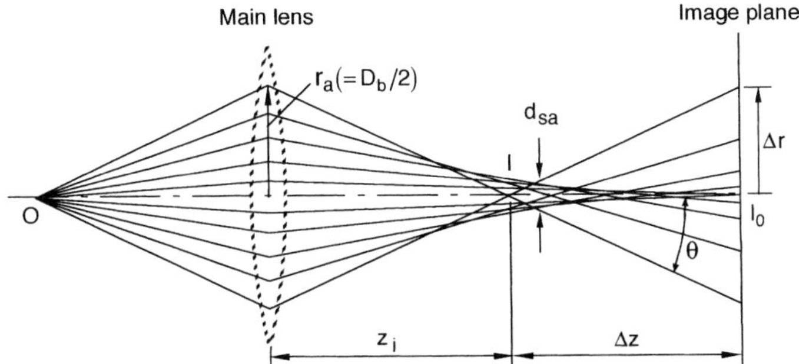

FIGURE 10. Spherical aberration of the main lens causing outer rays originating at object point O to intersect the axis at I before the Gaussian image point I_0 where the axis is intercepted by paraxial rays.

aberration of the main lens is usually the largest, its effect will be considered in the following in some detail.

As shown in Fig. 10, due to the spherical aberration of the main lens (shown as a thin equivalent lens), the outer rays of the beam, which originates at the object point O, will after convergence by the main lens always intersect the axis closer to the lens than the image plane at which the paraxial rays cross (Gaussian image point I_0). The longitudinal aberration is defined by the distance between point I at which a nonparaxial ray intersects the axis and the Gaussian image point I_0 as $\Delta z = I_0 - I$.

The longitudinal aberration Δz may be expressed in the form of power series of the radial position r_a of the ray at the main lens. In the power series only even terms appear (unlike that of the radial image coordinate of r_i already mentioned here), because the lens field is assumed to be axially symmetric and the value of Δz (a symmetrical function of r_a) has the same value for positive and negative values of r_a. Thus, we obtain

$$\Delta z = c_2 r_a^2 + c_4 r_a^4 + c_6 r_a^6 + \text{Res.} \tag{53}$$

where c_2, c_4 and c_6 are constants and Res. represents residual terms higher than the sixth-order.

Instead of the longitudinal aberration Δz, it is often more convenient to consider the radial or transverse aberration Δr as already mentioned. From Fig. 10, Δr is related to Δz as

$$\Delta r = \Delta z \theta \tag{54}$$

where θ is the slope angle of the ray and sufficiently small so that $\tan\theta \approx \theta$. The slope angle θ is expressed by

$$\theta = r_a / z_i \tag{55}$$

where z_i is the distance on the image side from the main lens to the point I where the ray intersects the axis.

Substitution of Eqs. (53) and (55) into Eq. (54), Δr can be written as

$$\Delta r = (c_2/z_i)r_a^3 + (c_4/z_i)r_a^5 + \text{Res.} \tag{56}$$

where Res. represents residual terms higher than the fifth-order. Assuming that r_a is not too large, we neglect the fifth- and higher-order terms in Eq. (56) resulting in

$$\Delta r = c_2' r_a^3 \tag{57}$$

where $c_2'(= c_2/z_i)$ is in effect a constant since the slope angle θ is assumed to be small. Thus, the cubic dependence of the spherical aberration Δr on the radial position r_a of the rays at the main lens is derived as a direct consequence of the symmetry of the lens system.

In the following, for the sake of convenience, instead of the radial position r_a, the beam diameter $D_b(= 2\,r_a)$ at the main lens will be used. The resultant circle of aberration $2\Delta r$ at the Gaussian image plane can then be expressed as a cubic dependence of the beam diameter D_b normalized by the lens diameter D_L. Thus, we obtain

$$2\,\Delta r = c_{sa}(D_b/D_L)^3 \tag{58}$$

where c_{sa} is the coefficient of the third-order spherical aberration. As shown in Fig. 10, at a distance in front of the Gaussian image point I_0 a sharply defined cross section of the beam is produced with a minimum diameter d_{sa}. This cross section is called the disc of least confusion. Normally beam focusing is adjusted so that the disc of least confusion falls on the phosphor screen, resulting in the smallest spot size. Since the disc of least confusion is as small as one-fourth the aberration circle $2\,\Delta r$ already mentioned here (Klemperer and Barnett, 1971d), we can write as follows:

$$d_{sa} = \frac{c_{sa}}{4}(D_b/D_L)^3 \equiv C_{sa}\,D_b^3 \tag{59}$$

where C_{sa} is the coefficient of the third-order spherical aberration defined by $C_{sa} = c_{sa}/4D_L^3$.

There are a large number of publications describing different ways of determining the spherical aberration coefficient (see, e.g., Ramberg, 1942; El-Kareh and El-Kareh, 1970b; Saito *et al.*, 1973, 1979). To calculate the spherical aberration coefficient computer programs are generally used. For example, Fig. 11 gives spherical aberration coefficients obtained by trajectory calculations in a Laplace electric field for a typical bipotential lens consisting of equidiameter cylindrical electrodes G_3 and G_4 as shown in Fig. 1 (see also Section III.B) whose lens diameter is D_L and spacing between electrodes is

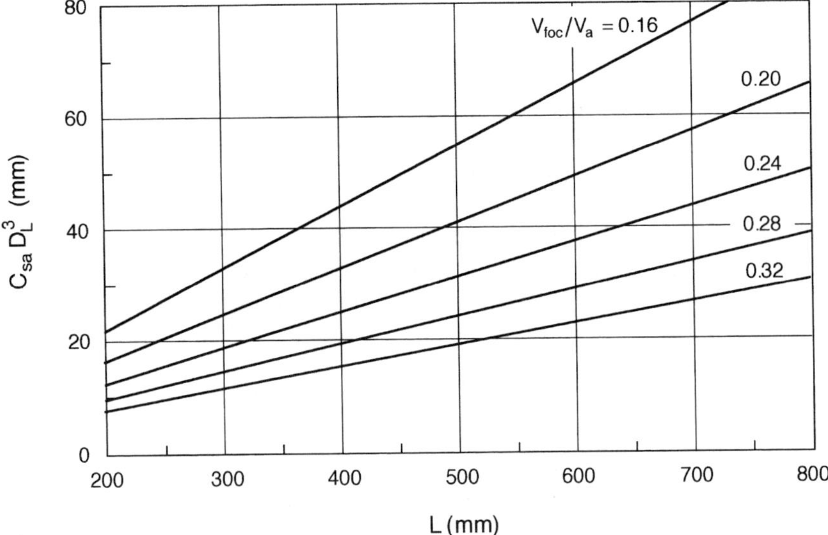

FIGURE 11. Spherical aberration coefficient C_{sa} multiplied by the cube of lens diameter D_L^3 for an equidiameter bipotential lens as a function of the lens to screen distance L for different values of the focus voltage to anode voltage ratio V_{foc}/V_a.

one-tenth that of the electrode diameter. In the figure the spherical aberration coefficient C_{sa} multiplied by D_L^3 is plotted as a function of the distance L from the main lens to the screen. As shown, separate curves are drawn for different values of the focus voltage to anode voltage ratio V_{foc}/V_a. The values of the ratio from 0.16 to 0.32 as shown are sufficient to cover the range used in most practical gun designs. This figure shows that for the bipotential lens the third-order spherical aberration is smaller at a higher focus voltage to anode voltage ratio with the bipotential lens.

It should be noted that in the case of a large beam diameter at the main lens it is necessary to consider the fifth- and higher-order terms in addition to the third-order aberration already mentioned here. However, in most practical gun designs where the beam diameter D_b is normally held at less than a half of the lens diameter D_L, it is usually sufficient to take only the third-order aberration into account.

4. *Spot Size Resulting from Combining the Three Limiting Factors*

To obtain the final spot size d_s actually observed at the screen it is necessary to take into account the effects of all limiting factors, namely, d_{th} (caused by the thermal velocity spread), d_{sc} (caused by the mutual repulsion of the electrons), and d_{sa} (caused by the spherical aberration of the main lens). However, since the mechanisms of the three factors are quite different, combining them becomes extremely complicated. For example, while thermal emission effects produce a spot with a Gaussian current density distribution, space charge and spherical aberration each produce different current density distributions.

To predict the exact intensity distribution of the spot as actually observed at the screen a sophisticated computer simulation of the trajectories of individual electrons is essential, taking into account all the forementioned effects (van den Broek, 1985; MacGregor, 1986; Aalders *et al.*, 1989; Bechis *et al.*, 1989; Oku *et al.*, 1992; Alig and Fields, 1997). However, for practical lens design purposes some simple methods are desirable, as discussed in what follows, which can combine these three fundamental factors to give the approximate overall spot size at the screen.

a. Models for Combining All Three Limiting Factors. Different approaches to this problem have been suggested by various research workers and designers of electron guns. In early analyses it was assumed for simplicity that the square of the resulting spot size at the screen was equal to the sum of the squares of the spot size resulting from each of the different mechanisms. In one approach of this type a quadratic combination was suggested for obtaining the final spot size by adding the Gaussian image size d_{th} with that of the aberration disc d_{sa}, with space-charge repulsion omitted (Mulvey, 1967). This approach was adopted in the design of a color CRT gun to optimize the overall spot size (Hughes and Chen, 1979). The overall spot size d_s is then expressed by

$$d_s = (d_{th}^2 + d_{sa}^2)^{1/2} \tag{60}$$

In another suggestion the Gaussian thermal spot size d_{th} was combined quadratically with the space-charge-limited spot size d_{sc} to obtain the overall spot size d_s for a narrow-angle beam where the spherical aberration of the lenses is negligible (Moss, 1968c), thus giving

$$d_s = (d_{th}^2 + d_{sc}^2)^{1/2} \tag{61}$$

This approach was used in a color CRT gun design (Wilson, 1975) in which the sum of the Gaussian thermal spot size d_{th} and the disc of least confusion d_{sa} due to spherical aberration of the main lens were combined quadratically with the space-charge-limited spot size d_{sc} as expressed by

$$d_s = \{(d_{th} + d_{sa})^2 + d_{sc}^2\}^{1/2} \tag{62}$$

In another approach the quadratic sum of the Gaussian thermal spot size d_{th} and the space-charge-limited spot size d_{sc} was combined linearly with the disc of least confusion d_{sa} of the spherical aberration in accordance with the following (Alig, 1980):

$$d_s = (d_{th}^2 + d_{sc}^2)^{1/2} + d_{sa} \tag{63}$$

Paralleling these approaches, a theoretical analysis was made taking into account the thermal velocity effect and space-charge repulsion. This resulted in a differential equation determining the radius of the beam in the absence of lens aberration (Weber, 1964). Calculations using this equation indicate that the overall spot size at the screen is given by a simple addition of the Gaussian thermal spot size d_{th} and the space-charge-limited spot size d_{sc} (rather than their quadratic combination) when the two spot sizes are of the same order of magnitude.

Since the aberration disc may be approximated by a uniform density distribution in the absence of thermal effects and space-charge repulsion, Barten (1984) suggested that the Gaussian distribution of the thermal spot should be combined with the spherical aberration of the main lens by a convolution over the disc of least confusion of diameter d_{sa}. This results in a bell-shaped distribution of the overall spot, which is further enlarged by the space-charge repulsion effect. As the latter does not require a convolution process, the spot size d_{sc} due to space charge can be added to the spot size obtained by a convolution of the Gaussian thermal spot and spherical aberration. The resultant overall spot size d_s was then expressed by

$$d_s = 0.7\,d_{th} + \sqrt{(0.3\,d_{th})^2 + d_{sa}^2} + d_{sc} \tag{64}$$

b. Spot Size Obtained by a Linear Combination of All Three Limiting Factors. The simplest approach, however, is to assume that the spot size d_s at the screen is determined by a linear combination of the three limiting factors. Although this assumption may not provide an exact result it is very simple and still provides useful insight into the effect of the various factors on the spot size. In fact, it can be shown that the overall spot size determined in this manner is very close (within 10%) to that obtained from Eq. (64).

As already discussed, both the thermal spot size d_{th} and the space-charge-limited spot size d_{sc} are inversely proportional to the beam diameter D_b in the main lens, whereas the disc of least confusion d_{sa} due to the spherical aberration of the main lens increases with the third power of this diameter. This results in a minimum value for the overall spot size at an optimum valuc of the beam diameter D_b at the main lens as shown here. By making use of Eqs. (45), (51) and (59), the overall spot size d_s can be written

$$d_s = d_{th} + d_{sc} + d_{sa} = (C_{th} + C_{sc})/D_b + C_{sa}\,D_b^3 \tag{65}$$

Differentiating Eq. (65) with respect to the beam diameter D_b and setting it to zero to obtain the minimum spot diameter yields

$$\frac{d\,d_s}{d\,D_b} = \frac{-(C_{th} + C_{sc}) + 3\,C_{sa}\,D_b^4}{D_b^2} = 0 \tag{66}$$

The optimum beam diameter at which the overall spot size is minimum is thus expressed by

$$D_b = \{(C_{th} + C_{sc})/3\,C_{sa}\}^{1/4} \tag{67}$$

By substituting this value into Eq. (65) the minimum spot size obtained is expressed by

$$d_s = 4\{(C_{th} + C_{sc})/3\}^{3/4}\,C_{sa}^{1/4} \tag{68}$$

Since the contribution of the spherical aberration of the main lens to the minimum spot size is given by the second term of the right-hand side of Eq. (65), by a substitution of Eq. (67) the contribution of the spherical aberration d_{sa} is expressed as

$$d_{sa} = \{(C_{th} + C_{sc})/3\}^{3/4}\,C_{sa}^{1/4} \tag{69}$$

A comparison of Eq. (68) with Eq. (69) indicates that under conditions of minimum spot size the contribution of the spherical aberration of the main lens (d_{sa}) accounts for one-fourth of the overall spot size d_s while the combined effect of the magnified crossover and space-charge repulsion accounts for three-fourths of the spot size. From Eq. (68), it can also be seen that a reduction of the spherical aberration coefficient of the main lens C_{sa} by a factor of n reduces the total value of the minimum spot size d_s by a factor of $n^{1/4}$.

To illustrate the foregoing, a plot of the overall spot size d_s as a function of the beam diameter D_b is shown by the thick-line curve in Fig. 12 for a high beam current of 4 mA (of a single beam). Also shown by thin-line curves are the thermal spot size d_{th}, space-charge-limited spot size d_{sc} and the disc of least confusion d_{sa} due to the spherical aberration of the main lens. The small black circle on the thick-line curve indicates the minimum point of the overall spot size d_s. It can be seen from the figure that the greatest factor determining the minimum overall spot size is the space-charge-limited spot size d_{sc}, followed by the thermal spot size d_{th} and the disc of least confusion d_{sa}, each contributing about the same amount.

The gun used in this calculation is of the bipotential type (see Section III.B) with a main lens diameter D_L of 8.0 mm, main lens to screen distance L of 400 mm, anode voltage V_a of 30 kV and focus voltage V_{foc} of 8.4 kV $(V_{foc}/V_a = 0.28)$. The spherical aberration coefficient C_{sa} of the main lens is

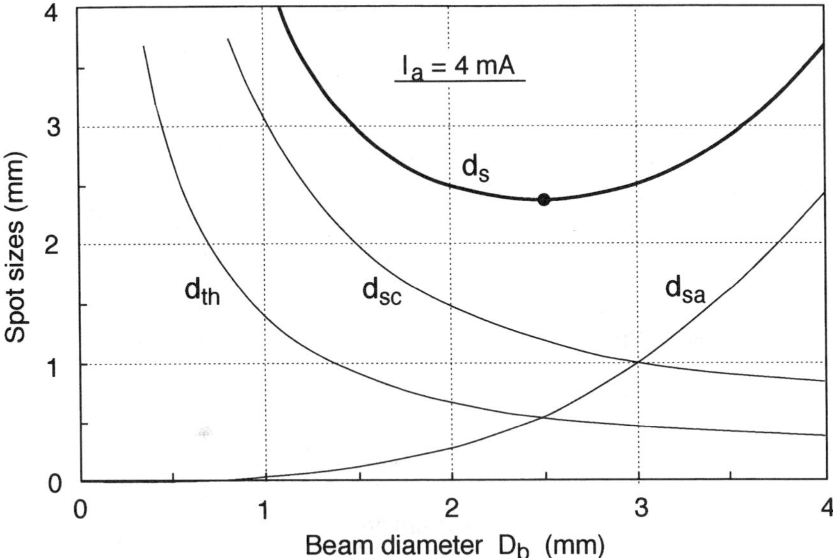

FIGURE 12. Overall spot size d_s (thick-line) as a function of beam diameter D_b for 4 mA single beam. Thermal spot size d_{th}, space-charge-limited spot size d_{sc} and disc of least confusion d_{sa} due to spherical aberration, indicated by thin-lines.

obtained as 3.8×10^{-2} mm^{-1} from Fig. 11. The beam current of 4 mA is almost the peak current used in TV applications for which a cathode loading j_{c0} of 3.7 A/cm^2 is assumed as a typical value in a practical design.

As the beam current is decreased the minimum value of the overall spot size d_s decreases due to a reduction in the space-charge-limited spot size d_{sc} and the thermal spot size d_{th}. Figure 13 shows a set of curves similar to those of Fig. 12, in which a beam current of 0.5 mA (close to the long-term average operating current of TV) corresponds to a cathode loading j_{c0} of 1.4 A/cm^2. As can be seen, the greatest factor determining the minimum overall spot size is now the thermal spot size d_{th} followed by the space-charge-limited spot size d_{sc} and the disc of least confusion d_{sa}, the latter two each contributing about the same amount.

As can be seen from Figs. 12 and 13, the space-charge-limited spot size d_{sc} decreases more rapidly with decreasing beam current than the thermal spot size d_{th}. This is explained as follows: As indicated in Eq. (51), the space-charge-limited spot size d_{sc} varies linearly with the beam current I_a (of a single beam) at a constant beam diameter D_b. On the other hand, as indicated in Eq. (45), the thermal spot size d_{th} varies as the root-mean-square (rms) of I_a/j_{c0}. Considering the fact that the cathode loading j_{c0} varies as $I_a^{3/5}$

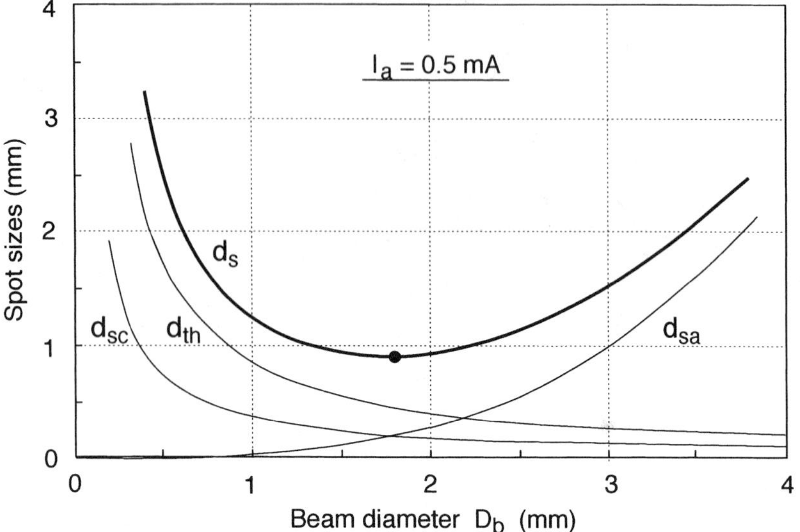

FIGURE 13. Overall spot size d_s (thick-line) as a function of beam diameter D_b for 0.5 mA of single beam. Thermal spot size d_{th}, space-charge-limited spot size d_{sc} and disc of least confusion d_{sa} due to spherical aberration, indicated by thin-lines.

(Francken, 1959/60), it follows that the thermal spot size d_{th} varies as $I_a^{1/5}$, being less dependent on the beam current.

Figure 14 shows the overall spot size d_s as a function of the beam diameter D_b for various beam currents ranging from 4 mA down to 0.1 mA. As can be seen, the minimum point indicated by the small black circle on each curve takes place at a reduced value of D_b as the beam current is decreased. This occurs because the disc of least confusion d_{sa} due to the spherical aberration of the main lens is unaffected by the beam current at a given beam diameter while both the thermal spot size d_{th} and the space-charge-limited spot size d_{sc} decrease with decreasing beam current as mentioned here.

The dashed-line curve in Fig. 14 shows an example of the change in beam diameter in an actual gun as explained in the following: The minimum value of the overall spot size d_s as a function of the beam current I_a (of a single beam) is shown by the solid-line curve in Fig. 15. As mentioned, the minimum spot size can be obtained only at the optimum beam diameter for each value of beam current. In an actual gun, however, the optimum beam diameter is obtainable only at a certain beam current due to the limitations of the prefocus lens. This generally results in excessively small beam diameters at lower currents and excessively large beam diameters at higher currents than would be required for the minimum spot size to be obtained.

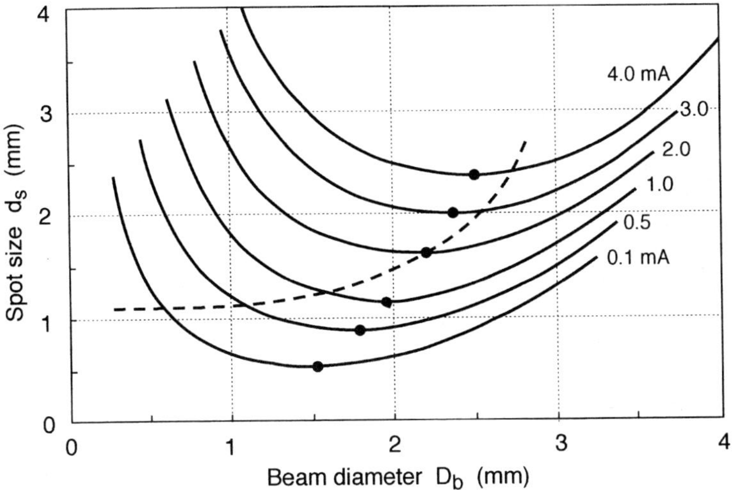

FIGURE 14. Overall spot size d_s for various currents as a function of beam diameter D_b. Optimum beam diameter (black circles) for minimum spot size decreases with decreasing beam current. Dashed-line curve shows an actually obtainable beam diameter characteristic.

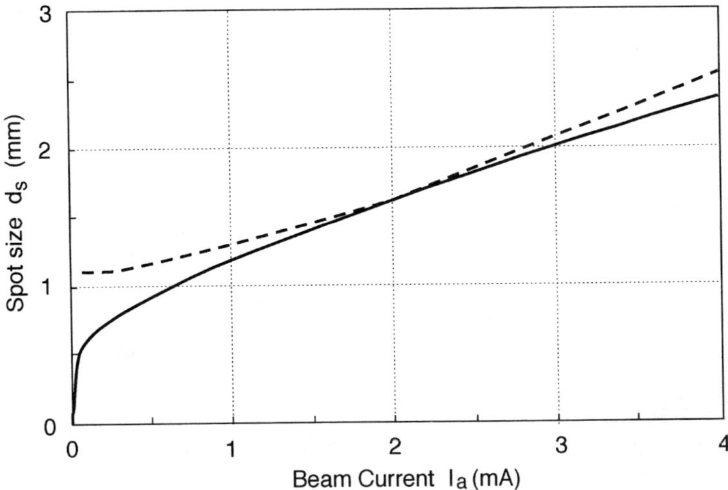

FIGURE 15. Minimum value of overall spot size d_s (solid-line) as a function of beam current I_a. Dashed line shows an actually obtainable spot size characteristic optimized at a high current of 2 mA.

The beam current at which the beam diameter is optimized depends on the particular application of the tube. For example, for TV applications the gun is normally designed in such a manner that the optimum beam diameter takes place at a relatively high current so that a small spot can be obtained at high-brightness levels. An example of the beam diameter D_b actually obtainable for various currents is shown by the dashed-line curve in Fig. 14, where the gun has been optimized at a current of 2 mA (of a single beam). As shown by the dashed-line curve of Fig. 15, this results in a spot size that is greater than the minimum attainable value both at lower currents and higher currents.

A further increase in the spot size may occur in practical guns due to the fact that the focus voltage V_{foc} at which a minimum spot is produced at the screen changes somewhat with a change in beam current whereas the gun usually operates at a fixed focus voltage. This may cause the spot to be out of focus when the beam current is varied. The change in focus voltage is caused by several factors such as space-charge repulsion, spherical aberrations of the lens system, and a shift in location of the crossover with the beam current.

E. Spot Growth Due to Magnetic Deflection

The discussions of spot size in the previous sections assume a beam that is not deflected. However, additional enlargement and distortion of the spot result from the magnetic deflection process. As discussed in what follows, aside from the fact that the deflected beam strikes the screen obliquely and also has an increased path length, two other major factors are involved: a) a difference in curvature between the imaging surface of the deflected beam and the screen surface; and b) astigmatism resulting in a difference in curvature of the imaging surface between the horizontal and vertical directions. Even a yoke with a relatively uniform magnetic field such as used in the older delta gun arrangement causes these effects. However, the situation is exacerbated with the currently used self-converging yoke whose magnetic field is highly astigmatic.

1. *Defocusing and Astigmatism of Spot*

The magnetic field of the yoke is usually located a short distance beyond the high-voltage anode of the gun. A uniform magnetic field is considered first that has a magnetic induction of B directed toward the reader (y-direction) as shown in Fig. 16, which deflects the beam in the horizontal direction (x-direction). Four electron rays are used to represent the boundary of the beam. As shown at the left of the figure, rays 1 and 3 are in the x-z plane and rays 2 and 4 are in the y-z plane. In the absence of the magnetic field the beam, shown by the dashed lines, is assumed to converge at the screen center C.

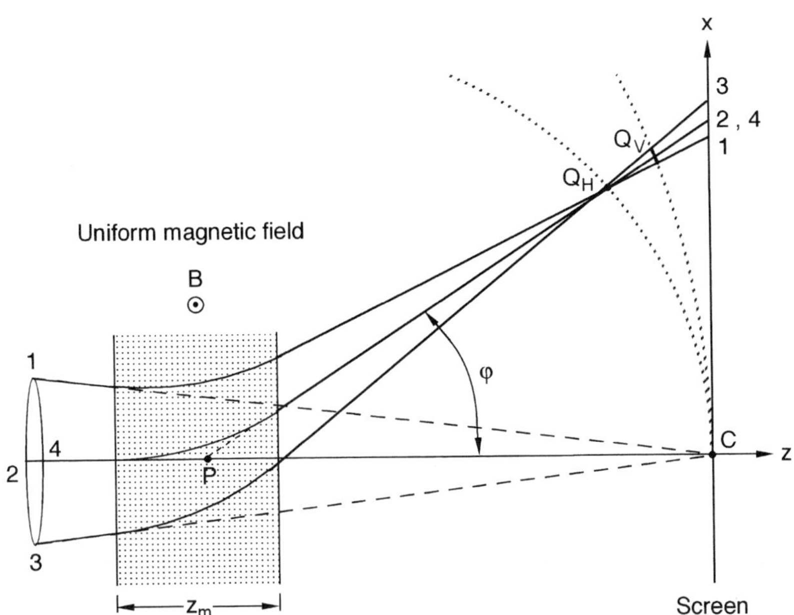

FIGURE 16. Horizontal deflection of a beam in x-direction by uniform magnetic field (magnetic induction B in y-direction) causes horizontal rays (1 and 3) to converge at Q_H while vertical rays (2 and 4) converge at Q_V such that $P - Q_H < P - Q_V \approx P - C$.

When the beam is deflected, rays 1 and 3 in the x-z plane converge at the point Q_H, closer to the deflection yoke than the undeflected beam. As shown in the figure the distance P–Q_H from the effective deflection point P to the point Q_H is less than the distance P–C from P to the screen center C. The effective deflection point P represents the intersection between the extension of the beam center and the z-axis.

On the other hand, rays 2 and 4 in the plane perpendicular to the x-z plane will converge at Q_V whose distance P–Q_V from P more nearly equals the distance P–C, resulting in P–Q_H < P–Q_V. This means that a greater converging action occurs in the horizontal (x-z) plane because of the magnetic field as will be discussed here. After passing their converging points, the rays diverge before reaching the screen where an elliptical beam spot will be produced whose shape is elongated in the direction of deflection.

The stronger converging action in the horizontal plane can be explained as follows: Electrons entering the magnetic field (assumed to be uniform and with sharp boundaries) move along a circular path whose radius R is given by

$$R = \frac{mv}{eB} \tag{70}$$

where m is the mass of the electron 9.107×10^{-31} kg, e is the electronic charge 1.602×10^{-19} C, and v is the velocity of electrons given by $\sqrt{2(e/m)V_a}$ in which V_a is the anode potential (see, e.g., Klemperer and Barnett, 1971e). The angle of deflection φ subtended by the beam center and the z-axis can be expressed as

$$\tan\varphi = z_m \Big/ \sqrt{R^2 - z_m^2} \tag{71}$$

where z_m is the length of the uniform magnetic field along the axis. If we assume $R \gg z_m$, this reduces to

$$\tan\varphi \approx z_m/R \tag{72}$$

Different deflection angles result for rays 1 and 3 due to their oblique incidence into the magnetic field region. In the figure ray 1 enters the field tilted downward toward the axis while ray 3 enters tilted upward. As a result, ray 3 has a longer path length in the field than ray 1 and also a longer path length than rays 2 and 4. Because of this ray 3 is deflected more than ray 1, according to Eq. (72), resulting in an increase in the converging action in the horizontal direction. On the other hand, the converging action in the y-z plane is unaffected because rays 2 and 4 enter the magnetic field under identical conditions, thus being subjected to the same converging action as the undeflected beam.

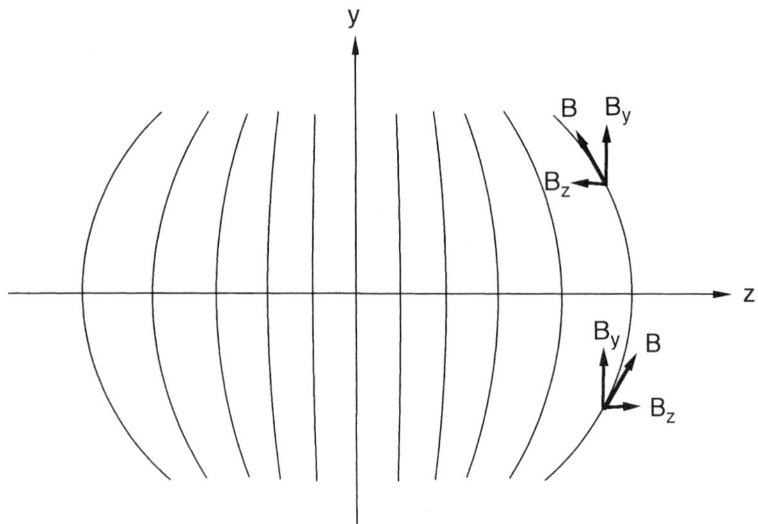

FIGURE 17. Actual magnetic field in y-z plane for horizontal deflection. B_z results from bending of lines of force of the barrel-shaped field z-component.

The preceding discussion, however, only applies to an ideal uniform magnetic field with sharp boundaries. In a practical case the field always begins and ends gradually in the axial direction with the result that at the boundary region the field is barrel-shaped in the y-z plane as shown in Fig. 17. Such a magnetic field by itself may produce a further converging action in the direction of deflection. In addition, it produces an increased converging action perpendicular to the deflection direction. As shown in Fig. 17, due to the curvature of the barrel-shaped field, mainly the horizontally deflecting field B_y, a component of the field B_z is created in the z-direction. This acts on a horizontally deflected beam such that the upper half of the beam above the x-z plane undergoes a downward force while the lower half undergoes an upward force, resulting in a converging effect on the beam in the vertical direction. In general, the horizontal deflection field has a converging effect, which is stronger in the vertical direction, thus resulting in $P - Q_H > P - Q_V$.

This situation is illustrated in Fig. 18. As shown, the horizontally deflected beam first converges at Q_V to a horizontal focus line, and then converges at Q_H to a vertical focus line. The resulting loci of Q_V and Q_H (shown by the broken lines) constitute the vertical and horizontal image surfaces, respectively. The difference in location of these surfaces produces astigmatism of the beam. Located between these surfaces is the circle of least confusion denoted by Q_{LC}, where the deflected beam has its minimum cross section. As

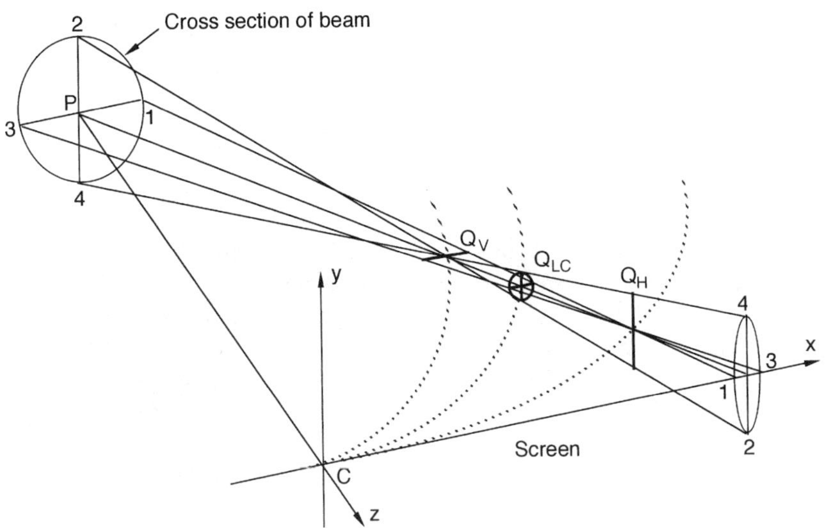

FIGURE 18. Horizontal deflection of a beam in x-direction by actual magnetic field results in location of Q_H and Q_V such that $P - C > P - Q_H > P - Q_V$. Astigmatism is given by $Q_H - Q_V$ with the circle of least confusion located at Q_{LC}.

a result of the foregoing, the beam is overfocused when it reaches the screen. As the overfocusing is stronger in the vertical direction, that is, in the direction perpendicular to the deflection, the resulting spot is vertically elongated and has an elliptical shape as shown in the figure.

In the case of the vertical deflection with a similar magnetic field, the spot will be horizontally elongated since the overfocusing is stronger in the horizontal direction, that is, in the direction perpendicular to the deflection. These situations also occur in older types of color CRTs using a delta gun arrangement and a relatively uniform magnetic field (Boekhorst and Stolk, 1962).

2. *Astigmatic Spot Defocusing in Self-Converging System*

As discussed in the accompanying article by Yamazaki (Design and Performance of Shadow-Mask Color Cathode Ray Tubes), modern color CRTs almost always use a self-converging deflection yoke in combination with in-line electron guns. In the self-converging system the three in-line beams are maintained converged at the screen during deflection by the specially designed astigmatic magnetic field of the yoke, thus avoiding the need for the dynamic convergence system which had to be added in early delta color CRTs. In the self-converging yoke the horizontal deflection field is pincushion-shaped and the vertical field is barrel-shaped. This produces the correct amount of diverging action on the in-line beams in the horizontal direction so that they remain converged everywhere on the screen. Although such magnetic fields produce, in addition, a converging action in the vertical direction, this does not disturb the beam convergence in this direction since they are all on a common horizontal plane (see, e.g., Heijnemans *et al.*, 1980; Dasgupta, 1992).

Due to its nonuniform magnetic field, a self-converging yoke causes a considerable degree of distortion of the cross section of the individual beams. Figure 19a shows the right-hand half of the horizontally deflecting field (as viewed from the screen side), which is pincushion-shaped; and Fig. 19b the upper half of the vertically deflecting field, which is barrel-shaped, assuming in both a beam that is circular when undeflected. As shown in Fig. 19a, since the horizontal field component B_y of the pincushion-shaped field becomes greater farther from the center, the deflected beam is subject to a diverging action in the x-direction and a simultaneous converging action in the vertical direction by the x-component of the magnetic field B_x. This results in a beam with an elliptical cross section that is horizontally elongated and vertically compressed. A similarly shaped cross section also occurs in the case of the vertical deflection as shown in Fig. 19b. As the vertical field component B_x of the barrel-shaped field is weaker away from the center of the screen, the beam

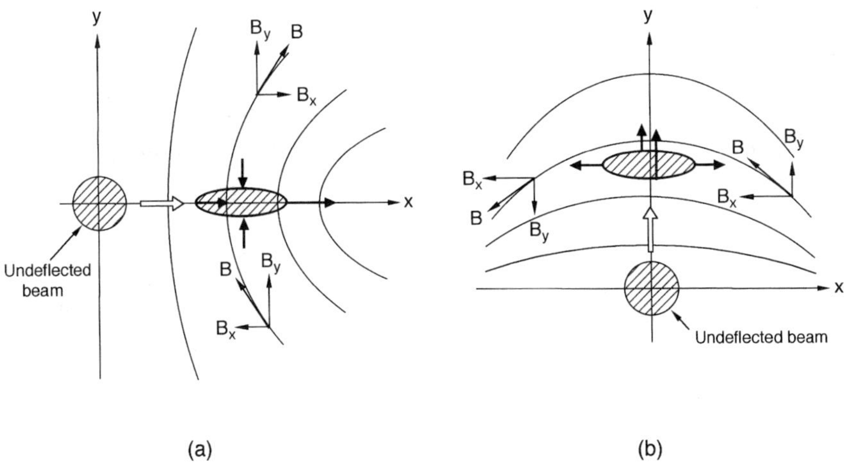

FIGURE 19. Self-converging magnetic field causes a distortion of the beam cross section (as viewed from the screen side); (a) pincushion-shaped horizontal deflection field; (b) barrel-shaped vertical deflection field, each in the x-y plane.

is subject to a converging action in the vertical direction with a simultaneous horizontal diverging action caused by the y-component of the field B_y because of its curvature.

The resultant effect of the deflection fields is to keep the spot in focus in the horizontal direction at every point of the screen, similar to the way the three in-line beams are kept converged. However, the spot is overfocused in the vertical direction. Referring to Fig. 18, this is equivalent to saying that the horizontal image surface Q_H falls entirely on the screen while the vertical image surface Q_V moves farther back from the screen. This results in an increase in astigmatism given by the difference in location $Q_H - Q_V$. Since the vertical overfocusing of the deflected spot is increased, a vertical haze is observed above and below the core of the spot, an effect which is most pronounced at the edges and corners of the screen as shown in Fig. 20. A computer simulation of these spot shapes is discussed by Lucchesi and Carpenter (1979).

Instead of determining the difference $Q_H - Q_V$ as already mentioned here, the astigmatism is often determined by means of the difference V_{ast} between the horizontal focusing voltage V_{foc}^H and the vertical focusing voltage V_{foc}^V of the spot expressed by

$$V_{\text{ast}} = V_{\text{foc}}^H - V_{\text{foc}}^V \tag{73}$$

This method is more convenient in the practical design of the gun and yoke. The value of V_{ast}, sometimes called the "spot astigmatism," depends on the

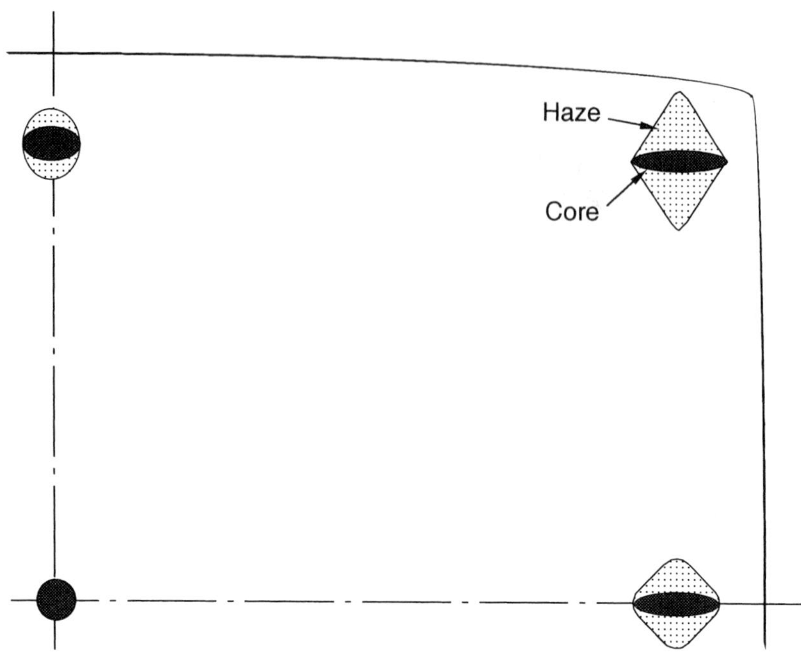

FIGURE 20. Typical spot shapes at various locations of the 1st quadrant of screen resulting from astigmatic deflection defocusing with a self-converging yoke.

type of gun being used. In a self-converging system the horizontal focusing voltage V_{foc}^{H} remains nearly constant during deflection since the spot remains in focus in the horizontal direction. As already mentioned, the spot is vertically overfocused so that $V_{\text{foc}}^{H} < V_{\text{foc}}^{V}$, making the spot astigmatism V_{ast} always negative assuming that the beam was originally circular.

Modern color CRTs use wide deflection angles such as 90°, 100° and 110° to reduce the tube length as much as possible and also use a relatively flat screen face (Adachi *et al.*, 1991). In these tubes the astigmatic deflection defocusing is a major factor limiting the spot size attainable at screen areas away from the center. In general, to minimize the spot defocusing due to magnetic deflection, it is desirable that the beam diameter in the region of the deflection yoke be as small as possible. This conflicts with the requirement of a large beam diameter at the main lens to minimize the effects of thermal velocities and space-charge repulsion on spot size. Some compromise on beam diameter is thus required.

To compensate for the vertical overfocusing of the deflected spot, quadruple elements are often added to the lens system of guns to produce a positive astigmatism that counteracts the negative astigmatism caused by the magnetic field. These gun designs will be discussed in Section III.E. Although the preceding discussion was concerned only with the center beam of the three in-line beams, the outer beams may in addition be affected in practice by the coma effect of the magnetic field. However, this can be counteracted by proper design of the deflection yoke (Iwasaki and Konosu, 1987; Kitagawa *et al.*, 1987).

F. Examples of Limits Achieved in Currently Available Tubes

Although a small spot size can be achieved with increasing anode voltage V_a, this requires an increase in deflection power. A limit on the anode voltage is also set by X-ray emission and the need for X-ray absorbing thick faceplates. A further problem is arcing between gun electrodes, which to a limited degree can be prevented by improved tube design and manufacturing procedures as discussed in Section V. Taking account of these factors, typical anode voltages for present tubes are in the 25–28 kV range for data display tubes and 29–32 kV range for large-screen TV and HDTV tubes. As discussed in Section II.D, to obtain a sufficiently high current in the beam spot a cathode must be used that is capable of supplying a sufficiently large emission current density (see also Section IV).

As already mentioned above, some design requirements are contradictory with respect to each other so that compromises must be made among them in any practical gun design, depending on the application of the tube. Three important applications of color CRTs are considered.

For conventional TV a wide deflection angle of 110° is almost always used with a narrow neck tube whose external diameter is 29 mm (Morrell, 1974). In some cases even a smaller neck diameter of 22.5 mm is employed (Hamano *et al.*, 1979). Usually, the gun is optimized at a relatively high beam current of 2–3 mA (of a single beam). This allows a sufficiently small spot size to be obtained at high brightness levels where the peak current may be as high as 4 mA.

For high-definition TV (HDTV) roughly twice as many scanning lines are required than in the conventional NTSC (National Television System Committee) system requiring more than twice the number of dots in the horizontal direction (Fujio, 1985; Lechner, 1985). To be capable of such a high-resolution, the tube design must meet various requirements (Yamazaki, 1988; Friedman *et al.*, 1991; van Raalte, 1992; Gorog, 1994). In addition to reducing the spot size a sufficient brightness must be maintained so that the HDTV display can be viewed under normal ambient light. Although in the early stages a large neck diameter of 37.5 mm was used with a deflection angle of 110° (Ohsawa *et al.*, 1990; Sakuraya *et al.*, 1991), a narrower neck diameter of 32.5 mm has become more common to obtain lower deflection power (Barbin *et al.*, 1990; Mitsuda *et al.*, 1991; Sayama *et al.*, 1991).

For data display tubes, uniformity of spot size and small raster distortion over the entire screen are essential for legible character display as well as for accurate graphic representation. Because of this a deflection angle of 90° is predominantly used since the deflection defocusing is much less than with 110° deflection (Yamada *et al.*, 1980; Barbin *et al.*, 1982; Eccles *et al.*, 1993). Although a neck diameter of 29 mm is mostly used, a smaller neck diameter has also been used recently in some cases (Shirai *et al.*, 1994). Since data display tubes are generally used under a low ambient light, the operating current is relatively low, the peak value being usually of the order of 0.3 mA (of a single beam), although this may be increased somewhat in large screen size tubes (Suzuki *et al.*, 1990; Nose *et al.*, 1995).

Typical spot-size data (horizontal and vertical widths at a 5% intensity level) measured at the screen center and corner of some widely used present-day tubes for the three major CRT applications mentioned here are given in Table I together with the corresponding anode current I_a shown under the spot data. The anode current I_a represents the value of the beam current (of a single beam) for which the gun design was optimized (see Section II.D.4.*b*). Also shown in the table are the operating anode voltage V_a, the average screen brightness (at a glass transmission of 50%), the corresponding long-term current (averaged over hours) of the three beams. In HDTV and data display

TABLE I
ELECTRON BEAM CHARACTERISTICS OF COLOR CRTS IN THREE MAJOR APPLICATIONS

Applications	TV	HDTV	Data display
Tube types diagonal size/ deflection angle/ neck diameter	68 cm/110°/29 mm	86 cm/106°/32.5 mm	51 cm/90°/29 mm
Anode voltage V_a (kv)	30	32	27
Average brightness (cd/m^2) (Glass transmission : 50%)	120	100	80
Average current of 3 beams (mA)	1.50	1.65	0.60
5% spot size (mm) horizontal × vertical			
Center	1.8 × 2.2	1.5 × 1.5	0.60 × 0.60
Corner	3.6 × 3.0	3.0 × 1.5	0.72 × 0.60
I_a (1 beam) for the spot size above (mA)	2	2	0.3

tubes dynamic astigmatism and focus correction is also used to compensate for the astigmatic deflection defocusing of the spot caused by the self-converging yoke as discussed in Section III.E.3.

The 5% spot size given here can be obtained from the measurement of the light intensity distribution of a static spot (without scan) on the phosphor screen by a microphotosensor such as a CCD camera. However, because of phosphor saturation effects the light intensity generally is not proportional to the current density of the beam spot. This effect can be reduced considerably, however, by switching the beam on with short pulses of the order of 0.01 μs (Suzuki *et al.*, 1976).

For an approximate estimate of the current density of the spot, it is assumed that the effect of phosphor saturation is negligible and the light intensity distribution of the spot is Gaussian. This is approximately valid for anode currents less than 0.5 mA. Assuming a Gaussian distribution, using Eq. (40) and the fact that the 5% spot width d_s is close to $\sqrt{3}$ times the 1/e width ($2\,r_{sp}$), the peak current density j_{sp0} can be expressed as

$$j_{sp0} = 0.38\,I_a/d_s^2 \quad [\mathrm{A/cm^2}] \tag{74}$$

where the anode current I_a (of a single beam) is given in milliamperes and the 5% spot size d_s is in millimeters.

By applying this formula to the center spot of the data display tube given in Table I ($d_s = 0.6$ mm, $I_a = 0.3$ mA), the peak current density of the spot j_{sp0} is found to be 0.32 A/cm^2. Similarly, for the center spot of the HDTV tube ($d_s = 1.5$ mm, $I_a = 2.0$ mA), the peak current density is found to be 0.34 A/cm^2. However, in the latter case the spot is normally much more bell-shaped than Gaussian and the actual peak current density of the spot may be somewhat lower.

III. Electron Gun Designs

A. *Introduction*

In the usual shadow-mask tube approximately 80% of the beam current delivered by the guns is intercepted by the shadow-mask, thus reducing the screen brightness by a corresponding amount. In part this loss is compensated for by the fact that three electron beams, converged to a common spot, are used in place of the single beam of black-and-white tubes. However, because the three guns have to fit inside a neck of limited diameter to minimize deflection power (especially in the case of tubes with wide deflection angles, e.g., 110°), individual guns with small diameter are required, resulting in an increase in lens aberrations. In addition, to obtain a sufficient brightness, relatively high beam currents are required; the result is an increase in space-charge repulsion. These lens aberrations and space-charge repulsion, in turn, cause an increase in spot size compared to that of a black-and-white tube using a single larger-diameter gun and smaller beam currents.

The main objective in color CRT gun design is thus to obtain a high-current beam whose spot diameter is sufficiently small. To achieve such a gun design it is essential that its lens system have a sufficiently reduced spherical aberration. Since a large degree of spherical aberration occurs at the main focusing lens, a reduction in spherical aberration of this lens is of primary importance. In addition, the performance of the prefocus lens is also important because, as will be discussed, it serves to reduce the diameter of the beam in the region of the main lens.

Unlike early color tubes, which employed three separate guns with cylindrical electrodes in a delta arrangement, present-day tubes usually employ an in-line unitized gun structure consisting of an assembly of metal plates, each with three apertures aligned with the other plates, to produce the three electron beams. Such structures result in more precise alignment of the lens systems and greater dimensional stability. However, because of the required close spacing of the beams the diameter of each of the three main lenses is reduced, causing increased spherical aberrations. However, by means of three-dimensional computer designs lenses with highly asymmetric aperture shapes have been developed. These have reduced spherical aberrations, allowing high-current beams with small spots to be obtained.

An alternative method for reducing the aberrations was developed by Sony. In this method, instead of using a separate main lens for each gun, a single large-aperture circular main lens was used. In this case the beams from the three guns are directed toward the center of this common main lens. Since they cross over close to the tube axis the effects of spherical aberration on the beams are minimized. After diverging from this crossover the three beams are then reconverged to a common spot at the screen by electrostatic deflectors mounted at the end of the gun system.

Deflection defocusing adds another constraint on the gun design of shadow-mask tubes. Although a self-converging deflection yoke can maintain the three beams converged at the screen, as already discussed, it requires an astigmatic magnetic field. This causes a strong overfocusing of the deflected beams in the vertical direction at the periphery of the screen. To reduce this deflection defocusing electrostatic quadrupole lenses have been developed in recent years. These, also using noncircular apertures, have been added to the guns.

A useful analytic tool for evaluating the spot-size performance of guns is the phase-space diagram. The use of such diagrams is also discussed in a separate section that follows.

B. *Types of Electrostatic Main Lenses*

1. *Bipotential Lenses*

The bipotential lens is perhaps the most basic form of electron lens used for the main lens of color CRT guns. The electron-optical properties of the lens can be derived from the potential distribution on the axis of symmetry as described in general references (e.g., Zworykin *et al.*, 1945b; Spangenberg, 1948d; El-Kareh and El-Kareh, 1970a). Guns with such a main lens are sometimes called bipotential-focus (BPF) guns. As shown in Fig. 21, the bipotential lenses consist of two adjacent coaxial cylindrical electrodes maintained at different potentials. (In practice, the edges of the electrodes facing each other are generally curled back to prevent high-voltage arcing.) Usually, the full anode voltage V_a is maintained on G_4 and the lower voltage V_{foc} (focus voltage) applied to G_3 is varied to adjust the strength of the lens and to focus the beam spot precisely at the screen. In the simplest case, as shown, the diameters of the cylinders are equal. Bipotential lenses in one form

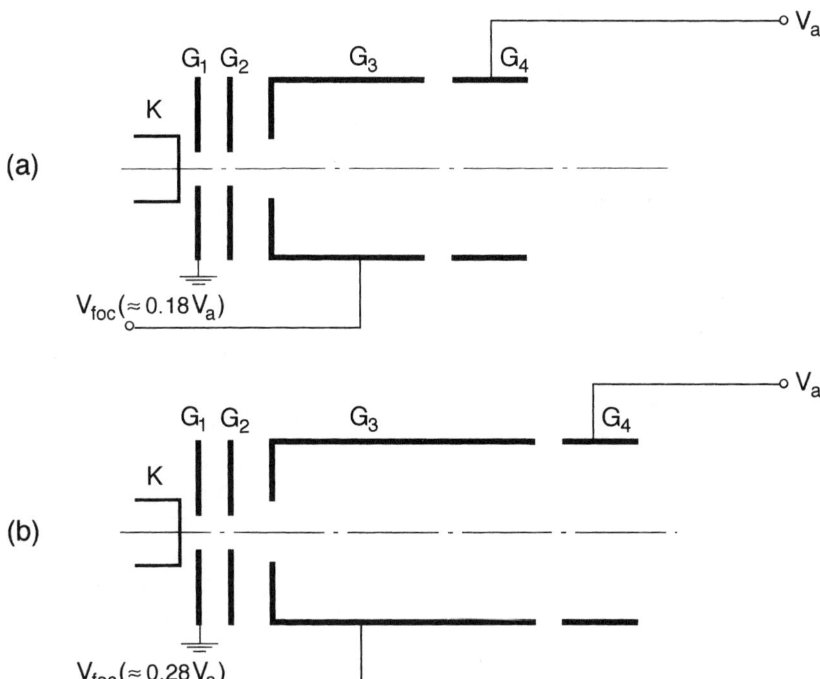

FIGURE 21. Bipotential-focus (BPF) electron guns; (a) with normal bipotential lens; (b) with hi-bipotential (hi-bi) lens.

or another are used extensively as the main lens of guns in shadow-mask tubes.

For some time after the development of the shadow-mask color tube, bipotential lenses with a ratio of focus voltage to anode voltage V_{foc}/V_a of approximately 18% as shown in Fig. 21a were widely used with a typical anode voltage of 25 kV (Morrell *et al.*, 1974a). With such a low-voltage ratio the focusing action is strong. This allows the cathode and its adjacent crossover (see Section II.B) to be positioned a short distance from the main lens, resulting in a short gun. However, a strong lens also results in a large spherical aberration, making it difficult to produce a small spot at the screen.

To reduce the spherical aberration of the main lens it is generally necessary to reduce its strength (Blacker *et al.*, 1976). This can be accomplished, for example, by increasing the focus voltage ratio V_{foc}/V_a from 18 to 28% of the anode voltage. However, this causes an increase in the focal length of the lens. This in turn requires that the G_3 electrode be lengthened as shown in Fig. 21b. Referring to Fig. 11 on p. 299, Section II, which shows the spherical aberration coefficient C_{sa} for an equidiameter bipotential lens, a 28% focus ratio results in a 50% reduction in spherical aberration compared to an 18% focus ratio. Since Eq. (68) shows that the minimum obtainable spot size is proportional to $C_{sa}^{1/4}$, the spot size with a 28% focus ratio will be $(0.5)^{1/4} = 84\%$ of the spot size with an 18% focus ratio. However, it should be noted that to obtain this spot-size reduction requires an optimum beam diameter at the main lens as discussed in Section II.D. To obtain such an optimum beam diameter, a suitable prefocus lens design is usually required.

The main lens of Fig. 21b is referred to as a high-bipotential (hi-bi) lens while the design of Fig. 21a with an 18% focus ratio is referred to as a low-bipotential (lo-bi) lens. From a practical point of view it should be noted that the increase in focus voltage from 18 to 28% still allows it to be supplied to the G_3 electrode of the gun through one of the pins of a normal tube base.

During the period from 1977 to 1979 the hi-bi type of the main lens came into use in the newly developed unitized gun designs using a 29 mm neck diameter (see Section III.D.2) such as Matsushita's Hi-BPF or hi-bi gun (Muranishi and Okamoto, 1977) and RCA's High-focus-voltage bipotential Precision In-line (HiPI) gun (Hughes and Chen, 1979). During the same period the hi-bi lens was introduced into the Philips 30AX system using a 36 mm neck diameter and conventional (separate) in-line guns (Barten and Kaashoek, 1978). Soon after, this type of gun design was adopted by other manufacturers because of its reliability and cost effectiveness as well as its simplicity and improved spot size compared to the conventional bipotential (lo-bi) gun with its lower focus voltage.

Using computer modeling and taking into account spherical aberration, the size of crossover and space-charge repulsion in the drift space between the main lens and screen, the spot size of one of the BPF guns of an in-line structure (assumed to fit into a standard 29 mm tube neck) was simulated. For a beam current of 4 mA, an anode voltage of 30 kV, and various focusing voltages, the analysis confirmed that spherical aberration of the main lens was reduced by an increase in focusing voltage. This also reduced the magnification of the crossover (magnified crossover), further reducing the spot size at the screen. In particular, the contribution of the spherical aberration to the spot size was shown to decrease in a near-linear inverse relationship with the focusing voltage in the voltage range from 5 to 9 kV (Say, 1978).

2. *Unipotential Lenses*

A typical unipotential lens, as shown in Fig. 22a, consists of three coaxial cylindrical electrodes G_3, G_4 and G_5 in tandem (see, e.g., Klemperer and Barnett, 1971f). Usually, the outer electrodes G_3 and G_5 are both maintained

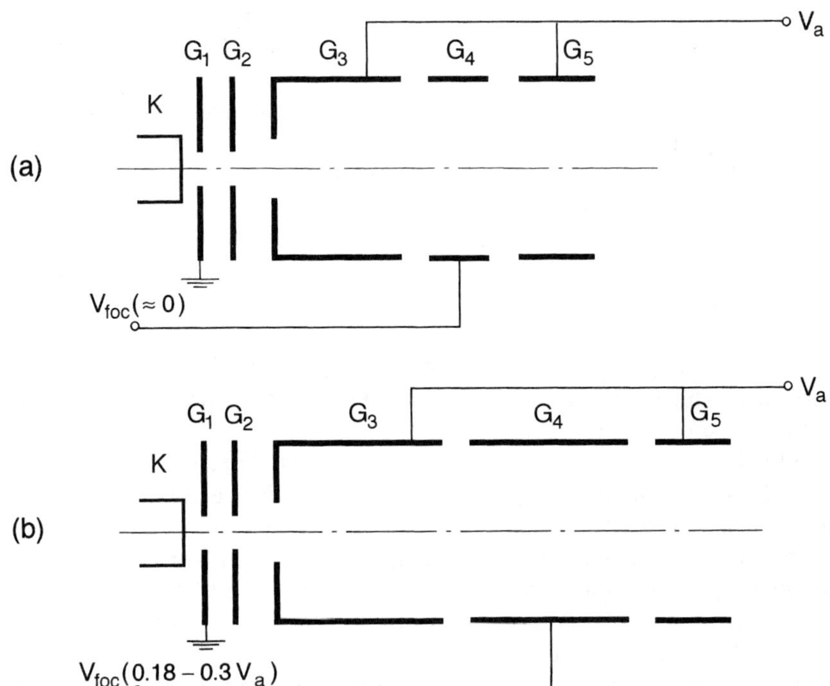

FIGURE 22. Unipotential-focus (UPF) electron guns; (a) with normal unipotential lens; (b) with hi-unipotential (hi-uni) lens.

at the full anode potential V_a while the relatively short central electrode G_4, used for focusing, is held close to ground potential. Guns with a unipotential main lens, sometimes called unipotential-focus (UPF) guns, thus have the advantage that the focusing voltage can be obtained from an existing low-voltage source within the TV set rather than requiring a separate high-voltage source as in the bipotential lens. However, because of the high potential gradients between the G_2-G_3 electrodes as well as between the G_3-G_4 and G_4-G_5 electrodes there is a greater tendency for arcing than in the BPF gun in which a similar high-potential gradient occurs only between the G_3-G_4 electrodes.

The high-potential gradient between G_2 (to which several hundred volts may be applied) and G_3 (to which the full anode voltage V_a of 20–30 kV is applied) produces a strong (prefocus) lens, which narrows the beam to a small diameter at the main lens. As a result of this, there is less variation in spot size versus beam current than in BPF guns. This is explained in Section II.D (see Fig. 14 on p. 305) where the relationship between spot size d_s and beam diameter D_b is shown for various currents. Also shown in Fig. 14 is a typical beam-diameter characteristic actually obtainable for a BPF gun, indicated by the dashed-line curve. As can be seen, the beam diameter for low currents, less than 2 mA in this case, is smaller than the optimum for which a minimum spot size is achieved. With a further reduced beam diameter, which may occur with a unipotential gun, the spot size will be increased more than that shown for a BPF gun. For beam currents higher than 2 mA, on the other hand, a reduced beam diameter with a UPF gun will be closer to an optimum, resulting in a smaller spot size.

It has been shown in fact that the spherical aberration of a unipotential lens whose G_4 length is equal to its diameter is nearly equal to that of a bipotential lens with the same lens diameter and the same object and image distances measured from the mid-plane of the main lens (van Gorkum and Spanjer, 1986). This supports the foregoing explanation about the difference in the spot-size characteristic attributed to the difference in the prefocus lens action between the BPF and UPF guns.

If the G_4 electrode is lengthened as shown in Fig. 22b and a higher voltage (4 to 5 kV, for example) is applied to it, the strength of a unipotential lens is weakened. A theoretical analysis shows that this results in a reduction of aberrations (Saito *et al.*, 1973). Based on this, a new UPF gun, the Hi-UPF or hi-uni gun, was developed in 1973 for 17 V (43 cm) and 19 V (48 cm) screen size, 114° deflection Trinitron tubes in which an increased G_4/G_3 voltage ratio of 18% was adopted (Yoshida *et al.*, 1973). A further increase in the voltage ratio to more than 30% was introduced in a subsequent UPF gun design (for unitized guns) resulting in an even further reduction in aberrations (Fukushima *et al.*, 1977).

3. *Other Lens Types*

In addition to the hi-bi and hi-uni gun designs already mentioned, other combinations of bipotential and unipotential lenses have been commercially used in large screen tubes with 90°, 100°, and 110° deflection angles. Such gun designs are sometimes called hybrid-focus or multistage guns (see also Takano, 1983). In one arrangement use is made of the fact that a unipotential lens can also operate with the center focus electrode at a voltage that is higher, instead of lower, than that of the outer electrodes. By providing an additional anode G_6, a unipotential lens can also be combined with a bipotential lens, thus allowing two types of hybrid lenses to be formed. These lens types may be referred to as the hi-uni-bi shown in Fig. 23a and the lo-uni-bi in Fig. 23b, respectively, depending on the voltage applied to the center focus electrode of the unipotential lens section.

In both cases the unipotential lens, consisting of G_3, G_4 and G_5, is combined with the bipotential lens (usually hi-bi design) created by the G_5

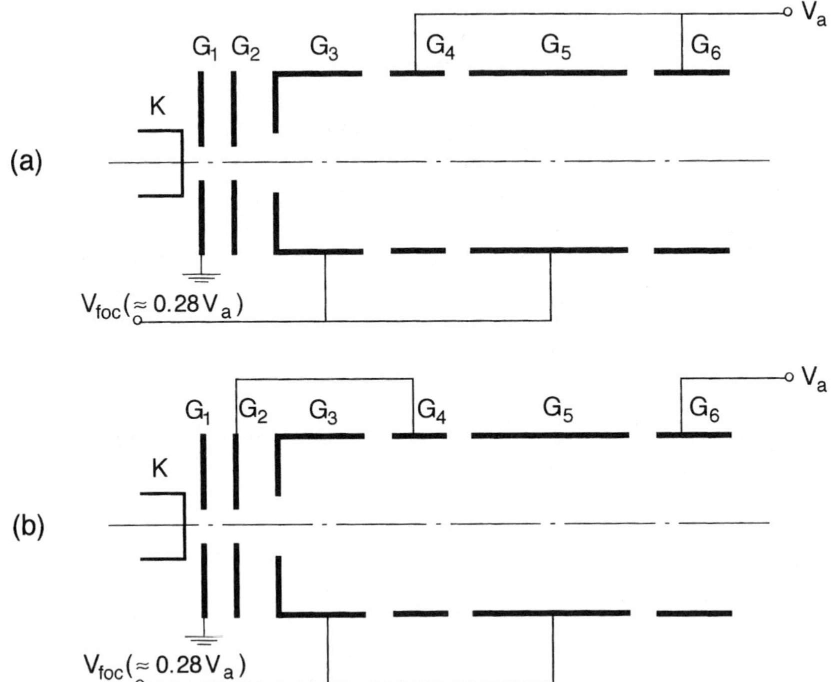

FIGURE 23. Hybrid-focus (unipotential plus bipotential) electron guns; (a) with hi-uni-bi lens which uses anode voltage connected to G_4 and (b) with lo-uni-bi lens which uses G_2 voltage V_{foc} connected to G_4.

and G_6 electrodes. Also, in both cases the variable focusing voltage V_{foc} is applied to G_3 and G_5, which are internally connected. Because a unipotential lens can operate with its center electrode at a higher potential than its outer electrodes, the center electrode G_4 can be internally connected to G_6 and operated at the full anode voltage V_a (hi-uni-bi) as shown in Fig. 23a. In the arrangement of the lo-uni-bi shown in Fig. 23b the center electrode G_4, which is internally connected to G_2 (or may be connected to G_1 at ground potential), is maintained at a lower potential than its adjacent electrodes as in the usual form of unipotential lens. Mitsubishi's Multistep-Focus gun (Washino *et al.*, 1979) and Hitachi's Bipotential Unipotential (B-U) gun (Yamaguchi *et al.*, 1979) employ the hi-uni-bi type lens, while Toshiba's Quadra Potential Focus (QPF) gun (Takenaka *et al.*, 1979) employs the low-uni-bi type lens.

Because of the dual lens nature of such hybrid gun designs they generally have a shorter focal length, thus allowing the total length of the gun to be reduced compared to the hi-bi gun design. However, fabrication of hybrid guns requires the alignment of seven electrodes compared to the five electrodes of the hi-bi gun. On the other hand, a computer analysis shows that the hi-bi design can have a smaller undeflected spot at high beam currents (Davis and Say, 1979). This is probably due to the fact that the hybrid system has greater spherical aberrations compared to a hi-bi system as a result of the added unipotential lens. It should be noted, however, that the spot size depends not only on the spherical aberration of the main focusing lens, but also on that of the prefocus lens, which may be different in guns with hybrid lens and hi-bi lens.

In the tripotential gun design shown in Fig. 24, introduced by Zenith (Blacker *et al.*, 1976), an intermediate voltage of about 40% of the anode voltage V_a is employed in addition to the focusing voltage of about 22% of

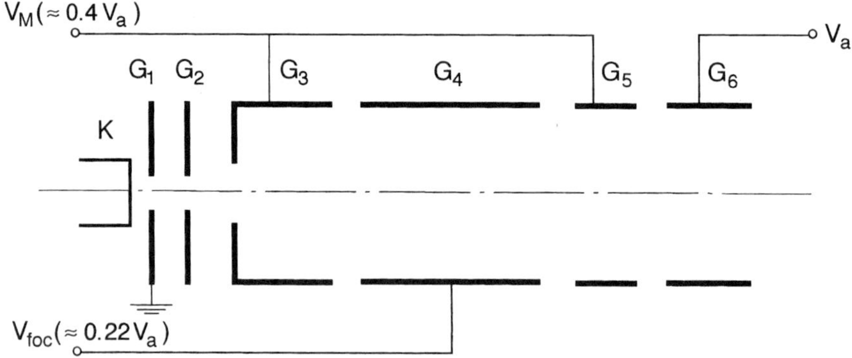

FIGURE 24. Tripotential-focus (TPF) electron gun which uses an intermediate voltage V_M connected to G_3 and G_5 with focus voltage V_{foc} connected to G_4, both supplied through the neck-base.

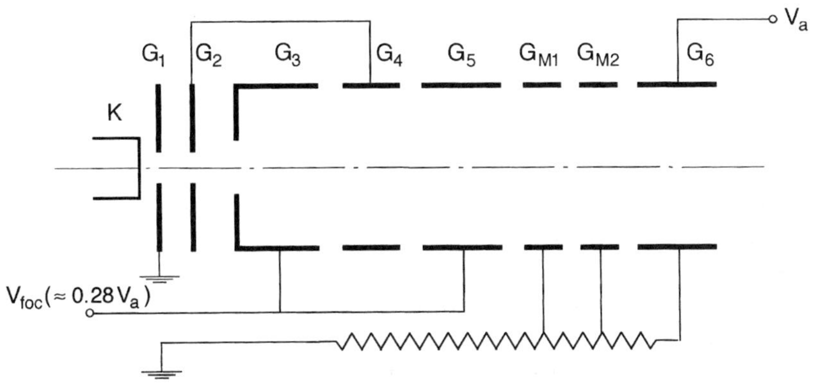

FIGURE 25. Hybrid-focus (lo-uni-bi) electron gun using two additional electrodes G_{M1} and G_{M2} between G_5 (focusing electrode) and G_6 (anode), with the intermediate potentials supplied to the additional electrodes by means of an internal voltage divider.

V_a. The intermediate voltage is applied to G_3 and G_5, which are internally connected and the focusing voltage applied to G_4. This results in a more gradual increase in potential along the tube axis in the main lens region from G_4 to G_6, thus producing a lens that is more extended than, for example, the hi-bi lens design. This is effectively the same as using a larger diameter lens whose spherical aberration is reduced. However, to apply the intermediate voltage through a pin of the tube base it was necessary to specially design the base to prevent high-voltage arcing and ensure low leakage.

A main focusing lens with an even more gradual increase in potential along the lens axis can be realized by inserting additional electrodes G_{M1} and G_{M2}, as shown in Fig. 25, in the lens region with progressively higher intermediate potentials applied to them from a voltage divider inside the tube neck. Such a lens was introduced by Toshiba in their 31 V (78 cm) screen size 110° deflection tubes (Yamazaki *et al.*, 1987). More recent developments of color tube guns of a similar type employ asymmetric main lens designs (see Section III.D.2.*c*) with a voltage divider inside the tube neck (Ueda *et al.*, 1996; Endo *et al.*, 1997). In these systems, however, a voltage divider with suitable stability is required.

C. *Performance of Prefocus Lenses*

1. *Reduction of Beam Diameter at Main Lens*

As already discussed in Section II, the prefocus lens PL between G_2 and G_3 causes electrons emerging from the triode section (consisting of K, G_1 and G_2) to enter the main lens with a reduced divergence angle (see Fig. 1). The

G_2 voltage is generally several hundred volts, while the G_3 voltage is several kilovolts (V_{foc}) in a BPF gun or it may be equal to the full anode voltage (V_a) of 20–30 kV in a UPF gun. It should be noted that the beam is always subject to a certain degree of converging action caused by the accelerating field near the exit of G_2. However, this can be enhanced by a proper design of the prefocus lens.

The optical equivalent of a lens system consisting of a prefocus lens PL and a main lens ML is shown in Fig. 26a. As discussed in Section II.D.1, the object O (crossover in the triode section) is imaged onto the screen as I (spot) by the combined action of the two lenses. In the ideal case, that is, in the absence of spherical aberration and space-charge repulsion, the spot size is given by the crossover size multiplied by the magnification of the dual lens system. The magnification M obtained from Lagrange's law, shown by Eq. (38) on p. 289, is repeated here:

$$M = \alpha_{cr}\sqrt{V_{cr}}/\alpha\sqrt{V_a} \tag{38}$$

Here α_{cr} and α are the half-angles subtended by the beam at the object O (crossover) and by the image I (spot) at the screen, while V_{cr} and V_a are the potentials in volts at the object and image, respectively. In accordance with

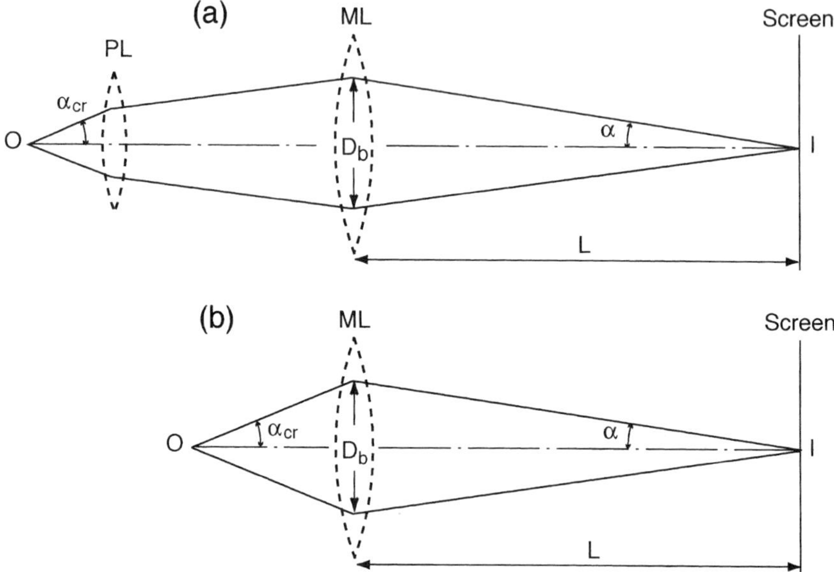

FIGURE 26. Optical equivalent of electron lens systems; (a) main lens ML combined with prefocus lens PL; (b) main lens ML only. With the same divergence angle α_{cr} at the object O (crossover), both lens systems have the same magnification if their convergence angle α at the image I (spot) is the same.

Eq. (38) the magnification of a lens system is determined only by the values of the parameters shown, irrespective of the lens type used to focus the beam and whether it is single or multistage.

For a fixed distance L between the main lens and the screen, the convergence angle α of the beam at the screen is determined by the beam diameter D_b at the main lens. Thus, with a stronger prefocus lens, which results in smaller values of D_b and α, the magnification M of the lens system is larger, and vice versa, assuming α_{cr}, V_{cr} and V_a are kept constant.

However, for a fixed beam diameter D_b at the main lens, the magnification of the lens system is the same whether or not a prefocus lens is present. For example, the lens system consisting of only an ML shown in Fig. 26b will thus have the same magnification as the lens system shown in Fig. 26a consisting of the ML and the PL. From the preceding discussions, in the ideal case where spherical aberration and space charge repulsion can be neglected (as in the case of a low current beam with a very narrow divergence angle), the addition of a prefocus lens results in a lengthening of the gun without affecting the spot size (Moss, 1968d).

In the case of color CRT guns for TV display, however, a wide range of beam currents from zero to a few milliamperes is commonly required. To obtain a large current the bias voltage between the cathode and G_1 must be reduced, allowing current to be drawn from cathode areas farther from its center. The convergence angle of the beam at the crossover is thus increased (although the crossover may also move somewhat farther from the cathode). This causes the beam to diverge from the triode section at a greater angle, expanding to a larger diameter at the main lens. Because the spherical aberration of the main lens increases as the cube of the beam diameter D_b, as discussed in Section II.D.3, the spot size at the screen may become considerably enlarged. A suitable prefocus lens to control the beam divergence entering the main lens is thus very important.

However, since a reduction in divergence angle by a prefocus lens results in a lens system with greater magnification and an increase in spot size, as mentioned here, a compromise must be made in the divergence angle of the beam. Usually, in gun designs for TV tubes the divergence angle of the beam is chosen so that at a high current, such as 2 mA, an optimum beam diameter at the main lens is achieved and a minimum spot size is obtained (see Fig. 14). This results, however, in an increase in spot size at low currents that is more than the minimum value obtainable. For data display tubes the divergence angle of the beam is chosen for lower currents such as 0.3 mA at which the gun is normally operated. In addition, since deflection defocusing is roughly proportional to the beam diameter in the deflection yoke (which is positioned a short distance beyond the main lens), there is a further reason for using a prefocus lens to reduce the beam diameter at the main lens.

Because the prefocus lens itself adds spherical aberration to the lens system it is very important that its aberrations also be minimized. One method of accomplishing this is to position the prefocus lens close to the crossover so that the beam diameter at the prefocus lens is as small as possible. This can be accomplished by reducing the distance between G_2 and G_3 and increasing the thickness of G_2. Such a prefocus lens design has been used in the RCA's HiPI bipotential gun design (Hughes and Chen, 1979).

In some designs of the prefocus lens additional electrodes may be used. For example, adding a G_{2S} electrode immediately after G_2 and maintaining its potential lower than G_2 creates a much stronger prefocus lens between the G_{2S} and G_3 electrodes in bipotential guns (Kawakami *et al.*, 1975). Another form of prefocus lens can be produced by positioning the lo-uni-bi lens (G_3-G_4-G_5) as shown in Fig. 23b very close to the triode section, creating in effect a multistage prefocus lens (Suzuki *et al.*, 1990). In these types of prefocus lenses, the lens strength can be controlled by the potential externally applied respectively, to G_{2S} or to the G_4 of the lo-uni-bi lens.

In a UPF gun a strong prefocus lens is more easily obtained than in a BPF gun due to the stronger accelerating field between G_2 and G_3 as already discussed. This frequently causes an excessive prefocusing effect on a low-current beam, making it difficult to obtain a sufficiently small spot size at such currents. To optimize the prefocus lens characteristic of the UPF gun of Trinitron tubes, an additional electrode G_M operating at a potential intermediate between that of G_2 and G_3 has also been used (Ichida *et al.*, 1987).

As already mentioned, in addition to the spot size achievable at the screen center, deflection defocusing has to be taken into account in considering the optimum beam divergence at the exit of the prefocus lens. This is particularly the case in a gun designed to operate with a self-converging yoke whose magnetic fields are strongly astigmatic, making the deflected beam subject to strong overfocusing in the vertical direction. To minimize this the beam diameter within the deflection field is frequently reduced in the vertical direction by introducing an astigmatic lens at the prefocus lens (G_2-quadrupole) or at the cathode lens (G_1-quadrupole) as will be discussed in Section III.E.2. Although this may result in a larger vertical spot size at the screen center, such a compromise is often made to enable a satisfactory spot size to be obtained over the entire screen.

2. *Cancellation of Spherical Aberrations*

In addition to the spherical aberration in the main lens discussed in Section II.D.3, such aberration also exists in the prefocus lens as well as in the cathode lens of the triode section of the gun (see, e.g., van Gorkum, 1985). These

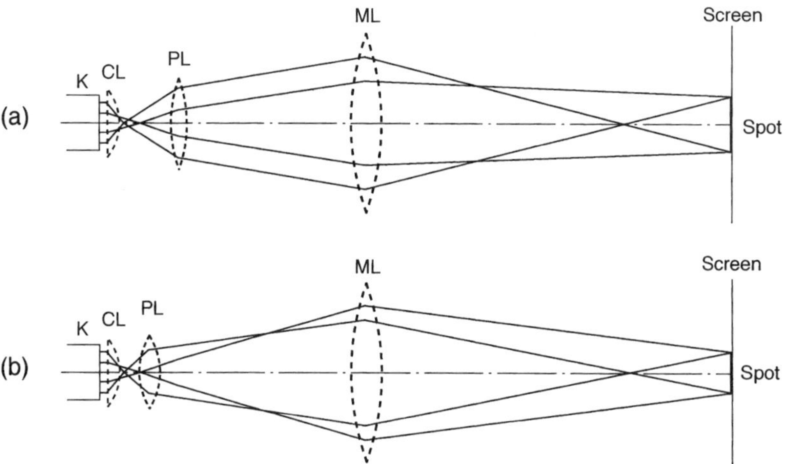

FIGURE 27. Schematic diagrams of gun lens systems; (a) non-ART gun, and (b) ART gun. In the latter, due to a strong prefocus lens PL, the position of the outer rays is interchanged with the inner rays before reaching the main lens (ML); CL represents the cathode lens.

aberrations generally contribute in an additive way to an increase in the spot size at the screen. More specifically, the effect of these aberrations is to produce a greater bending or focusing action on the outer rays of the beam compared to the inner rays, preventing these two types of rays from being brought to a common focus at the screen. In practice, focusing is adjusted so that the minimum diameter of the beam, referred to as the circle of least confusion, falls on the screen as shown in Fig. 27a. (For simplicity, thermal velocity effects and space-charge repulsion are neglected.)

It is possible, however, to reduce the total amount of spherical aberration of a gun by using a special prefocus lens [referred to as an aberration-reducing triode (ART)] (Gerritsen and Barten, 1987; Kitagawa *et al.*, 1987; Spanjer et al., 1987). In this arrangement, as shown in Fig. 27b, a prefocus lens PL with increased strength is placed close to the crossover. This causes rays in the space between the prefocus lens and the main lens to cross over so that those rays initially closer to the axis arrive at the main lens farther from the axis as shown. Because the aberration of the main lens tends to reverse this effect, it partly cancels the spherical aberration of the beam acquired in the cathode lens, reducing the total spherical aberration as well as the spot size. This process is more efficient at an increased beam current where electrons are drawn from areas of the cathode farther from its center and subject to a greater spherical aberration in the cathode lens.

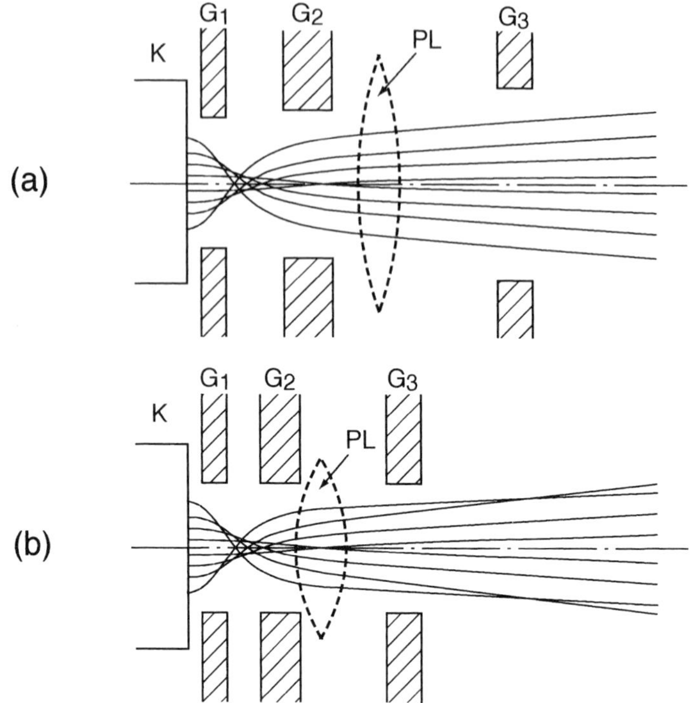

FIGURE 28. Electrode arrangements in triode (K-G_1-G_2) and prefocus lens section together with beam trajectories for a high current; (a) non-ART gun; (b) ART gun.

To create a strong prefocus lens close to the crossover, as required for the ART effect, the distance between G_2 and G_3 is reduced as shown in Fig. 28b and the aperture of G_2 and the aperture at the entrance of G_3 are made smaller. A proper thickness for G_2 also must be selected. In contrast to this, the more conventional beam-forming region shown in Fig. 28a produces a weaker prefocus lens farther away from the crossover. It should be noted that in both arrangements, due to the strong spherical aberration of the cathode lens the outer rays emitted from the cathode cross the axis closer to the cathode than the inner rays, thus resulting in a crossover which itself becomes more aberrant with increasing beam current. As shown in Fig. 28b, because the inner rays cross the axis at a position near the center of the ART prefocus lens, they undergo a reduced convergence despite an increased strength of the lens. On the other hand, the outer rays, which cross the axis closer to the cathode and maintain a certain distance to the prefocus lens, undergo an increased convergence due to the strong prefocus lens.

The ART system also has the advantage of reducing space-charge repulsion as the beam passes through the field-free drift space between the main lens and the screen. In conventional non-ART guns the current-density distribution of the beam in the drift region is highly nonuniform, having a bell-shaped cross section with its maximum at the center of the beam. However, in the ART system the low current-density outer parts of the beam are folded inwards as already discussed here, resulting in an increased current density of the outer portions of the beam. This significantly reduces the spot enlargement due to space-charge repulsion, especially in large tubes with their long gun-to-screen distance (long transit time) and high beam current. In the case of large 43 V (109 cm) screen size color tubes, for example, a more than 20% reduction in spot size was obtained at high beam currents using the ART design (Ashizaki *et al.*, 1988).

D. *In-Line Gun Systems*

As mentioned, almost all present-day color tubes employ an in-line gun arrangement and self-converging yoke. However, for a tube neck of given diameter this arrangement reduces the allowable lens diameter of the guns by approximately 30% compared to that of the older delta-gun arrangement. This results in an increase in lens aberrations and spot size, aside from increased deflection defocusing caused by the required nonuniform magnetic field of the self-converging yoke. To cope with the increased aberrations, new types of guns with specially shaped main focusing lenses have been developed as discussed in what follows (Section III.D.2.*c*).

1. *Separate Gun Systems*

The first shadow-mask tubes with in-line guns were introduced by General Electric in 1965 (Buchsbaum, 1966). These produced an 11-inch picture with a 70° deflection angle and employed a shadow-mask with the same hexagonal array of holes used in the delta system. The individual guns, constructed from cylindrical electrodes, were similar to those employed in earlier tubes that used the delta gun arrangement. Although dynamic convergence was obtained with the in-line guns by means of specially designed electromagnets placed around the tube neck coupled to pole-pieces attached to the end of the gun, the convergence circuits and their adjustments were significantly simplified compared to those required for the delta-gun tubes.

A few years later an in-line arrangement with three separate cylindrical guns using a 36 mm tube neck was introduced by Toshiba for their 19 V (48 cm) screen size and 110° deflection rectangular cone in-line system

(RIS). These tubes employed a slotted shadow-mask in place of the hexagonal hole arrangement (Ohta *et al.*, 1972). Also, simplified dynamic convergence, as in the earlier General Electric system, was used in conjunction with a magnetic field-controller positioned at the end of the gun to compensate for coma errors in convergence (seen as a mismatch in raster size between the outer and center beams). In-line arrangements with three separate cylindrical guns also came into use in Europe. Some examples of these were the Philips 20AX (Barten, 1974; Kaashoek, 1974) and 30AX systems (Barten and Kaashoek, 1978), both of which used 110° deflection and a 36 mm neck diameter.

2. *Unitized Gun Structures*

a. Early Arrangements. In 1972, RCA introduced a self-converging in-line color display system that did not require any dynamic convergence means. This employed a shadow-mask with an array of elongated holes or slots, which had a screen size of up to 19 V (48 cm) and a deflection angle of 90°. Using a precision static toroid PST yoke consisting of precisely wound toroidal coils, a strongly astigmatic magnetic field was created, which produced self-convergence (Barbin and Hughes, 1972). (See also review articles, e.g., Morrell *et al.*, 1974b; Branton *et al.*, 1979). Although its principle had been proposed earlier (Haantjes and Lubben, 1959) this was the first practical realization of the self-converging color CRT system. The neck diameter of these tubes was also reduced to 29 mm from the 36 mm used in previous systems. In addition, the gun design of these tubes represented a major improvement in that the three guns were integrated into a single structure, referred to as the unitized gun, employing common electrodes for the three beams. The in-line self-converging system was then used for large-screen (67 cm), wide-deflection (110°) tubes (Barkow and Gross, 1974; Morrell, 1974).

The unitized gun consisted of an assembly of metal plates, each provided with three apertures for the beams in registry with those of the other plates. By properly choosing the spacing between adjacent plates and applying appropriate potentials to these plate electrodes electron-optical lenses were obtained equivalent to those obtained with the adjacent cylindrical electrodes as already shown in Section III.B. Using three separate cathodes to which the respective video signals were applied, the currents of the individual beams could be separately modulated, with each beam successively traversing one of the apertures of the common electrodes. Since the beams could also be positioned closer to each other, with this unitized gun structure a reduction in neck diameter was made possible, thereby reducing the problems of maintaining the beams converged.

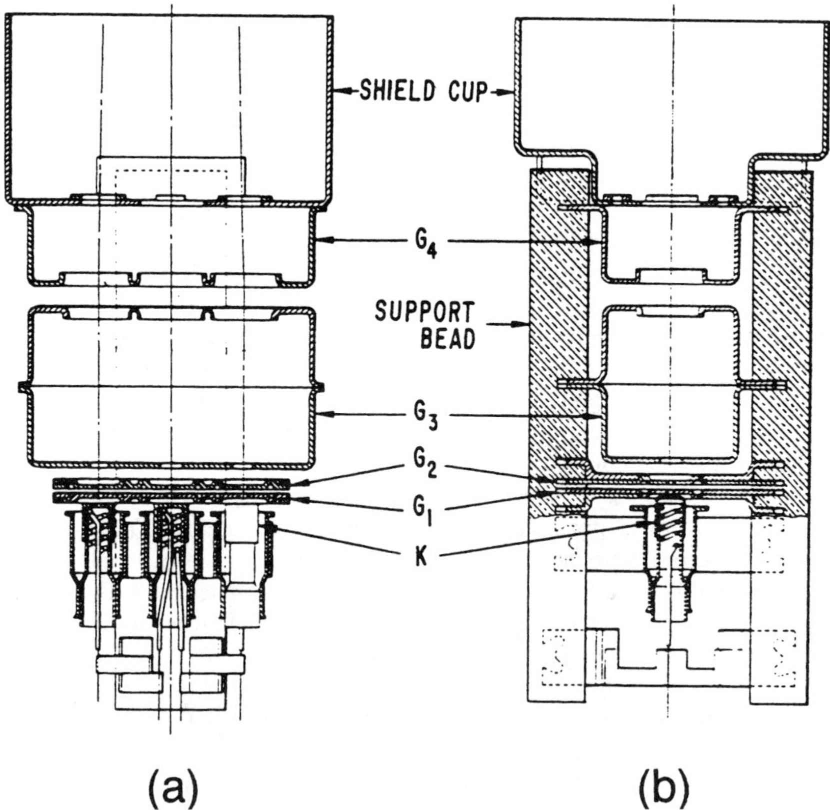

FIGURE 29. Structure of in-line 3-beam unitized gun; (a) horizontal section; (b) vertical section with supporting beads (multiform glass rods) [reproduced by permission from Morrell, A. M., *et al.* (1974). Color picture tubes, in *Advances in Image Pickup and Display*, Suppl. 1, p. 99].

The unitized gun structure with its bipotential main lenses is shown in Figs. 29a and b, representing a horizontal section and a vertical section, respectively. As indicated, G_1 and G_2 are planar electrodes. The focus electrode G_3 is made up of two pieces pressed together to form a box-like enclosure whose end plate at the side facing G_4 has three circular apertures. These apertures, together with the corresponding ones on G_4, create the main focusing lens fields when a potential difference is applied between G_3 and G_4. To avoid arcing due to the high potential gradient, the edges of the lens apertures of these electrodes are bent inward as shown, that is, away from the high-field region. The screen end of G_4 is provided with a shield cup (sometimes called the convergence cup) whose bottom plate may have small magnetic pieces placed around its apertures to compensate for coma errors resulting from the self-converging deflection field (see the article by Yamazaki, "Design and Performance of Shadow-Mask Color Cathode Ray Tubes" that precedes this article).

An important advantage of the unitized gun structure is the greater mechanical accuracy of its assembly compared to three cylindrical separate guns. In the case of separate guns, whose individual electrodes must be assembled with the aid of a jig, any misalignment of the electrodes will result in the electron beam emerging from the gun in an off-axis direction. In addition the three guns must be carefully aligned with respect to each other so that three beams are precisely centered in the deflection field with a precise spacing maintained between them.

The increased accuracy of the unitized gun is due mainly to the fact that the three lens apertures of each electrode are created by a mechanical press, ensuring that their position is accurate and reproducible. The multiple-aperture electrodes are then assembled into a unitized structure with the aid of a jig. The unitized gun also reduces the drift in convergence angle that results from the thermal expansion of the supporting elements holding the electrodes together in the older (separate) gun structures (Morrell, 1974).

Following the announcement of the precision in-line system by RCA, various manufacturers adopted this system with their own variations. Since about 1975 the unitized in-line gun structure with a 29 mm neck tube and its associated self-converging yoke has been the standard design for shadow-mask tubes. In Europe, however, separate in-line guns with a 36 mm neck tube were used with self-converging yokes until the mid-1980s in such systems as the Philips 20AX and 30AX already mentioned here.

b. Obtaining G_2-G_3 Static Convergence. Static convergence (i.e., convergence of the beams in the absence of a deflection field) requires tilting the two outer beams of the unitized gun toward the tube axis at an appropriate angle (typically about 1°) as they pass through their respective main lenses. This can be accomplished, as shown in Fig. 30a, by moving the center of the G_4 apertures of the two outer beams slightly farther away from the tube axis. However, the tilt angle of the outer beams changes with a change in the focus voltage applied to G_3, resulting in a shift in position of their spots at the screen. This makes an exact adjustment of focus and beam convergence difficult. In addition, because of the outward position of the G_4 apertures the shape of the outer beam spots is distorted; this causes a comet-like haze, referred to as spot coma or core-to-haze asymmetry, on the side of the beam spots opposite to the bending direction, resulting in a loss of resolution.

To reduce this problem another method of static convergence has been employed in which the outer beams are bent inward at the G_2-G_3 prefocus lens instead of at the G_3-G_4 main lens (Hosokoshi *et al.*, 1983; Gerritsen and Barten, 1987). This is accomplished, as shown in Fig. 30b, by shifting the center of the G_3 apertures of the prefocus lens of the outer beams farther from the tube axis. This causes the outer beams to be directed toward the center of their respective G_3-G_4 main lenses that are now axially symmetric. As a result of this, the core-to-haze asymmetry of the outer beam spots is almost

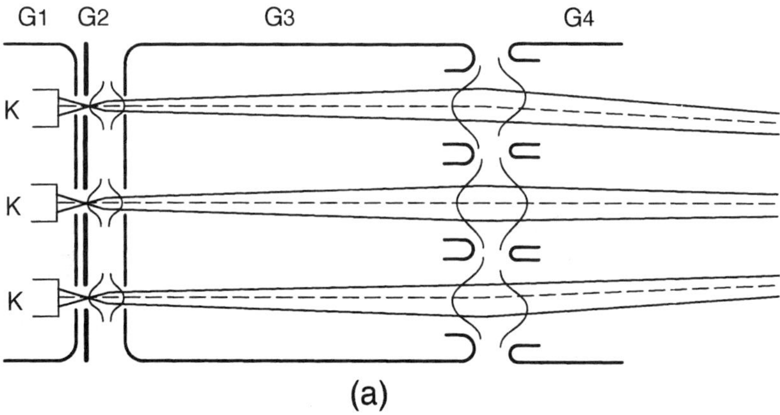

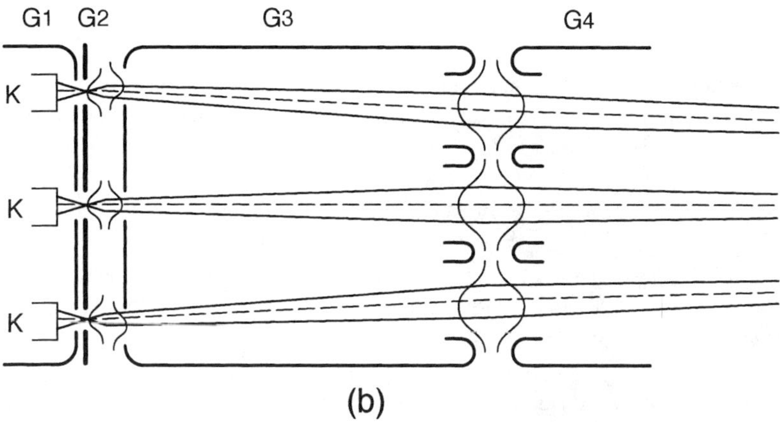

FIGURE 30. Methods of obtaining static convergence in a unitized gun; (a) displacement of G_3-G_4 main lens apertures; (b) displacement of G_2-G_3 prefocus lens apertures.

entirely avoided. Also, as there is no offset between the G_3 and G_4 apertures, identical parts can be used for both the G_3 and G_4 aperture plates. In this static-convergence method the bending of the outer beams occurs very close to their crossover, which serves as the object of the main lens. It can be shown that because of this the shift of the spot with a change in focus voltage applied to G_3 is significantly reduced.

To reduce further the effect of changes in the focus voltage on the static convergence, it is possible to offset the apertures of both the G_2-G_3 and G_3-G_4 lenses since the change in beam bending with a change in focus voltage is just opposite for these two lenses (Chen, 1985).

In addition to tilting of the beams by the gun structures as already mentioned, any remaining errors in static convergence can be corrected by means of magnetic multipole rings mounted outside the tube neck (see the article by Yamazaki, "Design and Performance of Shadow-Mask Color Cathode Ray Tubes" that precedes this article).

c. Asymmetric Lens Designs. In a conventional unitized in-line gun such as shown in Fig. 29, the main lenses are created by circular apertures provided on the end plates of the G_3 and G_4 electrodes. The front view of one of these aperture electrodes is shown in Fig. 31a and its horizontal (sectional) view is shown in Fig. 31b, where the diameter of the holes is designated by D_L, the center-to-center spacing of the holes (beam separation) by S, and the spacing between the neck glass and gun by a.

For a small spot size to be obtained at the screen, the lens diameter D_L should be as large as possible to reduce the spherical aberration as discussed in Section II.D. In a practical gun design, however, other factors such as beam convergence and high-voltage arcing must also be taken into account. For better beam convergence when deflected, a smaller beam separation S is preferable since convergence errors are roughly proportional to the beam separation. For this purpose a smaller lens diameter D_L is required. To ensure reliable operation at high anode voltage, the spacing a between the glass neck and the gun should be as large as possible, also requiring a smaller beam separation S.

In practice a compromise must be made among these three factors, namely, the lens diameter D_L, the beam separation S, and the spacing a between the neck glass and gun. In a typical tube with a 29 mm neck diameter (approximately 24 mm inside diameter), the best compromise may be obtained at a value of 5.5 mm for the beam separation S. For practical purposes, however, this leaves only 4.5 mm for the lens diameter D_L.

An important method for reducing the lens aberration without increasing the beam spacing S was introduced by Matsushita (Hosokoshi *et al.*, 1980) using a hi-bi unitized in-line gun. This method is based on the concept of the

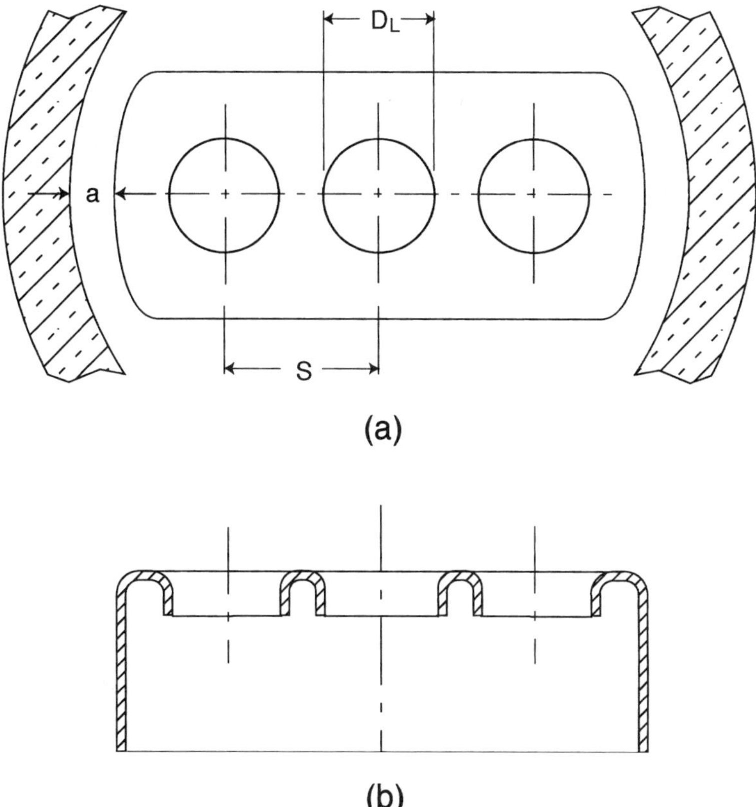

FIGURE 31. Electrode structure of a conventional main lens of unitized in-line guns with three separated circular apertures; (a) front view; (b) sectional (horizontal) view.

so-called overlapping field lens (OLF). This is illustrated in Fig. 32a and b, showing the end plate of the G_3 (focus) electrode and its horizontal section, respectively. As indicated, circular lens openings of diameter D are cut into the end plate of G_3. Since this diameter D is greater than the center-to-center spacing S, this results in the removal of some material between the aperture holes. Similar overlapping aperture holes are also cut into the end plate of the G_4 (anode) facing the G_3 electrode.

The removal of material between the apertures causes a large reduction in the horizontal focusing action of the G_3-G_4 lens. To compensate for this a pair of metal plates (called the field-forming electrode) are mounted behind the G_3 aperture plate and a similar pair mounted behind the G_4 aperture plate. As shown in Fig. 32b, each field-forming electrode pair is positioned a

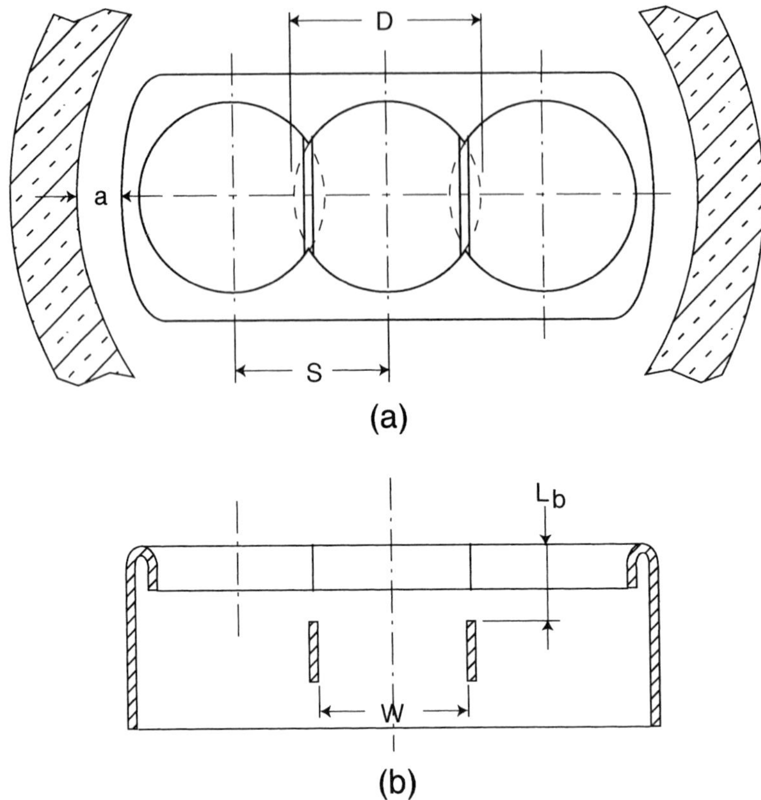

FIGURE 32. Concept of overlapping field (OLF) main lens of Matsushita; (a) front view; (b) sectional view.

distance L_b behind the aperture plate. These field-forming electrodes are maintained at the potential of the respective electrodes G_3 and G_4.

The horizontal and vertical sections of the OLF lens are shown in Fig. 33a and b, respectively, together with the electron trajectories to illustrate the lens action. Some of the equipotential surfaces between G_3 and G_4 are also shown. As indicated, the horizontal lens opening is defined by the spacing W between the field-forming electrodes, and is narrower than the vertical opening, which equals the diameter D shown in Fig. 32. Although the narrower opening produces a stronger lens action in the horizontal direction, this can be compensated for by increasing the axial distance between the field-forming electrodes of G_3 and G_4, which are positioned a distance (L_b) back from the curled ends of G_3 and G_4. In this way, identical lens action can be obtained in both the horizontal and vertical directions, which is close to the lens action

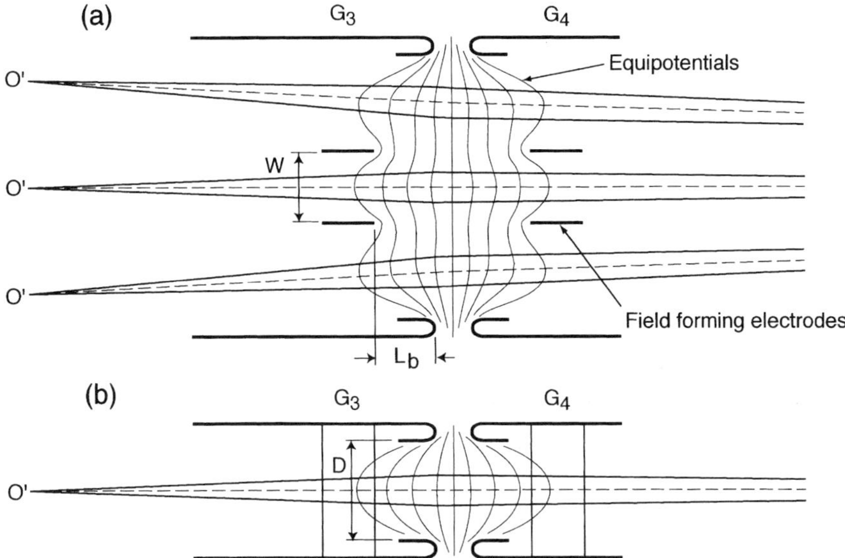

FIGURE 33. (a) Horizontal; (b) vertical section of OLF main lens consisting of G_3 (focusing electrode) and G_4 (anode) together with equipotentials and trajectories.

produced by a circular lens with a diameter D. The lens action for the side beams is very similar to that of the center lens except that the spherical aberration differs somewhat between the left and right halves of the horizontal section of each side beam. A 3D cut-away view of the G_3 and G_4 electrodes and the field-forming electrodes on both sides of the lens region is provided in Fig. 34.

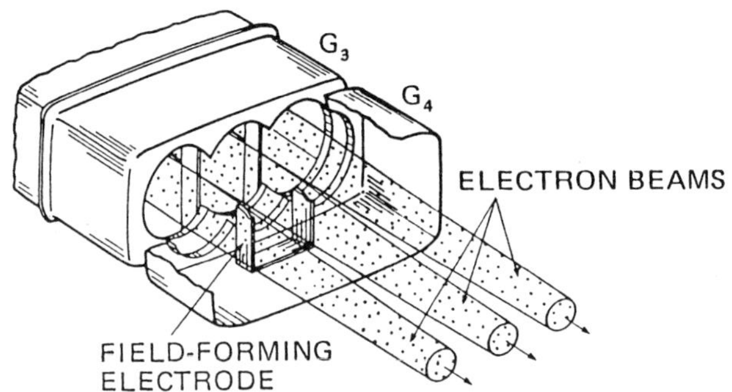

FIGURE 34. Complete structure of OLF main lens showing G_3 and G_4, each provided with field-forming electrodes [reproduced by permission from Hosokoshi, K., *et al.* (1983). Improved OLF in-line gun system, *Proc. 3rd Intern. Disp. Res. Conf. (Japan Disp. '83)*, pp. 272–275; Courtesy Society for Information Display].

The lens characteristics of the OLF lens, which has a nonsymmetrical field about its axis, were examined with 3D computer calculations and compared with those of a conventional circular lens having the same focus-voltage-to-anode-voltage ratio V_{foc}/V_a and the same distance L from the main lens to the screen (Hosokoshi *et al.*, 1980). These results showed that the magnification of the OLF lens was equivalent to that of a 7.6 mm circular lens in both the horizontal and vertical directions. The spherical aberration of the OLF lens, which was assumed to follow the same 3rd power function of the beam diameter D_b at the main lens as defined by Eq. (59) on p. 298 in Section II.D.3, was found equivalent to that of a 7.6 mm circular lens in the horizontal direction and to that of an 8 mm circular lens in the vertical direction. If 7.6 mm is taken as the equivalent (circular) lens diameter for the OLF lens, it is effectively 1.7 times larger compared to the conventional 4.5 mm diameter circular lens shown in Fig. 31, and at the same time the original beam separation of 5.5 mm is retained.

The larger effective diameter of the OLF lens in terms of aberrations allows the use of wider beams at the main lens, which more nearly fill the lens openings. This requires a change in the design of the prefocus lens, as discussed in Section II.D.4, resulting in a decrease in the magnified image of the crossover (magnified crossover) as well as the space-charge repulsion, leading to a further reduction in spot size. Thus, despite some residual aberrations generated by the axially asymmetric field, OLF lenses have been highly successful in reducing the spot size, especially at high beam currents.

Spot profiles obtained with an OLF gun using main lenses equivalent to a 7.6 mm diameter circular lens are shown in Fig. 35 in comparison to those of a conventional gun with 4.5 mm diameter circular lenses. These curves were obtained by optical measurements of the spot at the center of a 19 V (48 cm) screen, 90° deflection tube using a spot photometer at beam currents of 0.1, 0.25, 1.0, 2.5 and 3.5 mA. As shown, higher peak levels and narrower spots are obtained for the OLF gun, indicating a higher current density. The spot sizes, defined as their width at 5% of their peak value, are plotted in Fig. 36 as a function of the beam current. As shown, a spot-size reduction of approximately 20% was obtained with the OLF gun at high beam currents over 1 mA.

After the development of the forementioned OLF lens a new form of OLF lens structure with somewhat simpler design was introduced by Matsushita together with the G_2-G_3 static convergence system also mentioned in the foregoing (Hosokoshi *et al.*, 1983). In this OLF lens structure the field-forming electrodes were retained, but the overlapping apertures of three holes were replaced by a single oval-shaped large aperture.

In 1982, RCA introduced the expanded lens (XL) unitized hi-bi gun for their combined optimum tube and yoke (COTY-29) tubes also using a 29 mm

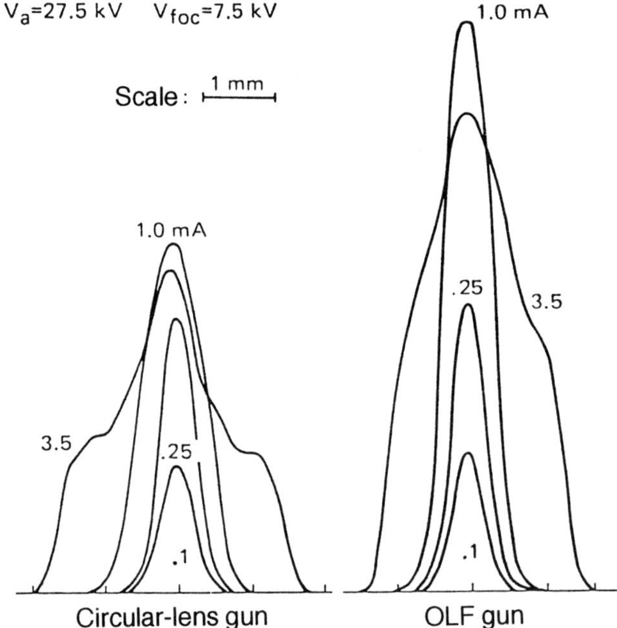

FIGURE 35. Spot profiles measured at screen center in 19 V 90° tubes for various beam currents indicated; OLF gun (right) and circular-lens gun (left) [reproduced by permisson from Hosokoshi, K., *et al.* (1980). A new approach to a high performance electron gun design for color picture tubes, *IEEE Trans. Consum. Electron.*, CE-2b (six), 452–458; © 1980 IEEE].

diameter neck (Morrell, 1982, 1983; Alig and Hughes, 1983). In the XL lens circular individual lens holes for the three beams are created on the flat adjacent end plates of the G_3 and G_4 electrodes. Unlike in the OLF lens, these holes (as shown in Fig. 37a) are limited in size and do not overlap. To avoid the high spherical aberration that would occur, each end plate is provided with a raised rim around its edge, as shown in Fig. 37b, forming an oval or racetrack-shaped aperture around the three circular beam apertures. As illustrated in Fig. 38a and b showing horizontal and vertical sections, respectively, the raised rims of the G_3 and G_4 electrodes extend toward each other, forming a large common lens region between them that encompasses the individual smaller lenses formed by the circular holes of the electrode plates. Superimposed on these electron lenses are the optical analogs of the large and small lenses shown by the broken lincs.

Because of the distance between the two aperture plates the strength of the small aperture lenses is reduced, lowering their spherical aberrations. However, the reduced strength of the aperture lenses is compensated for by the addition of the large common lens whose aberrations are also small

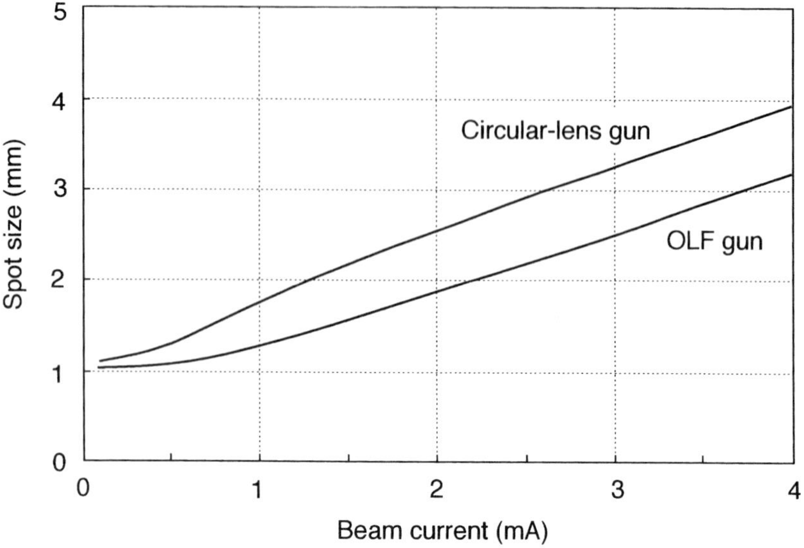

FIGURE 36. Spot size (width at 5% value) characteristics as a function of beam current for OLF gun and circular lens gun.

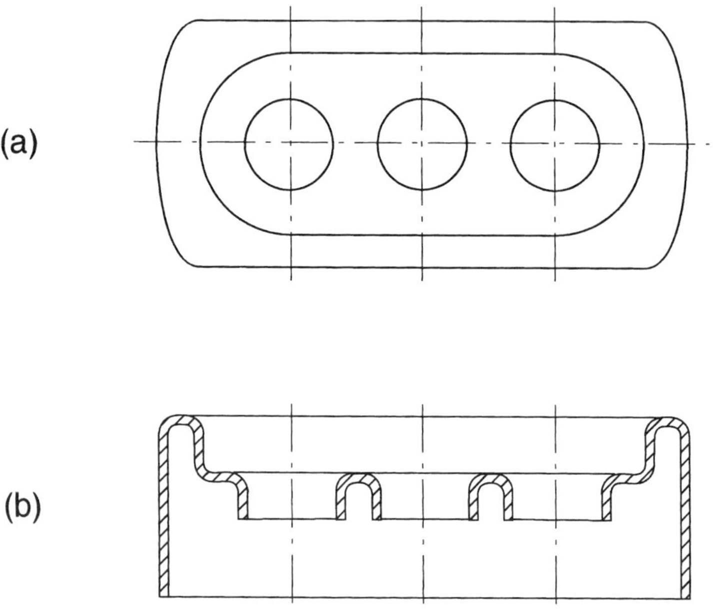

FIGURE 37. End-plate electrode of RCA XL main lens; (a) front view; (b) sectional view.

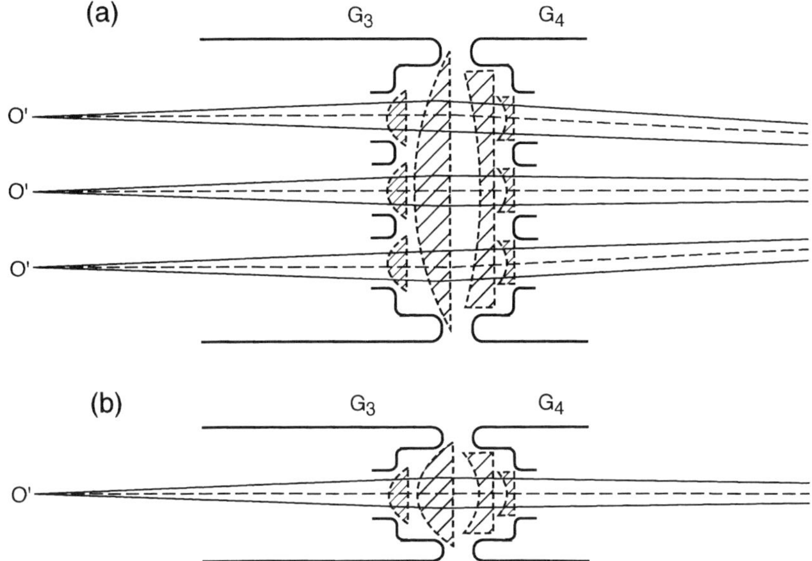

FIGURE 38. (a) Horizontal; (b) vertical section of RCA XL main lens consisting of G_3 (focusing electrode) and G_4 (anode) together with beam trajectories and superimposed optical analogs of the large and small lenses.

because of its greater size compared to conventional lenses such as shown in Fig. 31. Also the focus quality of this lens system is not limited by the beam separation as in the case of the OLF lens. Taking advantage of this basic characteristic, the beam separation of the COTY-29 system has been reduced from the 6.6 mm of a conventional system to 5.1 mm. The result is an improved beam convergence. (Although in the preceding discussions there are some differences in the explanation for the OLF and the XL lenses, their lens operation is very similar in principle.)

Because of the superior performance of guns of the preceding and other asymmetric main lens designs, other color tube manufacturers have adopted similar approaches with their own variations to increase the effective diameter and reduce the spherical aberrations of the main lenses. For example, in 1983 the conical field focus (CFF) lens of Philips ECG (Zmuda *et al.*, 1983) and the elliptical aperture (EA) lens of Hitachi (Shirai *et al.*, 1983) were introduced. The structure of the latter lens system is shown in Fig. 39. In 1986 the elliptical aperture, concave surface (ES) lens was announced by Hitachi (Miyazaki *et al.*, 1986), followed in 1989 by the polygon approach of Philips (Gerritsen and Himmelbauer, 1989) and in 1993 by the chain-link-shaped lens aperture of Chunghwa Picture Tube (Chen and Tsai, 1993). In one form or another the gun designs mentioned here are employed in most unitized in-line

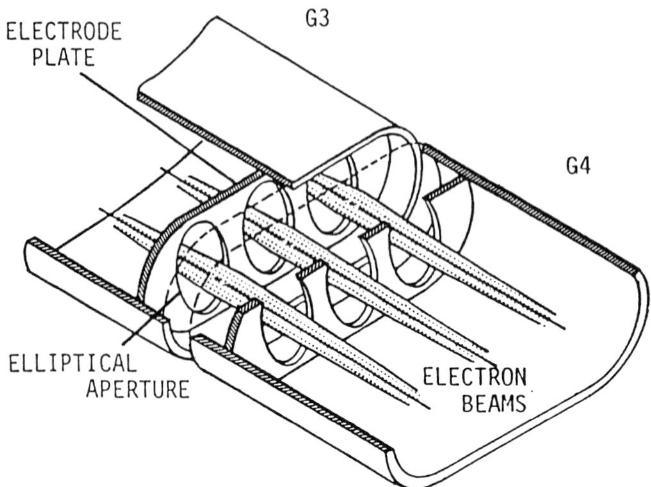

FIGURE 39. Elliptical aperture (EA) main lens of Hitachi consisting of G_3 (focusing electrode) and G_4 (anode), each provided with an electrode plate with three elliptical apertures [reproduced by permission from Shirai, S., *et al.* (1983). A rotationally asymmetric electron lens with elliptical apertures in color picture tubes, *Proc. 3rd Intern. Disp. Res. Conf.* (Japan Disp. '83), pp. 276–279; Courtesy Society for Information Display].

gun designs of present-day color tubes. The equivalent lens diameters, achieved through various asymmetric designs, are of the order of 8 mm for the 29 mm neck and 10 mm for the 32.5 mm neck used for large-screen tubes.

Asymmetric main lenses such as the OLF and XL already discussed here, especially where all three electron beams must be focused and converged simultaneously, are almost impossible to design by experimental means alone because of the many dimensional parameters involved. The development of such lenses has thus largely depended on the use of 3D computer simulations combined with experimental testing. For the determination of lens parameters, computer calculations of trajectories in the Laplace field, neglecting thermal effects of emerging electrons from the cathode and space-charge repulsion, can be very useful in many cases (Alig and Hughes, 1983; Hosokoshi *et al.*, 1983). However, an accurate prediction of the spot size requires that the computations include the thermal velocities and space-charge effects (MacGregor, 1983, 1986; Aalders *et al.*, 1989; Bechis *et al.*, 1989; Alig and Fields, 1997). The space-charge repulsion in particular strongly influences the beam trajectories in the beam-forming triode region where the current density is high and also in the long, field-free drift space between the main lens and the screen. Computations that ignore the space-charge repulsion predict spot sizes at high beam currents that may be smaller by a factor of two than the spot sizes actually measured.

It is important for commercially useful gun designs that they be relatively insensitive to dimensional variations that might occur in their assembly. For the purpose of assembly-tolerance analysis, simulations are made with misaligned, tilted, or incorrectly spaced grids to determine how much these errors degrade the spot from the ideal design performance. In this way designs can be adopted that are more manufacturable and cost-effective and minimize the factory scrap rates (Bechis *et al.*, 1989; Alig and Trinchero, 1995; Hellings and Baalbergen, 1997).

3. *Trinitron Gun System*

A somewhat different form of in-line gun structure was first introduced by Sony in 1968 for their 7-inch diagonal Trinitron color tubes (Yoshida *et al.*, 1968). In this system, the outer beams of three in-line beams enter the main lens obliquely so that all three beams cross over at its center as shown in Fig. 40. The main lens is a cylindrical large-diameter unipotential lens consisting of a low-voltage G_4 focus electrode positioned between the G_3 and G_5 electrodes, which are maintained at the full high-voltage anode potential. To direct the outer beams toward the axis their cathodes are inclined toward the tube axis as shown. In another arrangement the apertures of the outer prefocus lenses between G_2 and G_3 may be offset in the same way as for the G_2-G_3 static convergence, already discussed in the preceding.

After crossing over at the center of the main lens, the outer beams diverge and enter an electrostatic deflection system. With the two inner deflection plates maintained at the final anode potential, slightly lower potentials applied to the outer plates cause the outer beams to bend inward so that all three beams converge at the center of the screen (static convergence). Because the

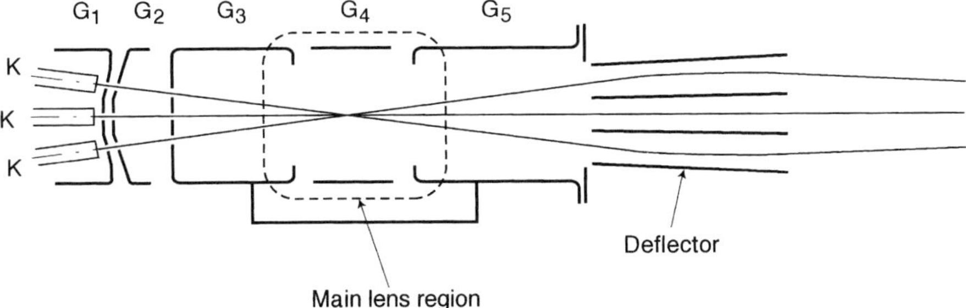

FIGURE 40. Horizontal section of Sony Trinitron gun system with the center beam and tilted outer beams crossing at the center of the large diameter unipotential main lens (G_4) and then directed toward the tube axis by electrostatic deflectors.

amount of deflection required to produce beam convergence is relatively small, the resulting effect on spot distortion is small. (Although not shown in Fig. 40, the voltages for the outer deflection plates may be obtained from a voltage divider provided inside the tube neck.) The same deflection system also allows dynamic convergence to be obtained by applying suitable time-varying voltages (with a parabolic waveform) to the outer electrodes synchronized with the currents applied to the magnetic deflection yoke.

It is important to note that since the main lens has a relatively large diameter compared to the individual main lenses of conventional systems that use separate or unitized in-line guns, the beams are subjected to a relatively small amount of spherical aberration. The diameter of the main lens for 29 mm neck guns may be as large as about 12 mm. In addition to this, the use of a unipotential-type main lens results in a smaller spot size at increased beam currents. In HDTV tubes, which require high brightness and high resolution, this is a great advantage—although the gun structure is somewhat longer and more complicated compared to the conventional unitized in-line gun systems.

Since the initial introduction of the Trinitron gun system it has been incorporated in many tubes developed by Sony such as, for example, wide deflection (114°) tubes of up to 30 V (76 cm) screen size (Yoshida *et al.*, 1973, 1974; Yoshida, 1977), data display tubes with 90° deflection of up to 19 V (48 cm) screen size (Yamada *et al.*, 1980), data display tubes with 20 V × 20 V (51 cm × 51 cm) screen size with a 1:1 aspect ratio (Sudo *et al.*, 1986), 32 V (81 cm) 110° deflection tubes with a flatter-face screen (Inoue *et al.*, 1992), and 34 V (86 cm) 110° deflection wider aspect ratio (16:9) tubes for HDTV (Sakuraya *et al.*, 1991). Recently, however, unitized in-line guns employing main lens diameters effectively enlarged by the asymmetric lens designs discussed in the foregoing are becoming more common than Trinitron guns for data display tubes operating at relatively low beam currents, for example, less than 0.5 mA per beam (see e.g., Kikuchi *et al.*, 1996).

E. Gun Designs for Reducing Deflection Defocusing

As discussed in Section II.E, the astigmatic magnetic field of a self-converging yoke causes a considerable amount of negative astigmatism of the individual beams. This results in a strong overfocusing in the vertical direction, causing the spots to have a large vertical haze, especially at the screen edges and corners. To compensate for such overfocusing, gun designs incorporating additional astigmatic elements (quadrupole lenses) with noncircular apertures are common. These elements produce positive astigmatism to counteract the negative astigmatism caused by the magnetic field. General properties of astigmatic electron-optics in CRT gun designs are described by Hasker (1971, 1973).

As will be discussed, the strength of the astigmatic (quadrupole) field produced may be constant or vary with the application of dynamic voltages in synchronism with the deflection, with the latter providing better compensation for the vertical overfocusing of the deflected spot. This allows a minimum spot size to be attained at every point on the screen.

1. *Quadrupole Lenses*

The ideal quadrupole lens consists of four elongated hyperbolic-shaped electrodes positioned around the z-axis, a cross section of which is shown in Fig. 41 (Klemperer and Barnett, 1971g). It has four planes of symmetry intersecting along the z-axis with an angle $\pi/4$ between them. The lens, centered at $z = 0$, extends in the z-direction. The aperture of the lens $2a$ is defined by the diameter of the hypothetical circle tangential to the four electrodes. We assume one pair of the opposing electrodes 1 and $1'$ to be held at a positive potential $+V$ with respect to ground while the other opposing pair 2 and $2'$ is at a negative potential $-V$. The resulting electric field lines are

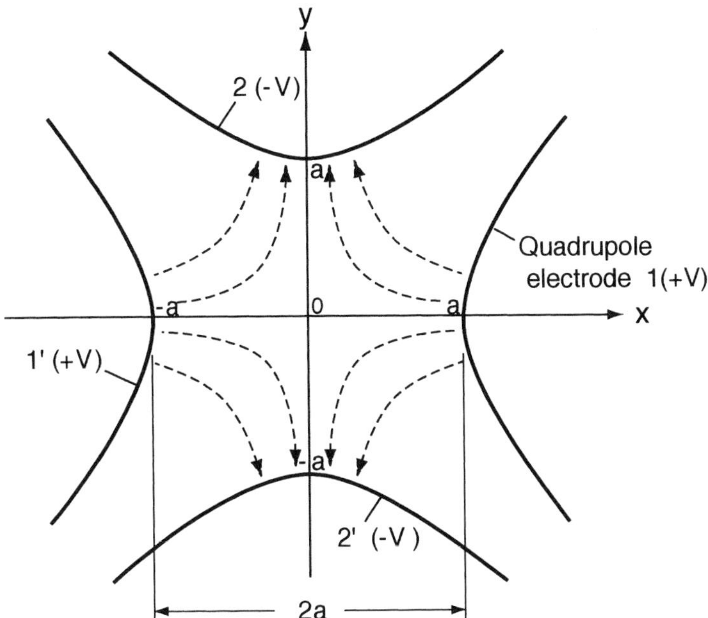

FIGURE 41. Quadrupole lens consisting of four hyperbolic electrodes (1, $1'$, 2, and $2'$) and resulting electric field shown by dashed lines.

shown by dashed lines. It is assumed here that electrons move in the direction of the reader (the positive z-direction).

Those electrons initially in the xz-plane will diverge toward the positive 1 and $1'$ electrodes. However, those electrons initially in the yz-plane will be repelled by the negative potentials on electrodes 2 and $2'$, converging toward the z-axis. Electrons incident in other than the xz- and yz-planes will follow skewed trajectories, moving away from the z-axis in the x-direction (diverging) while approaching the z-axis in the y-direction (converging). In the case of a circular beam this would result in a horizontally elongated and vertically reduced cross section of the beam, whose initial shape is circular. If one wants a converging action in every plane centered on the z-axis two sets of quadrupole lenses must be provided in tandem and oriented 90° with respect to each other so that the divergent plane of the first lens corresponds to the convergent plane of the second lens. However, for reducing deflection defocusing as will be discussed here, the astigmatic property of a single quadrupole lens is utilized.

Since the electrodes are assumed to be sufficiently long, the potential distribution in the quadrupole lens can be expressed as $V(x, y)$ in the xy-plane at any axial position. For electrodes of general shapes, the potential distribution can be expressed in the form of a power series expansion as follows:

$$\frac{V(x, y)}{V} = K_2\left(\frac{x^2 - y^2}{a^2}\right) + K_6\left(\frac{x^6 - \cdots}{a^6}\right) + \cdots \tag{75}$$

where coefficients K_{2n} $(n = 1, 3, \ldots)$ are constants depending upon the form of the cross section of the electrodes (Grivet, 1972b). For an ideal quadrupole lens with hyperbolic electrodes as already mentioned here, the constants K_2 and K_6 are thus 1 and 0, respectively, so that we obtain for this case

$$\frac{V(x, y)}{V} = \frac{x^2 - y^2}{a^2} \tag{76}$$

The field components E_x and E_y are then given by

$$\left.\begin{aligned} E_x &= -\frac{\partial V(x, y)}{\partial x} = -\frac{2x}{a^2}V \\ E_y &= -\frac{\partial V(x, y)}{\partial y} = \frac{2y}{a^2}V \end{aligned}\right\} \tag{77}$$

Thus the electric field strength E produced by the ideal quadrupole lens is proportional to the distance from the z-axis in the radial direction throughout the entire useful area of the lens.

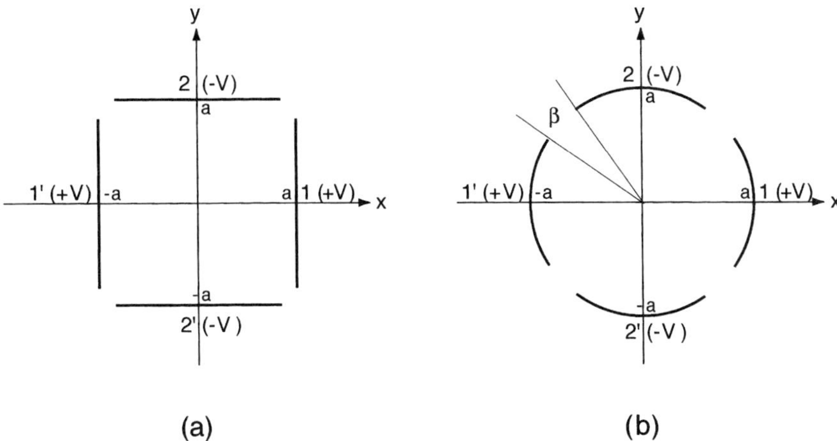

FIGURE 42. Quadrupole lens structures using: (a) planar electrodes; (b) circular electrodes.

In practice, however, electrodes with hyperbolic cross sections are difficult to fabricate and simpler shapes are generally used. Examples of simpler structures with planar and concave circular electrodes are shown in Fig. 42a and b, respectively. The potential distribution of these lenses, despite their changed electrode shape, is still very similar to that of the lens with hyperbolic electrodes of Fig. 41, which is considered ideal, particularly in the region near the z-axis. As already cited in Grivet (1972b), the lens with planar electrodes has 1.037 and 0.009 for the coefficients K_2 and K_6 in Eq. (75), respectively, while the lens with the circular electrodes has coefficients $K_2 = 1.27\ (\sin\beta)/\beta$ and $K_6 = 0.042\ (\sin 3\beta)/3\beta$, where β is the angle subtended by the gap between electrodes as shown in Fig. 42. If one chooses $\beta = \pi/3$, K_2 and K_6 are found to be 1.05 and 0, respectively.

Other forms of quadrupole lenses useful for CRT gun designs, such as shown in Fig. 43, can be constructed by using two opposing plates at a right angle to the z-axis with noncircular apertures on one or both of them and a potential difference applied between them. In the structure shown in Fig. 43a, for example, a vertical slot on the 1st plate is aligned with a circular aperture on the 2nd plate, while in Fig. 43b a vertical slot on the 1st plate is aligned with a horizontal slot on the 2nd plate, respectively. It should be noted that in these "quadrupole" lenses the fields produced in the lens region are similar to those produced by the lenses of Figs. 41 and 42 but without requiring four electrodes.

Actual quadrupole lenses used in CRT guns are usually very short in length. Because the required lens strength is relatively weak, a short lens can be

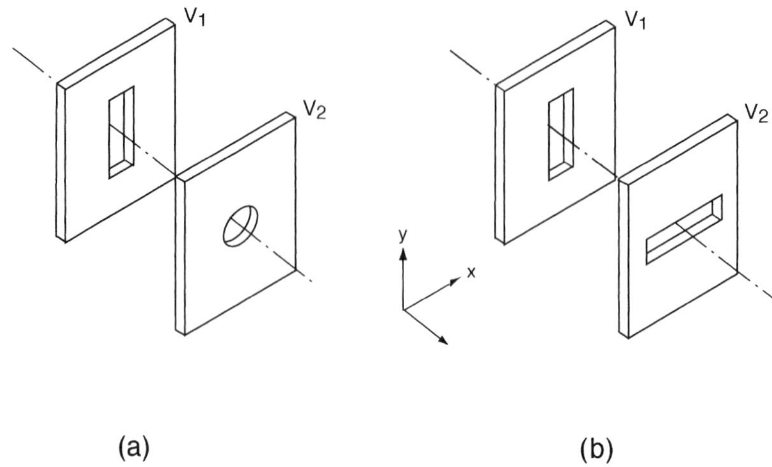

FIGURE 43. Structures for producing quadrupole field with two planar electrodes using: (a) vertical slot and circular aperture; (b) vertical slot and horizontal slot.

adequate, although the potential distribution as derived in the preceding may only be valid in the region near the lens center ($z = 0$). In addition, the principal plane of the lens coincides closely with the lens center and the absolute value of focal lengths for the convergent plane (f_x) and the divergent plane (f_y) are nearly equal. The focal lengths or lens strength of these quadrupole lenses can be determined from trajectory calculations using 3D computer programs (van Oostrum, 1985).

2. *Static Quadrupole Systems*

Quadrupole lenses with fixed electrode voltages whose strength is constant (referred to as static quadrupole lenses) can be made part of the main lens electrodes (already discussed here) and/or part of the triode electrodes to be discussed in what follows. The triode design with a quadrupole lens at G_1 is sometimes called the G_1-quadrupole and that at G_2 the G_2-quadruple. Although both systems are used to obtain smaller spot sizes by reducing the astigmatic deflection defocusing, they may cause somewhat different effects on the spot characteristic and their choice in practical gun design depends on the application of the tube.

a. G_1-Quadrupole. The G_1 quadrupole was first introduced into the 30AX tube of Philips (Barten and Kaashoek, 1978). Its structure is shown in Fig. 44 for one of the three electron beams. As shown, G_1 consists of two thin plates, G_{1A} facing the side of the cathode and G_{1B} facing G_2, the former having a square hole and the latter a vertical slot instead of the usual circular

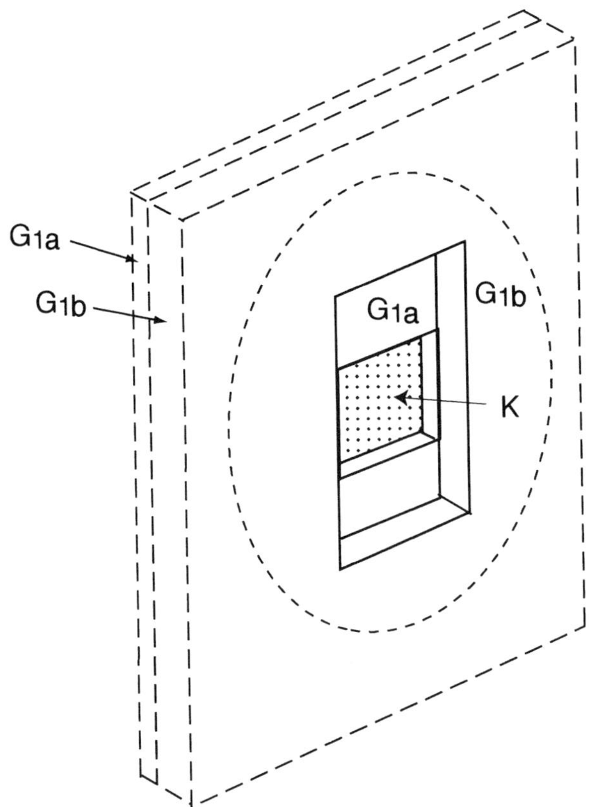

FIGURE 44. Structure of G_1-quadrupole (showing one of three G_1 apertures).

aperture. The sides of the G_{1A} aperture are approximately 0.7 mm long and its thickness is about 0.1 mm. The sides of the G_{1B} aperture are about 0.7 mm $\times$ 2 mm and its thickness is about 0.2 mm. The resulting nonsymmetrical G_1 structure forms a quadrupole lens field in front of the cathode which produces positive astigmatism of the beam while restricting the vertical dimension of the emitting area of the cathode; this is explained in what follows.

The horizontal and vertical sections of G_1 are shown in Fig. 45a and b, respectively, together with schematic beam trajectories (neglecting the prefocus lens between G_2 and G_3). As indicated, in the horizontal section, due to the greater thickness of G_1, the curvature of the equipotential surfaces caused by the penetration of the G_2 potential is greater compared to that in the vertical section. This causes two separate crossovers along the z-axis, first at

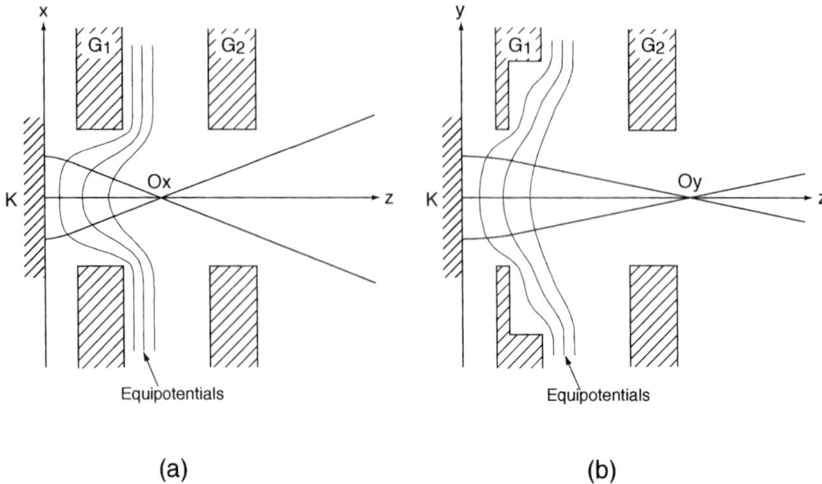

FIGURE 45. Equipotentials and beam trajectories in G_1-quadrupole (schematic); (a) horizontal; (b) vertical section. Prefocus lens is neglected.

O_x in the horizontal and then at O_y in the vertical plane. Thus a greater distance from the horizontal crossover O_x to the main lens (not shown) results. As a consequence, the beam in the horizontal plane is subject to a stronger convergence by the main lens, which is assumed to be nonastigmatic. Therefore, if the focus is adjusted for the horizontal direction at the screen center as is normally done in the self-converging system, the spot is underfocused in the vertical direction. This vertical underfocusing (positive astigmatism) serves to counteract the vertical overfocusing of the spot (negative astigmatism) during deflection caused by the magnetic field.

In addition, the beam-diverging angle after the crossover is less in the vertical plane than in the horizontal plane due to the greater distance of the vertical crossover O_y from the cathode. This results in a vertically reduced cross section of the beam while it traverses the magnetic field, reducing the vertical haze of the deflected spot. Another feature of the G_1-quadrupole is that as a result of its location close to the cathode the quadrupole lens action continues to be effective until a high beam current is reached. The same is not true for the G_2-quadrupole (mentioned in what follows), whose action decreases with increasing beam current. This characteristic makes the G_1-quadrupole suitable for use in TV tubes operating at high beam currents.

A modified G_1-quadrupole may be used to incorporate the ART effect (see Section III.C.2) where instead of a square hole a vertically shortened rectangular hole is used for G_{1A} (Gerritsen and Barten, 1987). This causes the split crossovers in the horizontal and vertical directions to coincide again on the axis. In the ART design the crossover must be placed at an axial position predetermined by the ART prefocus lens condition as already discussed above. Although the modified G_1-quadrupole still provides a vertically reduced cross section of the beam while it passes through the magnetic field of the yoke, it no longer produces a sufficient amount of positive astigmatism, thus making necessary, for instance, an additional positive astigmatism in the main lens design. In another gun design employing the G_1-quadrupole, an additional astigmatic element with noncircular holes may be used in the unipotential lens section of a lo-uni-bi type hybrid gun (Garnier *et al.*, 1994).

b. G_2-Quadruple. The commonly used G_2-quadrupole structure is shown in Fig. 46. On the surface of G_2 facing G_3, a small recessed area in the shape of a rectangle is provided around the circular G_2 aperture. The area of the recessed portion is of the order of 1.5 mm $\times$ 1 mm and its depth is 0.25 mm. The diameter of the G_2 aperture is 0.4 mm. Horizontal and vertical sections of G_2 are shown in Fig. 47a and b, respectively. With the vertical opening of the recessed portion narrower than the horizontal opening, this produces a stronger convergence of the prefocus lens in the y-direction than in the x-direction as a result of the potential difference between G_2 and G_3. The vertically reduced cross section of the beam in

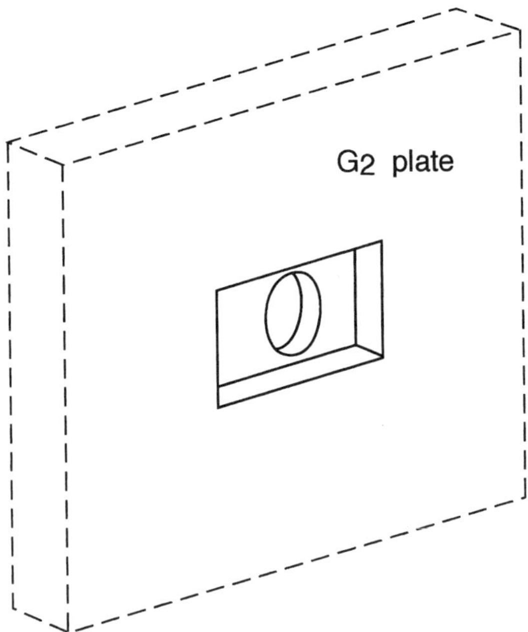

FIGURE 46. Structure of G_2-quadrupole (showing one of three G_2 apertures).

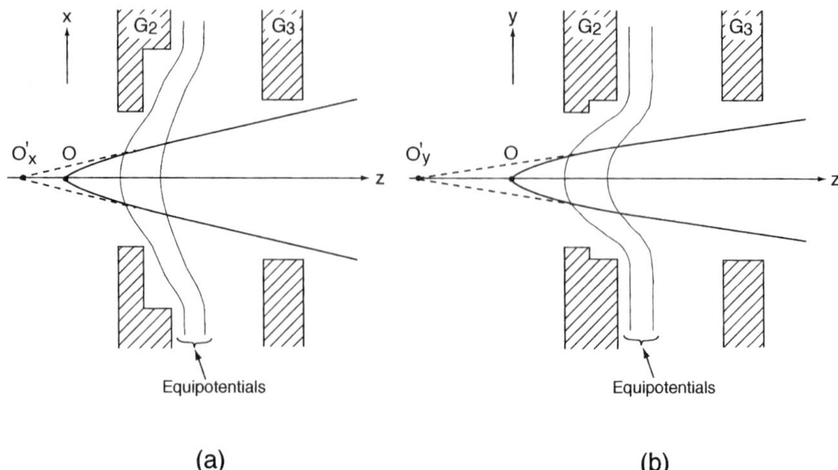

FIGURE 47. Equipotentials and beam trajectories in G_2-quadrupole (schematic); (a) horizontal; (b) vertical section.

turn reduces the vertical haze of the spot when the beam is magnetically deflected.

Unlike the G_1-quadrupole already mentioned, the G_2-quadrupole produces negative astigmatism of the beam, which is not desirable. As shown in Fig. 47, because of the astigmatic prefocus lens, the virtual crossover is split so that O'_y in the vertical plane is farther to the left of O than O'_x in the horizontal plane. This results in a vertically stronger convergence of the beam by the main lens and a vertically overfocused spot at the screen (negative astigmatism) because, as previously mentioned, the focus is adjusted for the horizontal direction. This negative astigmatism must be offset by the main lens design, usually by adding a certain amount of positive astigmatism.

The prefocus lens action generally decreases with increasing beam current since the crossover moves closer to the lens as the current is increased. For this reason, the G_2-quadrupole is limited to use in guns operating at low beam currents and requiring uniform spot sizes over the entire screen, although its fabrication is much easier than that of the G_1-quadrupole. The G_2-quadrupole is thus widely used in guns for data display tubes (Hirabayashi *et al.*, 1982; Chen, 1982; Suzuki *et al.*, 1983; Watanabe *et al.*, 1985; Nakanishi *et al.*, 1986).

Another possibility is to place the G_2-quadrupole lens on the G_1 side of G_2. In this way a similar effect can be obtained as with the G_1-quadrupole mentioned in the foregoing (Chen and Hughes, 1980; Yun *et al.*, 1994). However, the effect is somewhat weaker, especially when the beam current is increased.

In both the G_1- and G_2-quadrupole systems mentioned here, although the vertical dimension of the deflected spot can be reduced, the spot at the screen center remains underfocused in the vertical direction, causing it to be vertically elongated. In the design of the total gun system the spot size chosen represents a compromise between the center and corners to achieve the most uniform focus over the entire screen. It is thus not possible with a static quadrupole system to obtain the minimum spot size at each point of the screen.

3. *Dynamic Quadrupole Systems*

By modulating the quadrupole lens in synchronism with the deflection current of the yoke the correct amount of positive astigmatism of the beam can be produced at each point of the screen to neutralize the negative astigmatism caused by the magnetic field (Ishida *et al.*, 1986). In guns designed for this purpose, in addition to the quadrupole lens the main lens is also modulated in accordance with the deflection. This system is sometimes called the dynamic astigmatism and focus or DAF system. Although the principle has been known for some time (Barten, 1982), practical guns of this type with a unitized structure have been developed only relatively recently (Ashizaki *et al.*, 1986; Yamane *et al.*, 1986).

The DAF gun is basically a bipotential gun. However as shown in Fig. 48 the focusing electrode G_3 is split into two sections G_{3-1} and G_{3-2}. On their surfaces facing each other three rectangular apertures are provided; they are oriented vertically and horizontally, respectively. Each of the three opposing aperture pairs corresponds to the quadrupole lens shown in Fig. 43b, with a quadrupole field being produced when a potential difference is applied between G_{3-1} and G_{3-2} of the quadrupole lens (QL). The main lens ML created between G_{3-2} and G_4 usually produces a nonastigmatic field.

As shown in Fig. 48, the focusing voltage V_{foc} is applied to G_{3-1}, and the anode voltage V_a is applied to G_4. In operation the dynamic voltage v_{dyn} is superimposed on V_{foc} and applied to G_{3-2} in synchronism with the beam deflection. With an increasing positive potential on G_{3-2} with respect to G_{3-1}, the required quadrupole field produced at QL simultaneously weakens the ML. In what follows, the operation of the DAF gun is discussed in greater detail.

A schematic of the lens system of the DAF gun is shown in Fig. 49 where (a) is the horizontal and (b) is the vertical section. The beam starting from the object point O′ (virtual crossover) passes through the QL and then the ML. In the absence of any dynamic voltage the quadrupole lens is inactive, with the main lens focusing the beam at the screen center as shown by the solid lines. With the application of the dynamic voltage the quadrupole lens becomes

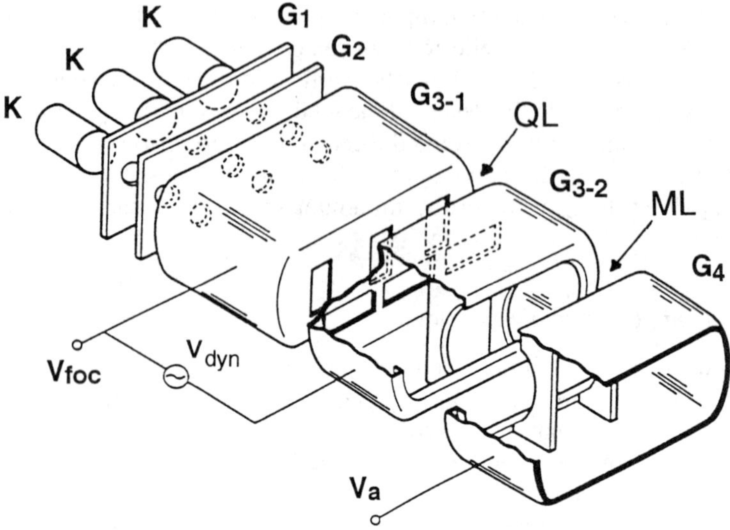

FIGURE 48. Structure of dynamic astigmatism and focus (DAF) gun system [reproduced by permission from Suzuki, H., *et al.* (1986). Progressive-scanned 33″ 110° flat-square color CRT, *Proc. SID*, **28**(4), 403–407; Courtesy Society for Information Display].

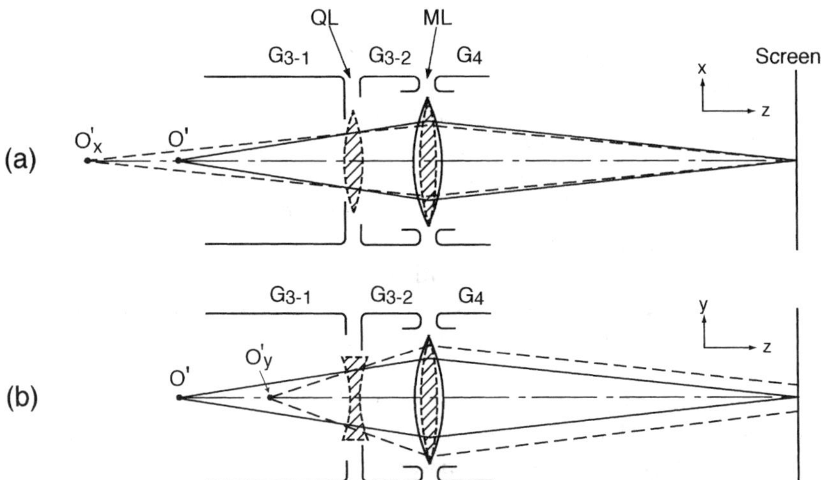

FIGURE 49. Optical models for DAF gun; (a) horizontal and (b) vertical section, where QL and ML represent the quadrupole lens and main lens, respectively. (The beam is assumed to reach the screen without deflection.)

active (converging the beam horizontally and diverging it vertically), while at the same time the main lens action is weakened. These lens actions are indicated by the dashed lines. In the figure, for the sake of simplicity, the effect of the magnetic deflection field on the beam is not considered.

Extending the rays back from QL to the axis as shown by the dashed lines, virtual images of the object O′ result; O'_x in the horizontal and O'_y in the vertical plane. These serve as new objects for ML. With increasing dynamic voltage and resultant increase in the quadrupole lens strength, O'_x moves back while O'_y moves forward. If the main lens strength were constant, an increasing object to main lens distance would cause a horizontally overfocused spot at the screen. In the DAF gun this is compensated for by a simultaneous weakening of the main lens so that the horizontal focusing is not disturbed by the dynamic voltage. As already discussed, in the self-converging system the horizontal focusing is maintained by the magnetic field. On the other hand, the spot is vertically underfocused (positive astigmatism) for two reasons: the reduced object to main lens distance and the weakening of the main lens. This serves to compensate for the vertical overfocusing of the deflected spot (negative astigmatism) caused by the magnetic field.

The required dynamic voltage v_{dyn} can be determined from the spot astigmatism V_{ast} caused by the magnetic field. As defined by Eq. (73) on p. 311 in Section II.E, the spot astigmatism can be defined by the difference between the horizontal and vertical focusing voltages, namely, $V^H_{foc} - V^V_{foc}$. If the astigmatism were fully compensated for, the resultant focusing voltage for the nonastigmatic beam would be nearly equal to its average value, since the beam would have a common image surface between the horizontal and vertical image surfaces. With the average focusing voltage increasing during deflection over the focusing voltage V_{foc} of the center spot due to deflection defocusing, this increase should give the required dynamic voltage and we obtain

$$v_{dyn} \approx \frac{V^H_{foc} + V^V_{foc}}{2} - V_{foc} \tag{78}$$

As mentioned, the horizontal focusing voltage is constant over the entire screen in the self-converging system so that V^H_{foc} equals V_{foc}. Consequently, we obtain

$$v_{dyn} \approx -V_{ast}/2 \tag{79}$$

Specifically, the required dynamic voltage equals approximately one-half the spot astigmatism with reversed sign.

The actual spot astigmatism depends on the deflection angle and type of gun used. In 90° tubes with a 28% focus hi-bipotential (rotation-symmetric) gun operating at an anode voltage of 25 kV, a maximum spot astigmatism of about −1200 V was measured at the screen corners, while with 110° deflection the same type of gun operated at 30 kV, a spot astigmatism as large as −4000 V was measured (Suzuki *et al.*, 1987). According to Eq. (79), the maximum required dynamic voltages for these tubes are found to be 600 V and 2000 V, respectively. In practice, however, about half these values are usually sufficient to obtain good focus over the entire screen. This allows parabolic waveforms in synchronism with the deflection with peak values of 300 V and 1000 V, respectively, to be used for these tubes.

The DAF system has proven very useful for reducing the astigmatic deflection defocusing while maintaining self-convergence. Various guns using this principle have become common in high-resolution color CRTs used for data display and HDTV (Shirai and Fukushima, 1987; Suzuki *et al.*, 1987; Katsuma *et al.*, 1988, Shirai *et al.*, 1989; Bloom and Hockings, 1989; Chen and Gorsky, 1989, 1990; Sugawara *et al.*, 1995). Although the spot performance achieved by these guns is basically very similar, the quadrupole lens structures vary considerably, depending on the manufacturer. Variations of the quadrupole lens structures shown in Figs. 42 and 43 are generally used. A dynamic astigmatism and focus correction system has also been introduced in the Sony Trinitron (Ichida *et al.*, 1988). Because the dynamic voltages required in these guns are still very high (from the point of view of circuit design), special gun designs allowing a reduced dynamic voltage have been used in some cases (Shirai *et al.*, 1991; Iguchi *et al.*, 1992).

Although a minimum spot size at each point of the screen can be achieved by the DAF gun, the spot shape still remains considerably distorted during deflection, horizontally elongated, and vertically shortened. Aside from a loss in horizontal resolution, this can cause undesirable moiré in the raster pattern (see the article by Yamazaki, "Design and Performance of Shadow-Mask Color Cathode Ray Tubes" preceding this article). The spot distortion is partly due to the fact that the magnification of the total system of the DAF gun and deflection yoke increases in the horizontal direction and decreases in the vertical direction with beam deflection (Gerritsen and Sluyterman, 1989). As already discussed, magnification is a major factor in determining the spot size at low currents (magnified crossover).

The spot shape can be made rounder by using additional quadrupole lenses adjacent to the triode section of the DAF gun and driving them with the same dynamic voltage. This quadrupole lens causes the divergence angle of the beam to be increased in the horizontal direction and reduced in the vertical direction before entering the DAF quadrupole lens, thus reducing the difference in total magnification between the horizontal and vertical directions. This gun, sometimes called the DQ-DAF system because of the use of double-stage quadrupole lenses, is useful for data display tubes that operate at low beam currents and require a rounder spot shape (Suzuki *et al.*, 1993; Konosu, 1995; Kikuchi *et al.*, 1996; Natori *et al.*, 1997). However, in the design of such guns, special consideration is required to prevent the dynamic voltage from affecting the beam convergence (Ueda *et al.*, 1995).

At increased beam currents (several milliamperes), however, it is very difficult to obtain a rounder shape of the deflected spot when a self-converging yoke is used (Alig, 1993). In wide-angle deflection tubes particularly, such as 110°, a considerable elongation of the spot in the horizontal direction (by nearly a factor of 2 with respect to the center, together with a reduced vertical size) occurs at screen edges and corners (Gerritsen and Sluyterman, 1989), which even the DQ-DAF system can not improve (Suzuki *et al.*, 1992).

It is of interest to note that the spot distortion is considerably less with a uniform-field (nonself-converging) yoke such as used in tubes with the earlier delta gun arrangement in conjunction with dynamic convergence correction. In view of this in some applications requiring higher brightness and higher resolution over the entire screen, uniform-field yokes with dynamic convergence may be used in combination with in-line guns to obtain better convergence (Ando *et al.*, 1985; Kuramoto *et al.*, 1989). As previously discussed, even a uniform-field yoke causes a certain amount of astigmatism of the deflected beam. However, this astigmatism can be fully compensated for by using the DAF principle together with specially designed dynamic convergence components (Gerritsen and Sluyterman, 1989; Suzuki *et al.*, 1990).

F. Use of Phase-Space Diagrams for Evaluating Spot Sizes

A useful method for studying the properties of an electron beam in a particular electron-optical system is to use a phase-space diagram. Based on knowledge of the beam characteristics at some particular point, such diagrams enable one to obtain a better understanding of the effects produced at some other point by changes in the electrode system (see also Washino *et al.*, 1979). For example, as discussed here, in the study of the triode system such as the ART discussed in Section III.C.2, the phase-space diagram allows one to understand the effect of the prefocus lens on the cancellation of spherical aberrations.

Such diagrams show the relation between the position coordinate and radial momentum of each electron of a beam as it passes through a plane perpendicular to the gun axis (z-axis). In an axially symmetric system the position of an electron can be expressed by its radial position r from the gun axis and its corresponding radial momentum $m\,dr/dt$, where m is the mass of the electron and dr/dt is its radial velocity. It is possible, however, to use the angle of divergence or slope $r'(= dr/dz)$ of the electron ray with respect to the gun axis instead of the momentum. This is legitimate in the regions of the gun where there is no axial acceleration of the electrons (field-free regions) because dr/dt can be expressed by $(dr/dz)\ dz/dt$ and dz/dt is a constant in such regions.

In the phase space diagram, the slope r' of each electron ray of the beam is plotted against its radial position r at a properly chosen plane perpendicular to the axis. The beam of electrons can thus be represented by a cloud of points on the diagram because the electrons passing through the plane at some position r from the axis may have various values for the slope r' with respect to the axis. This is due to the fact that the electrons may have an angle between 0 and 90° with respect to the normal due to the thermal spread in their transverse velocities. The entire electron beam can thus be represented by an area in the phase-space diagram called the emittance diagram whose contour contains information on the diameter and divergence angle of the beam including the effects of thermal velocity, space-charge repulsion, and geometrical aberration. Such an emittance diagram can be drawn at any plane perpendicular to the axis, having a particular shaped contour in accordance with the axial position.

As an example the phase-space diagram can be applied to the virtual image of the crossover (virtual crossover). As discussed in Section II.B, the electrons emitted from the cathode are first converged by the cathode lens (CL) to a crossover O, which is then partially converged by the prefocus lens (PL) as shown in Fig. 1 (p. 273). In the ideal case where the effects of thermal velocity spread, space-charge repulsion and spherical aberration in both the cathode and prefocus lenses are neglected, the virtual crossover would be a single point on the axis, as indicated by O′, from which electron rays diverge with various angles entering the field-free space of G_3. Shown by the emittance diagram such a virtual crossover would be a straight line along the r'-axis.

Taking into account the spherical aberration in both the cathode and prefocus lenses, the backward extensions of the rays (from the field-free space beyond the prefocus lens) do not converge to a single point but cross the axis at different points as shown by the dashed lines in Fig. 50a. This results in a circle of least confusion, which defines the virtual crossover due to spherical aberration. Therefore, the emittance diagram for the virtual crossover, that is, the slope r' versus position r relationship, is now a curved line, as shown in the graph of Fig. 50b, reflecting the spherical aberration. In the diagram the maximum value of r represents the beam radius and the maximum value of r' represents the beam divergence angle at the virtual crossover.

Aside from the emittance curve already mentioned, a phase-space acceptance curve for the main lens can be drawn as explained in what

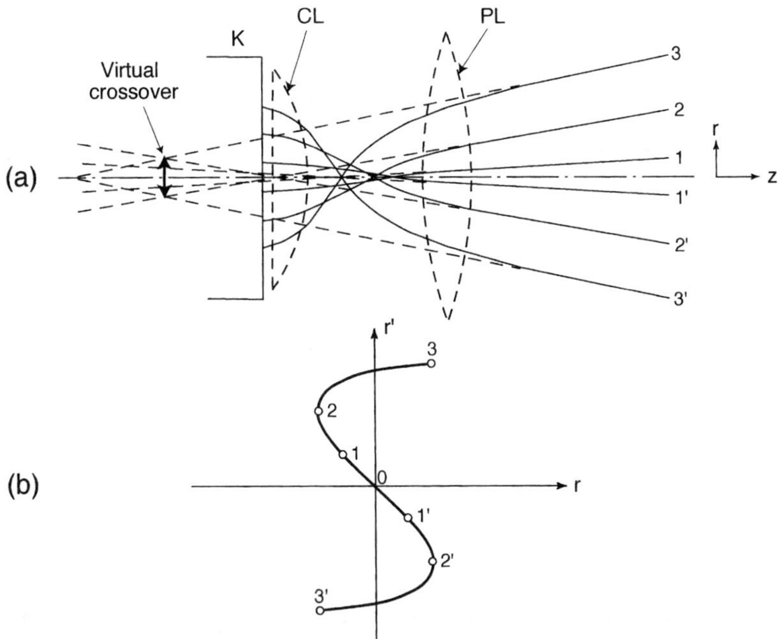

FIGURE 50. Emittance phase-space diagram at virtual crossover neglecting thermal and space-charge effects; (a) virtual crossover defined by the circle of least confusion; (b) S-shaped emittance curve relating slope r' with radial position r.

follows. By making use of the acceptance curve in conjunction with the emittance curve, the spot size at the screen can be determined, even though this spot size neglects the effect of space-charge repulsion in the drift space between the main lens and the screen.

Figure 51a shows the ML together with the electron trajectories. Assuming spherical aberration of the main lens, if the rays emerging from the main lens are to converge to a point at the screen center ($r = 0$), it is required that in the object space (before entering the main lens) the rays cross the axis at different points depending on their slopes. As shown, paraxial rays close to the gun axis cross the axis farthest from the main lens while wide angle rays (or marginal rays) cross the axis nearest to the main lens. We assume an object plane at a place beyond the paraxial point as indicated by a dashed line in the figure labeled object plane. In the object plane, the slope r' of each ray is plotted as a function of position r, resulting in a curve such as that shown in Fig. 51b. Alternatively, this curve could be considered the acceptance curve that the rays would have to fall on in order to converge to a single point at the screen.

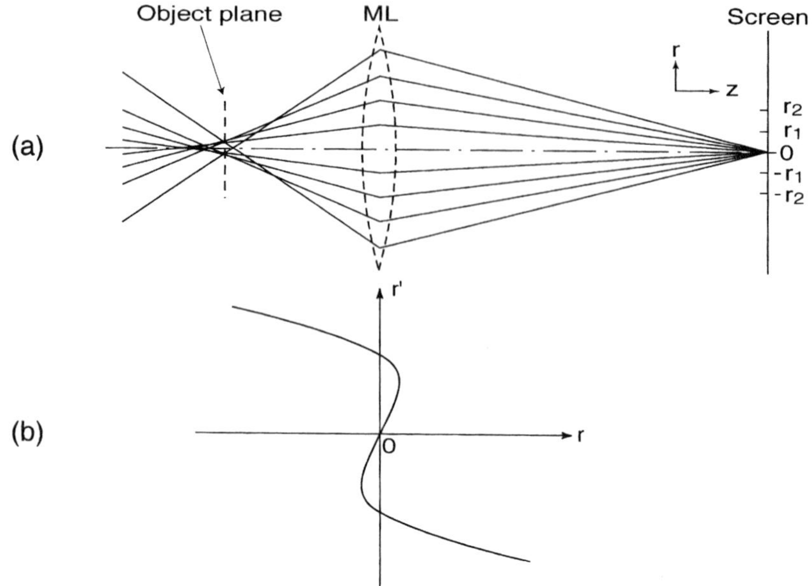

FIGURE 51. Acceptance phase-space diagram for the main lens; (a) electron rays converging at a point on the screen after focusing by main lens, and (b) acceptance curve obtained at object plane which relates slope r' with radial distance r.

The shape of the acceptance curve would depend on the axial position of the object plane chosen.

Similarly, acceptance curves can be obtained for other rays to converge at off-axis points on the screen such as r_1, $-r_1$, r_2 and $-r_2$ as indicated in Fig. 51a. These acceptance curves are shown in the graph of Fig. 52 by the thin-line curves using the same notation. The curve with the rays for the on-axis (0) point previously mentioned is also included. Although all the acceptance curves are very similar in shape they cross the r-axis of the phase-space diagram at different points.

Also shown in the same figure is the forementioned emittance curve indicated by the thick line. The emittance curve is just fitted into the space enclosed by a pair of the acceptance curves corresponding to r_1 and $-r_1$, being tangent to these acceptance curves both at its ends and intermediate parts where the radial position of the curve is maximum. This matching between the emittance and acceptance curves results from the fact that the object plane shown in Fig. 51a was properly positioned with respect to the virtual crossover where the emittance curve was plotted as shown in Fig. 50. As a result of this, all rays defined by the emittance curve fall, after focusing

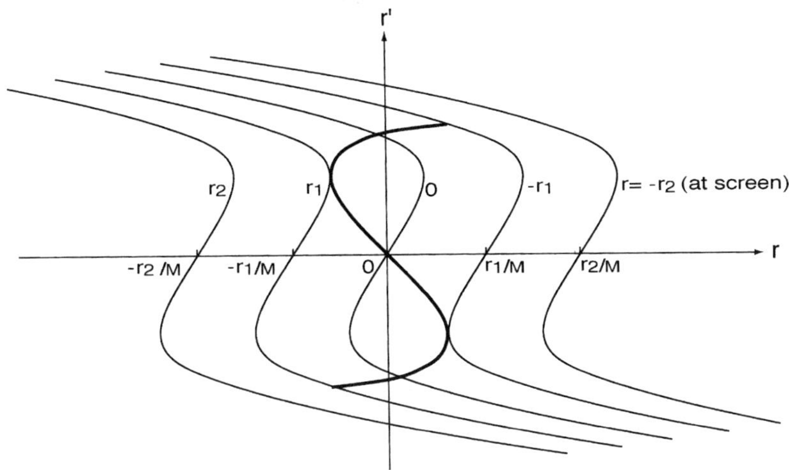

FIGURE 52. S-shaped emittance curve enclosed by the pair of acceptance curves for r_1 and $-r_1$, indicating that all rays on the emittance curve fall within the diameter $2r_1$ on the screen, resulting in a spot size of $2r_1$.

by the main lens, within the diameter $2r_1$ on the screen. This equals the circle of least confusion and the diameter of the minimum obtainable spot.

As discussed, by fitting the emittance curve within a pair of the acceptance curves that are as close as possible, the spot size can be evaluated. It should be noted that the dimensional scale on the phase-space diagram is $1/M$ of that on the screen where M is the magnification of the main lens. Therefore, the resulting spot size on the screen is thus given by the distance between the acceptance curves on the phase-space diagram multiplied by M.

Actual emittance diagrams include the effects of thermal velocity spread and space-charge repulsion in the triode section in addition to the spherical aberration. Such emittance diagrams can be obtained experimentally using special tubes with a diaphragm at the exit of the final anode of the guns where an array of a large number of pinholes is provided (Bijmer, 1981; Ikegami *et al.*, 1986). Acceptance diagrams of the main lens can be obtained by means of electron-optical ray-tracing programs in the Laplace field where the space-charge repulsion effect is neglected.

An example of the use of the phase-space diagrams is their application to the ART lens system discussed in Section III.C.2. Corresponding to the electron trajectories of a non-ART gun and an ART gun shown in Figs. 27a and b (p. 327), phase-space diagrams for a beam current of 4 mA are shown in Figs. 53a and b, respectively. Both the emittance and acceptance diagrams shown here are those drawn for a plane perpendicular to the gun axis close to the virtual crossover, which serves as the object of the main lens, so that an

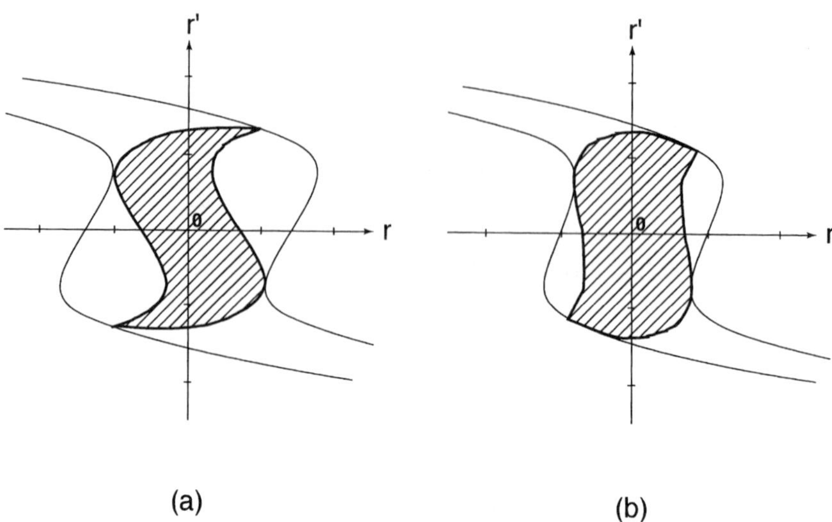

FIGURE 53. Examples of actual emittance diagrams; (a) non-ART gun; (b) ART gun. The latter has a curved shape fitting within a pair of acceptance curves closer together and resulting in a smaller spot size.

optimum-focused spot size can be obtained at the screen. The emittance curves are results of the pinhole measurements and the acceptance curves are based on computer calculations of trajectories for the main lens without taking into account space-charge repulsion.

As indicated, the emittance diagrams of Fig. 53 are enclosed areas instead of the curved lines such as in Figs. 50 and 52. This results from the fact that experiments include the effects of thermal velocity spread and space-charge repulsion in the triode section. The S-shaped contour of the diagrams of Fig. 53 indicates the effect of spherical aberration from the cathode and prefocus lenses. It can be seen that the marginal rays and some intermediate rays of the contour come in contact with the acceptance curves of the main lens. This is the condition, as already mentioned, for a minimum spot size attainable at the screen. The spot size is then given by the distance on the r-axis between the two acceptance curves multiplied by the magnification of the main lens M. Thus, by fitting the emittance diagram within the closest possible pair of acceptance curves, the resulting spot size at the screen can be evaluated.

As shown in Fig. 53b, the action of the ART prefocus lens transforms the emittance diagram into a more curved shape that fits within a pair of the main lens acceptance curves which are closer together, indicating that the spot size at the screen is reduced. Moreover, the closest fit is no longer determined by the extreme rays but by more intermediate rays. This is another way of saying that the position of the inner and outer electron paths are interchanged as seen in the beam trajectories of Figs. 27b and 28b. Further detailed discussions of the ART effect based on the phase-space analysis are given in the references (see van Gorkum and van den Broek, 1985; Spanjer *et al.*, 1989, 1991).

IV. Cathodes

A. *Introduction*

Modern CRTs almost always employ conventional oxide-coated cathodes. These use an electron emitter consisting of the oxides of alkaline-earth metals such as BaO, SrO and CaO deposited on a supporting metal plate of high-purity nickel. Although the oxide cathode is now considered a mature technology, efforts to improve it continue with the goals of obtaining higher current density and longer life.

In recent years dispenser cathodes have come into use for tubes requiring an increased current density. Of these the impregnated cathodes (referred to as I-cathodes) are most common. The emitter of this cathode consists of a pellet of porous tungsten impregnated with a mixture of oxides such as BaO, CaO and Al_2O_3. Although this type of cathode can operate at a higher cathode loading and has longer life compared to conventional oxide cathodes, it is more complicated and expensive to manufacture and requires operation at a higher temperature.

Both types of cathodes will be discussed in detail in Sections IV.C and IV.D, respectively, which follow the general discussion of cathodes in Section IV.B, which follows. In Section IV.E, the cathode assembly, which integrates the cathode and its heater into the electron gun structure, will be discussed as well as its influence on the overall performance of the gun.

B. *General Description*

1. *Typical Structure*

Figure 54 shows the basic structure of a typical cathode (oxide cathode) used in CRTs. The emitter usually consists of a layer about 70 μm thick of the triple oxides (Ba,Sr,Ca)O, although double oxides (Ba,Sr)O may also be used. Of the oxides, barium oxide (BaO) plays an essential role in electron emission as discussed later. The cap-shaped metal base (substrate) is usually 1.0 to 2.0 mm in diameter, consisting of high purity Ni (99.99%), which contains a closely controlled small quantity of activator or reducing agents such as Mg, Si, and Al. The operating temperature of the oxide emitter is usually

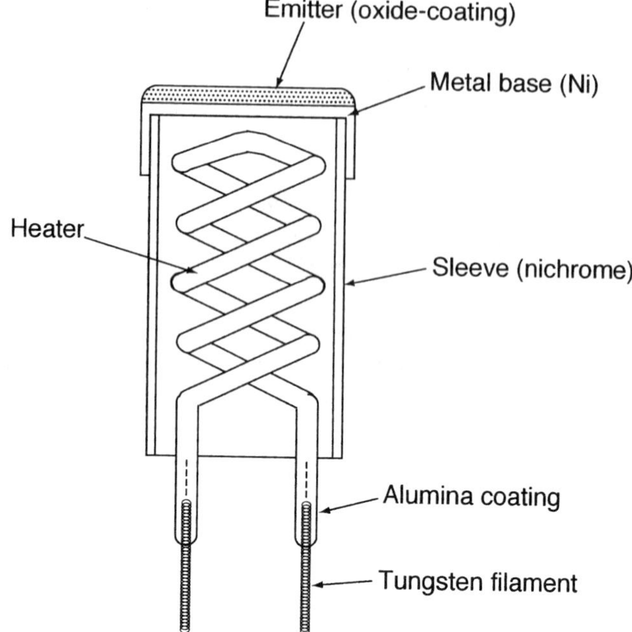

FIGURE 54. Typical structure of the cathode and its heater (oxide cathode).

maintained in the range of 700–800°C by means of a heating element in the form of an alumina-coated tungsten filament inserted in the nichrome cylinder (sleeve) whose top end is covered with the oxide-coated metal cap. The alumina coating of the filament serves to insulate the heating element from the remaining cathode structure electrically.

2. *Emission Properties*

The emission current density of electrons emitted from a thermionic cathode is given by the Richardson-Dushman equation. This was derived theoretically based on the Fermi-Dirac energy distribution of electrons in a metal (see, e.g., Herrmann and Wagener, 1951b). The Richardson-Dushman equation is of the form:

$$j_s = A_0 T^2 \exp(-11600\phi/T) \tag{80}$$

where j_s is the emission current density in A/cm^2; T is the temperature in °K; ϕ is the work function in electron volts (eV); and A_0 is a constant of proportionality. Although the theoretical value of A_0 is 120 A/cm^2 deg^2, in practice it is generally much smaller. Since it appears that little can be done to increase the constant A_0, if one wishes to achieve an increase in the current density j_s at a given temperature, the work function ϕ must be reduced.

The emission current density j_s indicates the maximum emission capability of a cathode and is referred to as the temperature-limited current density. It is also called the saturation current density since it does not increase above some sufficiently high anode voltage. In order to calculate the saturation current density as a function of temperature with the aid of Eq. (80), the values of both A_0 and ϕ are required for each particular emitter material. These quantities can be determined experimentally as described in the references (see, e.g., Herrmann and Wagener, 1951b; Beynar and Nikonov, 1964; Medicus, 1979; Cronin, 1981; Hasker and van Dorst, 1989).

In the emission equation the saturation current density j_s is influenced most by the value of the exponential term, making it a very sensitive function of the work function ϕ and the temperature T. For example, at a temperature of 1000°K only a 0.2 eV decrease in work function from 2.0 to 1.8 eV increases the emission 10 times. Similarly, for a work function of 2.0 eV, increasing the temperature from 1000 to 1100°K also increases the emission 10 times. Unfortunately, the higher the temperature, the higher the evaporation rate of the emitter material, resulting in a reduction in the life of the cathode. Therefore, a compromise must be found in the operating temperature between obtaining high emission and long life.

3. *Modes of Operation*

In most applications of electron tubes the current is space-charge-limited. In this mode of operation, as discussed in Section II.C.2.*d*, because of a potential minimum created in front of the cathode by the thermal velocities of the emitted electrons, only electrons with sufficiently large initial velocities able to pass through the minimum can serve as the beam current of an electron gun. The resulting current density j drawn from the cathode reaching the anode is given by Eq. (33) of Section II. As shown by this equation, the current density j reaching the anode of a planar diode varies only slightly with the cathode temperature T.

If the anode voltage is increased sufficiently so that the current density j reaching the anode becomes equal to the saturation current density j_s at a given cathode temperature, the potential minimum disappears and the field strength at the cathode surface becomes zero.

With further increase of the anode voltage, an accelerating field develops at the cathode surface, which serves to reduce the effective work function ϕ, resulting in an increase in the saturation current density j_s beyond that given

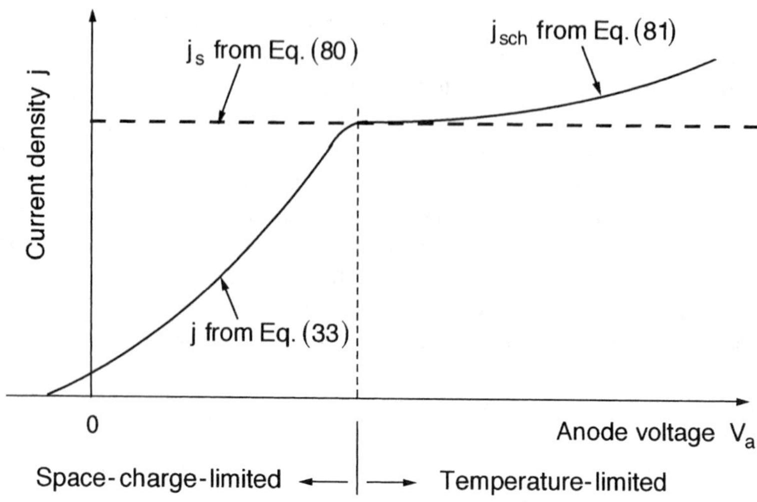

FIGURE 55. Schematic representation of current-voltage characteristic of planar diode.

by Eq. (80). This is known as the Schottky effect. The resulting current density, denoted by j_{sch}, is given by the following equation:

$$j_{\text{sch}} = j_s \exp(4.4 E_c^{1/2}/T) \tag{81}$$

where j_{sch} is in A/cm^2 and E_c is the accelerating field strength at the cathode surface in V/cm (see, e.g., Spangenberg, 1948a). The current density versus anode voltage relationship in the space-charge-limited and temperature-limited (saturation) modes are shown schematically in Fig. 55.

During operation the saturation current density j_s of a cathode frequently decreases due to a gradual increase in work function as will be discussed. If j_s decreases so that it is no longer sufficient to provide the necessary space-charge-limited current density j, the current flow will become saturated and temperature-limited. In a CRT gun this shift will generally occur at the central portion of the cathode where the current density drawn is a maximum. If this occurs, the current-density distribution, which normally is bell-shaped, as shown in Fig. 5b (p. 282) of Section II.C, will become truncated at the top. As a result the intensity distribution of the spot at the screen will be correspondingly changed and may affect the resolution of the CRT (Dallos *et al.*, 1981). The beam current also will be affected more directly by any change in temperature or surface condition of the cathode.

4. *Requirements and Limits*

a. Cathode Loading. As discussed in previous sections, a high cathode loading is required to produce a high-intensity beam spot at the screen for high resolution and large-screen color CRTs. The cathode loading is defined here by the central cathode loading j_{c0} given by Eq. (31) on p. 286 of Section II. With the conventional oxide cathode the maximum allowable value for j_{c0} is about 1.4 A/cm^2 under the condition of long-term-averaged beam current (Barten, 1989), for example, over a period of hours. Cathode loading greater than this value will considerably shorten the life of the cathode. However, for instantaneous peak currents such as occur for white peaks in a picture display, a cathode loading up to several amperes/cm^2 may be permitted. As will be discussed, an increased cathode loading of about 2 A/cm^2 is possible with recently developed high-performance oxide cathodes and, for a still higher cathode loading, an impregnated cathode can be used, providing up to about 4 A/cm^2 for a long-term-averaged beam current with a long life. For high-end applications of color CRTs in the future, a cathode operable at a cathode loading as high as 10 A/cm^2 with a long life may be required.

b. Operating Life. In all cases the saturation (temperature-limited) emission current of a cathode decreases with operating time. The rate of the decrease depends upon various factors such as the type of emitter, the process of activation, other materials used in the tube and the release of gases from them, the operating temperature, and the cathode loading at which the tube operates. Although there is no clear definition of the emission level at the end of life, for practical purposes this occurs when the emission has fallen to about half its initial value. With present color CRTs, a lifetime of 10,000 to 20,000 hours is generally obtained.

c. Operating Temperature. In general, a low operating temperature is desirable. Although the saturation current density rapidly increases with operating temperature as shown in Eq. (80), the evaporation of activators also increases with temperature, thus resulting in a shortening of the life. Another problem occurring at high temperatures is the evaporation of materials from the cathode onto G_1 and G_2. Electron emission from these electrodes (grid emission) may then cause stray emission of light at the screen and cause an error in the adjustment of the cutoff voltage of the tube. In addition, effective dimensional changes of these electrodes resulting from these deposits also can be a major factor in causing shifts in the cutoff voltage. Higher temperatures can also cause leakage between the cathode and its alumina-coated heater, resulting in circuit problems.

C. *Oxide Cathodes*

As already mentioned, the oxide cathode is the most widely used type of cathode for CRTs. It is easy to fabricate, operates at a relatively low temperature, and provides a useful level of emission current density with a relatively long life. To fabricate such a cathode three major processes are involved: preparation and deposition of materials; decomposition; and the activation (sometimes called aging). These processes are described in detail by Herrmann and Wagener (1951a). Since then, to allow mass production of monochrome and color CRTs, effort has been directed toward simplifying the cathode structure, reducing the activation time and the susceptibility to poisoning from various gases in the tube as well as increasing the stability in air before processing.

1. *Fabrication*

a. Preparation of Coating. The starting materials for the oxides are carbonates of alkaline-earth metals, the triple carbonates (Ba, Sr, Ca) CO_3 and double carbonates (Ba, Sr) CO_3, respectively, for the corresponding oxides mentioned in the preceding. The former consists of about equal mol ratios of barium and strontium carbonates with a small mol ratio of calcium carbonate, while the latter consists of equimolar proportions of each carbonate. Generally, these carbonates are obtained by co-precipitation from an aqueous solution of the corresponding nitrates to which sodium carbonate or ammonium carbonate is added. Sodium carbonate is more commonly used commercially, resulting in the following chemical reactions to produce barium carbonate:

$$\mathrm{Ba\,(NO_3)_2 + Na_2CO_3 = BaCO_3 + 2\,NaNO_3} \tag{82}$$

The triple or double carbonates thus produced are washed, dried, and ground in a ball mill together with a solution of a binder (usually nitrocellulose dissolved in a volatile organic solvent) to reduce their particle size to 2–5 μm. The resulting suspension is then sprayed onto the surface of the nickel cap of the cathode.

b. Decomposition. After the gun assembled with the cathode is mounted in the tube, it is heated for about 10 minutes by applying power to the filament to raise the cathode temperatures to about 1050°C while maintaining a vacuum of the order of 10^{-4} torr. In this process the cellulose binder decomposes first, after which the alkaline-earth carbonates decompose into oxides and carbon dioxide. The carbon dioxide is then pumped out, leaving a coating of pure oxides. The reaction of the decomposition for the case of the triple carbonates is written as follows:

$$(\mathrm{Ba,\ Sr,\ Ca})\,\mathrm{CO_3} = (\mathrm{Ba,\ Sr,\ Ca})\mathrm{O} + \mathrm{CO_2} \tag{83}$$

Because a porous layer of oxides is required (porosity of the order of 40%) for good emission as mentioned below, the decomposition process must be carried out with great care. If the temperature is raised too quickly or too high, the oxide coating might sinter and lose its porosity (see also Nakanishi, 1994). In some cases, depending on the manufacturer, the decomposition process is performed after the tube has been sealed off.

c. Activation. Immediately following decomposition the cathode produces a certain amount of emission current at the operating temperature. However, this emission is generally not fully stable and varies considerably from tube to tube. To obtain a sufficiently large and stable emission current that does not vary from tube to tube, a further process, in which the getter is flashed, is necessary after the tube is sealed off. This evaporates a Ba film on the inner walls of the tube, which adsorbs the residual gases and produces a much better vacuum, of the order of 10^{-5}–10^{-6} torr. Following this the activation process (sometimes called aging) takes place.

In the activation, the cathode is heated up to 900–1000°C, while increasing positive voltages with respect to the cathode are applied stepwise to G_1 and G_2. In this process a few to several tens of volts are applied to G_1 and 100–200 V applied to G_2. For a stable emission current to be reached, an activation time of approximately 60 minutes is usually required. Without the application of positive voltages, an emission slump could occur in a short time at the operating temperature. During the activation process the currents from the cathode drawn to G_1 and G_2 also serve to degas these electrodes by electron bombardment (Sato and Yamamoto, 1954). In addition, the electron flow can ionize residual gases in the tube, which are then more effectively adsorbed by the getter.

It is known from various investigations that free excess Ba atoms are created in the crystals of the oxide coating during the activation (see, e.g., Rittner, 1953). This is due to reduction of the barium oxide (BaO) at the boundary between the oxide layer and the nickel substrate, which contains small amounts of reducing agents such as Mg, Si and Al. (The nickel of a typical cathode contains the order of 0.05% of Mg.) Examples of the chemical reactions for reduction with these reducing agents are as follows:

$$\left.\begin{aligned} \mathrm{BaO} + \mathrm{Mg} &= \mathrm{MgO} + \mathrm{Ba} \\ 4\mathrm{BaO} + \mathrm{Si} &= \mathrm{Ba_2\,SiO_4} + 2\mathrm{Ba} \\ 4\mathrm{BaO} + 2\mathrm{Al} &= \mathrm{Ba\,Al_2O_4} + 3\mathrm{Ba} \end{aligned}\right\} \tag{84}$$

Although the quantity of free Ba atoms produced is very small (of the order of 0.01% of the oxide), this barium is essential for the emission of electrons as will be discussed. In order to obtain enough free Ba atoms, the activation must be carried out in a sufficiently high vacuum. As the quantity of free Ba atoms produced is highly influenced by the degassing of other electrodes, the activation is normally carried out in the sealed-off tube after the getter is flashed.

The resulting free Ba atoms diffuse into the grains of the oxide coating and are incorporated into the crystal lattice of the BaO. They finally migrate to the surface of the oxide grains (see Fig. 57), where they are gradually lost by evaporation. Because the reduction, diffusion, and evaporation all increase with increasing temperature, an optimum range of activation temperatures exists in which the reduction and diffusion are large enough but the evaporation is not too large. Another important factor to be taken into account is the evaporation of the BaO onto other electrodes such as G_1 and G_2, which then become electron-emissive and cause undesirable stray emission of light at the screen.

2. *Mechanism of Emission*

The mechanism of the oxide cathode is quite complicated and various explanations have been proposed. However, as proposed by Wilson (1931), it is now considered most likely that the oxide is an *n*-type semiconductor with excess Ba atoms as the donor. According to the semiconductor model, BaO plays an important role in electron emission. With a band gap of approximately 4 eV, BaO is a good insulator when absolutely pure and stoichiometric in composition, allowing no electron conduction. However, if free Ba atoms exist in the BaO crystal lattice, they are located a small distance (in the order of 1 eV) below the conduction band as shown in Fig. 56. As a result, if the temperature is increased, some of the electrons from the Ba donors can be raised to the conduction band, from which they can be thermally emitted from the surface of the grains into the vacuum.

The thermionic emission from the oxide cathode also follows a Richardson-Dushman type of equation given by Eq. (80). In this case the effective work function for the oxide cathode is the sum of two quantities, the external work function χ and the internal work function υ as shown in Fig. 56, given by $\phi = \chi + \upsilon$. In a well-activated oxide cathode, a value of the work function ϕ of the order of 1.0–1.2 eV is obtained at a temperature of 700–800°C. Further details of the semiconductor model of the oxide cathode are given elsewhere (see, e.g., Herrmann and Wagener, 1951d; Imai, 1962).

From the emission mechanism mentioned here, the role of the free Ba atoms is essential since the number of free Ba atoms determines the

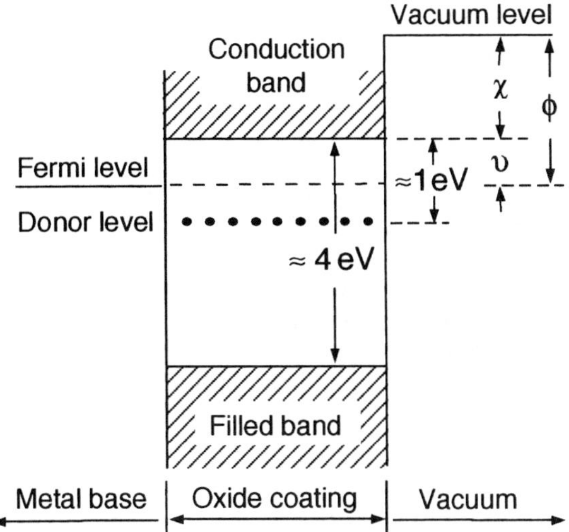

FIGURE 56. Potential diagram of BaO with free Ba. The Ba donor level is about 1 eV below the conduction band. The effective work function ϕ is equal to $\phi = \chi + \upsilon$ where υ is the internal work function and χ is the external work function.

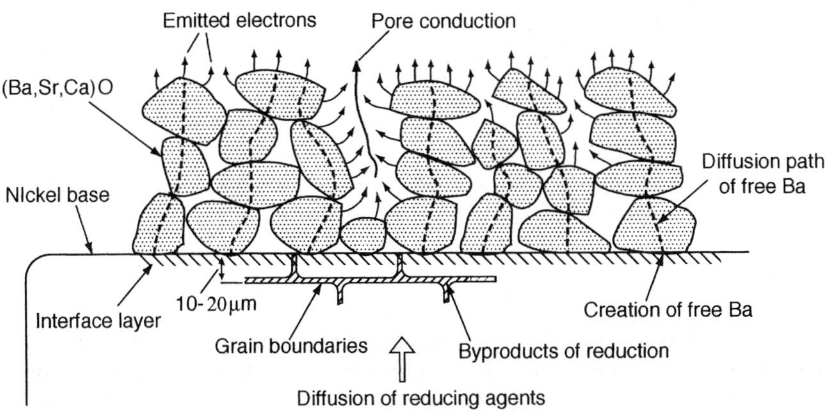

FIGURE 57. Oxide-coating layer on Ni metal base indicating the flow of emitted electrons from the oxide grain surface and the diffusion path of free Ba through the grains.

concentration of donors upon which the emission current depends. During operation, reducing agents continuously migrate by diffusion from within the nickel metal base to the boundary between the oxide layer and the nickel substrate, replenishing the supply of free Ba atoms that ultimately reach the BaO surface. In Fig. 57 a schematic representation is given for the path of the free Ba atoms and the electron emission from the grain surfaces.

During operation the free Ba atoms on the surface of the oxide grains are gradually lost by evaporation as well as poisoning by residual gases in the tube. Although the sealed-off tube is always maintained at a high vacuum of 10^{-7}–10^{-8} torr (with a vacuum of the order of 10^{-10} torr being reached after long operation by the gettering action of the Ba film evaporated on the inner wall of the tube), some gases still remain. Commonly found are H_2, H_2O, CH_4, CO, N_2, CO_2, O_2, Ar, C_nH_n, and so on (Nakanishi, 1994). The active surface of the emitter is thus subject to chemical reactions with these gases as well as bombardment by ions of these gases. Continuous replenishment of the Ba is, therefore, necessary. This is supplied by continuing reaction of the BaO with the reducing agents in the nickel base and the diffusion of the Ba to the surface, thus maintaining a state of equilibrium. An important factor in this process is thus the availability of appropriate reducing agents in the nickel base (Imai, 1962; Ouchi *et al.*, 1982; Nikaido *et al.*, 1982).

The role of the other oxides, SrO and CaO, is not yet clear in terms of the emission mechanism. It is believed that these oxides may play roles such as reducing both the evaporation of Ba and BaO and the work function. The presence of Ca may help to grow larger crystals of the carbonates during precipitation, which may be a factor in reducing the evaporation of Ba and BaO during operation (Shafer and Turnbull, 1981).

As already mentioned, because the oxide layer has many pores between its grains, electrons can flow through the open spaces between the grains, as shown in Fig. 57, as well as through the grains themselves. Based on such an electron flow through the pores, which has been confirmed by experiments, a model of electron pore conduction was proposed (Loosjes and Vink, 1949, 1950). This model shows that the shape of the oxide crystals and their packing are important factors in the design of the oxide cathode.

3. *Effects of Electrical Resistance*

When the electrons pass through the coating, the oxide grains become further heated because of their relatively high resistance. This causes them to grow in size, melt and evaporate, thus limiting the emission current density from any cathode area to about 5 A/cm^2 (Yamamoto, 1988). In an actual CRT cathode, as discussed in Section II.C, because of the radial variation in the external electric field at the cathode surface the current density distribution is bell-shaped, being highest at the center as expressed by Eq. (25) on p. 284 and shown in Fig. 5b on p. 282. Therefore, the average allowable current density over the entire emitting area in a CRT gun will be 5/2.5 or 2 A/cm^2 according to Eq. (31) on p. 286.

In addition to the resistance of the oxide grains, the current density is limited by the so-called "interface resistance," which results from the formation of high-resistance byproducts at portions of the interface between the oxide coating and the nickel substrate as shown in Fig. 57. These byproducts, such as MgO, Ba_2SiO_4 and Ba Al_2O_4, result from the reduction of BaO by agents in the nickel substrate such as Mg, Si and Al (Imai, 1962). Such interface products are produced also between grain boundaries of the nickel substrate just beneath the surface to a depth of 10–20 μm (Ouchi *et al.*, 1982) as indicated in Fig. 57. The formation of these interface products increases with operating time.

Aside from their high resistance, interface products impede the reducing agents from diffusing from inside the nickel substrate to its surface. This is another factor limiting the cathode life since it depends on the rate of reduction of BaO, which in turn is determined by the supply rate of the reducing agents.

4. *Improved Oxide Cathodes*

To satisfy the demand for higher current density and longer life of oxide cathodes, various proposals have been made (e.g., see Lemmens and Zalm, 1961/62; Jenkins, 1969; Rutter and Schimmel, 1979). Although these proposal were not used in a practical way, in recent years the incorporation of a small quantity of other oxide grains into the carbonates has been found effective in increasing the life as well as the cathode loading.

One such cathode is the Sc_2O_3-dispersed oxide cathode (Saito *et al.*, 1986, 1990, and 1995). This has an emissive layer consisting of a mixture of the ordinary triple carbonates (Ba, Sr, Ca)O and scandium oxide Sc_2O_3. In contrast to the maximum cathode loading of 1.4 A/cm^2 of conventional oxide cathodes, the new cathode can operate at 2 A/cm^2 at the same temperature as conventional oxide-cathodes, namely, approximately 760°C. An example of the percentage drop in beam current from its initial value versus operating time for the Sc_2O_3-dispersed oxide cathode compared to the conventional oxide cathode, measured in a color CRT, is shown in Fig. 58 (Nakanishi *et al.*, 1990). In an accelerated test under dc operation at 2 A/cm^2 the lifetime (to half-value) of the new cathode was found to exceed 30,000 hours.

Another type of oxide cathode capable of a higher current density and operation without increasing the temperature is the indium co-precipitated oxide cathode. In an accelerated test at high current density the lifetime of this cathode was found to be twice that of a conventional oxide cathode (Hara *et al.*, 1990).

The exact reason for the improvement in emission from cathodes using the additives mentioned is not yet fully understood. It has been suggested that the scandium oxide prevents interface products such as barium silicate (Ba_2SiO_4) from being formed when emission current is drawn (Nakanishi *et al.*, 1990).

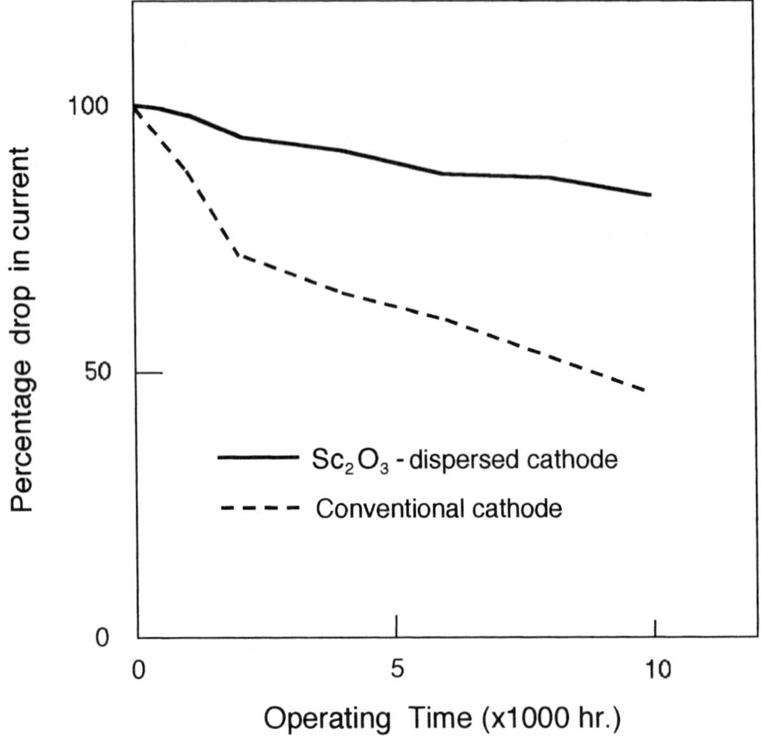

FIGURE 58. Percentage drop in current as a function of operating time for Sc_2O_3-dispersed oxide cathode and conventional oxide cathode.

The use of oxide cathodes in one form or another for color CRTs is expected to continue since their cost-competitiveness is very high and they can provide a current density that is still useful for many applications. However, to meet the increased current density requirement of high-end applications, such as HDTV and high-resolution display systems, a higher-performance type cathode such as the impregnated cathode which will be discussed below, is expected to be increasingly employed.

D. Dispenser Cathodes

1. L-Cathode

In the late 1940s, a new type of cathode called the L-cathode was invented at Philips, bearing the initial of its inventor (Lemmens *et al.*, 1950). As shown in Fig. 59a, the L-cathode consists of a porous tungsten disc covering a cavity

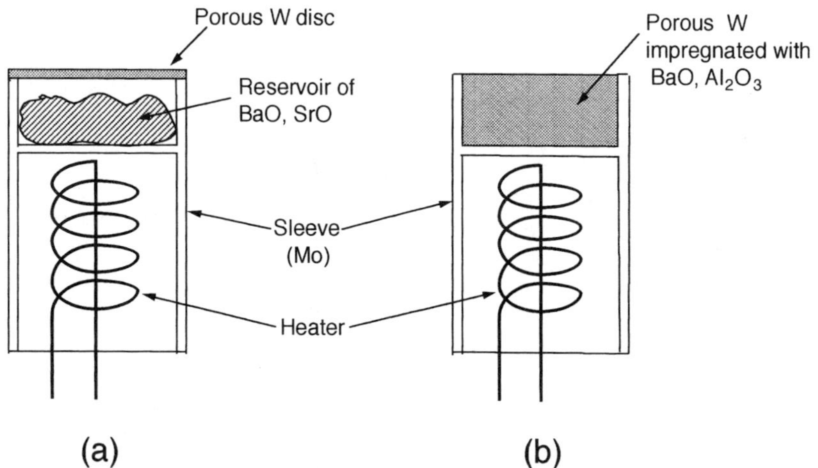

FIGURE 59. Structures of dispenser cathodes; (a) L-cathode; (b) impregnated cathode (A-type).

made of refractory metal such as molybdenum, which contains barium and strontium carbonates. When heated the carbonates decompose into the oxides at a temperature of about 1000°C, the resulting CO_2 escaping slowly through the pores of the tungsten disc usually over a period of an hour or so. The cathode can then be activated by further heating up to about 1200°C. This causes free Ba atoms to be produced due to the reducing action of the tungsten as follows:

$$6\,\mathrm{BaO} + \mathrm{W} = \mathrm{Ba_3WO_6} + 3\,\mathrm{Ba} \tag{85}$$

The liberated Ba atoms then evaporate onto the tungsten disc and migrate over its outer surface.

The electron emission mechanism of the L-cathode is significantly different from that of the oxide cathode. In the oxide cathode, as already mentioned, the coating layer of BaO is a semiconductor in which liberated free Ba atoms become the donors from which the electrons are emitted. In the L-cathode, on the other hand, free Ba atoms do not serve as donors but form an atomic monolayer on the surface of the porous tungsten substrate. It is generally believed that the Ba atoms of this monolayer combine with O atoms from within the pores or from residual gas in the tube before forming a monolayer of Ba-O, which in turn plays an essential role in the emission of electrons (see also, e.g., Jenkins, 1969).

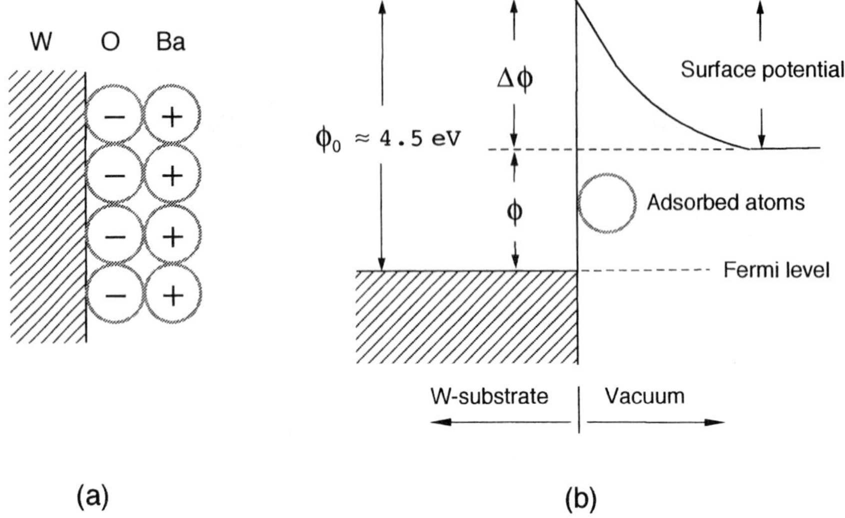

FIGURE 60. Potential diagram of W surface with the atomic Ba-O monolayer with resultant dipole effect shown in (a) and effective work function ϕ reduced to $\phi = \phi_0 - \Delta\phi$ shown in (b).

As shown in Fig. 60a, the Ba-O monolayer creates an electric dipole on the surface with the positive charge of the barium facing outward. This reduces the surface potential of the monolayer, which is indicated by $\Delta\phi$ in Fig. 60b. Since the value of $\Delta\phi$ is about 2.5 eV, the resulting work function is reduced to approximately 2.0 eV from the 4.5 eV of pure tungsten (Yamamoto, 1988). Because the value of the work function is still higher than that of the oxide cathode (1.0–1.2 eV) the operating temperature of the L-cathode must be about 300°C higher than that of the oxide cathode. The mechanism of the formation of the electric dipole is very complicated. Recent research shows that this involves charge transfer processes between Ba, O and W atoms at the surface (Mueller, 1989).

During operation the emitter surface of the L-cathode is continuously supplied with the BaO from the reservoir. This compensates for the Ba atoms lost from the emitter surface by evaporation. For this reason, the name "dispenser cathode" was given to such cathodes. In addition, because the emitter is metallic, its ohmic resistance is very low. For these reasons, the L-cathode can operate at a very high current density for a very long time. However, it also has many drawbacks—most of which are associated with its complicated structure and manufacturing processes. In addition, such cathodes are difficult to outgas, have a long activation time, higher heater power, and high cost, all of which have been major deterrents to its use.

2. *Impregnated Cathodes*

To overcome the problems of the L-cathode, another type of dispenser cathode was developed at Philips, referred to as the impregnated cathode or I-cathode (A-type). As shown in Fig. 59b the emissive materials, consisting of a mixture of BaO and Al_2O_3, are impregnated into a porous tungsten pellet instead of being held in reserve in a cavity in the form of carbonates (Levi, 1955, 1957/58). The impregnation is carried out by melting the mixture either in a vacuum or in a hydrogen atmosphere to prevent the tungsten pellet from oxidizing and allowing the mixture to soak into the porous tungsten. After impregnation, the mixture of barium oxide and alumina is converted to an oxide compound of barium and aluminate (barium aluminate). This activation is similar to that used for the L-cathode. It seems likely that the free barium diffuses along the pores to activate the surface of the tungsten in accordance with the following reaction:

$$6(\mathrm{BaO})_3\,\mathrm{Al_2O_3} + \mathrm{W} = \mathrm{Ba_3WO_6} + 6(\mathrm{BaO})_2\,\mathrm{Al_2O_3} + 3\,\mathrm{Ba} \qquad (86)$$

With the impregnation of the emitter material into the porous tungsten pellet, many of the drawbacks of the L-cathode were considerably reduced, although the work function of the A-type cathode was approximately 2.3 eV higher than the 2.0 eV of the L-cathode. Later, in order to reduce the evaporation of Ba atoms and also improve the emission, calcium carbonate was added to the emissive material of the A-type cathode (Levi, 1965; see also Cronin, 1981). The modified cathodes are referred to as the B-type and S-type impregnated cathodes since they have different molar ratios of the BaO, CaO and Al_2O_3 (5:3:2 and 4:1:1, respectively). Both cathodes have a work function of approximately 2.1 eV.

It was subsequently found at Philips that electron emission is greatly enhanced if the emitter surface of the impregnated cathodes (B- and S-types) is coated with a thin film of Os (Zalm and van Stratum, 1966). This reduced the work function from about 2.1 eV to 1.8 eV, allowing either a decrease of operating temperature of about 100°C for the same emission current or an increase of current density by approximately 10 times at the same temperature. Such coated cathodes are referred to as Magic or M-type impregnated cathodes in view of their remarkable improvement. However, since the osmium oxide (OsO_4) is toxic in nature an alloy of Os and Ru is commonly used today instead of Os. In some cases Ir is used for the coating material instead of the Os-Ru alloy (Kimura *et al.*, 1989). A schematic cross section of the tungsten pellet itself of the M-type cathode is shown in Fig. 61. After preparation the pellet is fixed to the top of a sleeve made of a refractory metal such as molybdenum into which an alumina-coated tungsten filament is inserted as in the oxide cathode.

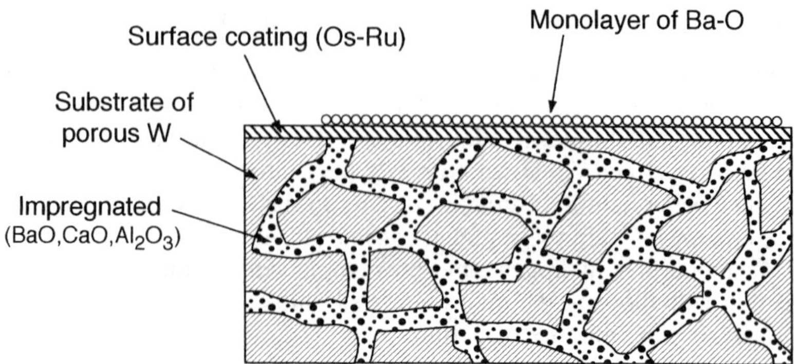

FIGURE 61. Cross-sectional view of the porous tungsten pellet for M-type impregnated cathode.

As already mentioned, the impregnated cathode can operate at a considerably higher current density compared to the conventional oxide cathode. However, despite extensive research and development, impregnated cathodes still have various drawbacks such as higher operating temperature, longer activation time, and higher cost compared to the oxide cathode (Falce, 1983; Nakanishi, 1987, 1994). [An overview of various types of dispenser cathodes is given by Thomas *et al.* (1990).]

However, in view of the demand for higher resolution color CRTs and the necessary increase in cathode loading, the application of impregnated cathodes into high-resolution color CRTs has moved forward in spite of their increased cost and more complicated tube fabrication processes compared to the oxide cathode (Falce and Breeze, 1991; Falce, 1992; Farina *et al.*, 1992).

3. *Choice of Cathode in CRTs*

As discussed in the previous sections, it is difficult to achieve both high resolution and high brightness at the same time. Thus, in practical designs, a compromise must be found between these two factors. In color CRT applications requiring both high brightness and high resolution such as HDTV, it is believed that a maximum cathode loading at the center j_{c0} of approximately 10 A/cm^2 may be required. In effect the cathode must have a temperature-limited saturation current density j_s exceeding this value.

A saturation current density j_s of about 10 A/cm^2 can be obtained with the M-type impregnated cathode using BaO, CaO and Al_2O_3 (4:1:1) with a sputtered top coating of Os-Ru of the order of 1 μm thick when operating at a temperature of approximately 1000°C. (Endo *et al.*, 1990; van der Heide, 1992). It has been shown that this cathode can operate for more than 10,000 hours in color CRTs at a cathode loading j_{c0} of about 4 A/cm^2 maintaining a virtually constant beam current. Similar results of cathode performance have been obtained with the M-type impregnated cathode, which uses a top coating of Ir instead of Os-Ru (Kimura *et al.*, 1990).

However, the higher operating temperature of the impregnated cathode compared to the oxide cathodes may cause increased stray electron emission from the G_1 and G_2 electrodes onto which active barium compounds are evaporated from the cathode (Lee *et al.*, 1994). It may also cause the drift in the cutoff voltage to be more pronounced, as discussed later. In addition, the correspondingly higher heater temperature (approximately 1150°C) may cause a short life due to leakage through the insulation of the alumina coating of the filament. To keep the heater temperature as low as possible, the cathode structure must be designed so that there is more efficient transfer of heat from the heater to the cathode. An effective means for accomplishing this is to blacken the alumina-coated surface of the heater wire as well as the inner surface of the molybdenum sleeve surrounding the heater.

The fabrication process of tubes with impregnated cathodes may be similar to that of conventional oxide cathodes. Following outgassing, the cathode is activated by gradually increasing the temperature to about 1200°C over a period of approximately 30 minutes (Endo *et al.*, 1990; Kimura *et al.*, 1990).

The M-type impregnated cathode with an Os-Ru coating has been used in guns designed for high-resolution color CRTs (30-inch and 40-inch types) developed for an HDTV system (Kubo, 1982; Ashizaki and Suzuki, 1984). Since these use twice as many scan-lines as tubes for the NTSC system, the guns employ a reduced G_1 aperture (0.4 mm in diameter) to obtain the small spot size required. This requires a cathode loading j_{c0} exceeding 3 A/cm^2 for the long-term-average beam current, a value which is well over the limit of about 1.4 A/cm^2 cathode loading of conventional oxide cathodes.

Large-screen (43-inch) color CRTs for conventional NTSC TV have also adopted the impregnated cathode (Ashizaki *et al.*, 1988). Also, because of the trend in large color CRTs of increasing the picture contrast by reducing the transmission of the glass faceplate (to about 36%) an increase in beam current is required, resulting in a higher cathode loading (Adachi *et al.*, 1991).

Although data display or computer monitor tubes operate at a relatively low beam current, they use a G_1 aperture whose diameter is sometimes as small as 0.35 mm to obtain a reduced spot size for high resolution. In some data display tubes whose screen size is larger than 20 inches, impregnated cathodes have been introduced (Suzuki *et al.*, 1990; Nose *et al.*, 1995).

Since the high operating temperature of the M-type impregnated cathode may cause problems, it is vitally important to obtain a cathode with a current density capability of about 10 A/cm^2, which can operate at a lower temperature. Of the various types of dispenser cathodes, the scandate-impregnated cathode, which has a layer of W and Sc_2O_3 on the surface, is considered a promising candidate to replace the M-type cathode in the future (Hasker and Stoffelen, 1985; Hasker *et al.*, 1989; Yamamoto *et al.*, 1986, 1989; Gibson *et al.*, 1989; Koganezawa and Yamamoto, 1991).

E. The Cathode Assembly

1. *Typical Structure*

Although high emission and long life are essential, the thermomechanical characteristics of the cathode are also very important because they have a large influence on the gun performance (Corson, 1974). For example, a variation in beam current resulting from thermal expansions in the cathode and other electrodes of the triode section (see the G_1-doming discussion that follows) becomes increasingly difficult to reduce when a smaller G_1 aperture is used to obtain a smaller spot size. To minimize this problem the design of the cathode assembly is very important.

Although various types of cathode assembly are used, depending on the manufacturer, a typical structure is illustrated in Fig. 62. In this structure the cathode sleeve is supported at its lower end by ribbons fixed to the upper end of a metal cylinder used for heat shielding, which in turn is surrounded by a ring-shaped ceramic insulator. This ceramic ring is then fixed to the inside of the cap-shaped G_1 electrode which, together with other gun electrodes, is supported by the bead glass rods (not shown).

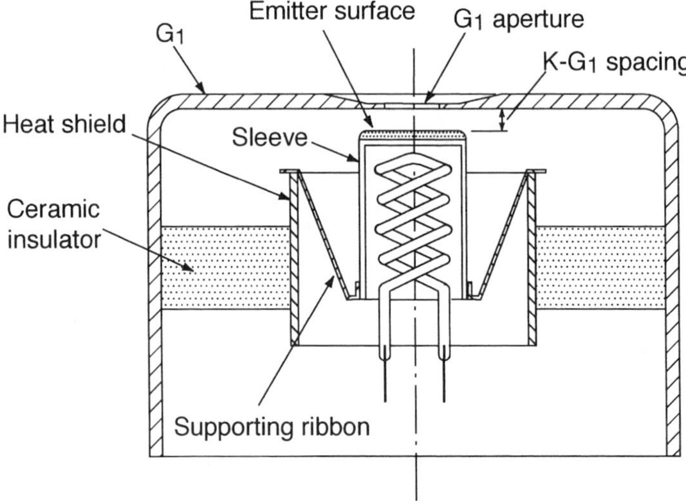

FIGURE 62. An example of typical cathode assembly structure.

The characteristics of the gun are very sensitive to the cathode to G_1 distance (or K-G_1 spacing), which is of the order of only 0.1 mm. In the illustrated structure, the influence of the thermal expansion of the cathode on the K-G_1 spacing is reduced by the supporting ribbons whose expansion counterbalances the expansion of the cathode sleeve.

2. *Important Design Factors*

a. Power Consumption. The power consumption of each cathode is typically 1.2–1.5 W with a nominal heater voltage rating of 6.3 V resulting in a total cathode power consumption per tube of 3.6–4.5 W. To save energy, there is a trend to use cathodes with a reduced power consumption of 0.5–0.7 W. These lower-wattage cathodes have a smaller emitter diameter of 1–1.5 mm compared to about 2.0 mm for the usual cathode. Low-wattage impregnated cathodes also have been developed that consume about 0.65 W (van der Heide, 1992). The design of a low-wattage cathode is described in Almer and Kuiper (1961/62).

A lower-wattage cathode reduces evaporation of active barium compounds onto G_1 and G_2 due to its reduced emitting area, thus decreasing spurious emission from these electrodes. In addition, the G_1 and G_2 temperatures are kept lower because of reduced radiation from a smaller emitting area of the cathode, thus minimizing the electron emission from their surfaces.

b. Input Capacitance. Because the cathode capacitance with respect to G_1 (and other electrodes) determines the impedance seen by the video amplifier, it limits the bandwidth of the signal. In conventional TV tubes, where the video frequencies extend only to about 4–5 MHz, the cathode capacitance plays a relatively small part and does not limit the bandwidth. However, in high-resolution applications requiring higher-frequency video signals, a reduction in this capacitance is very important.

The cathode capacitance depends on the size and structure of the cathode assembly including the lead wires. Generally, the smaller the size of the assembly, the lower the capacitance. For example, the capacitance of a 1.4 W cathode is approximately 5 pF, while for a 0.65 W cathode the capacitance is reduced by 20% to about 4 pF due to its smaller size. In general, to obtain a low capacitance the dimensions of the individual parts such as the cathode sleeve, the heater, and the G_1 structure must be reduced (Sudo *et al.*, 1986).

c. Warm-Up Time. The time after switching-on the cathode until the picture appears is referred to as the warm-up time. This time depends on the thermal capacity of the cathode and the effective heater power. The lower the capacity and the higher the power, the faster the warm-up. To reduce the thermal capacity, the size of the cathode and its support structure must be minimized without reducing the size of the emitter surface. To increase the effective heater power, a more densely wound filament is used in combination with a black coating (as previously mentioned) on the inside surface of the sleeve to improve the heat transfer from the filament. Such quick-heating cathodes are commonly used in modern color CRTs, having a warm-up time that is generally 5 to 10 seconds.

d. G_1-Doming. In addition to the warm-up, several minutes are generally required after warm-up before the picture becomes fully stabilized in brightness and color balance. This is caused by thermal transient phenomena in various parts of the cathode assembly and electrodes of the triode section. Because of their close proximity to the cathode, G_1 and G_2 heat up by conduction as well as radiation. For oxide cathodes their temperatures become approximately 250°C and 160°C, respectively, while for impregnated cathodes the temperatures are 360°C and 220°C, respectively (Nakanishi *et al.*, 1989).

The resulting thermal expansions cause a variation in the K-G_1 and G_1-G_2 spacings, resulting in a change in beam current with time before a final state of thermal equilibrium is reached. This is referred to as the "G_1-doming" because of the domed shape of G_1 after expansion and due to the greater effect on the K-G_1 gap. The actual change in beam current with time is very complicated, generally differing for each of the three beams because of mechanical differences in their position and electrode support structure.

The K-G_1 and G_1-G_2 spacings are usually of the order of 0.1 mm and 0.2 mm, respectively. With a reduced G_1 aperture diameter these spacings are further reduced. For example, in the case of a high-resolution tube with a G_1 aperture diameter of 0.35 mm, the spacings will be halved. With such small spacings, the G_1-doming becomes more pronounced. To reduce this doming, increased thermomechanical stability of the cathode assembly and triode structure is required as well as a lower cathode operating temperature.

e. Drift of Cutoff Voltage. During operation over a long period, the cutoff voltage varies gradually, generally decreasing and causing a decrease in brightness in the absence of any readjustment. This voltage drift occurs both with oxide and impregnated cathodes, although the mechanisms are different.

In the oxide cathode the drift can occur due to an increase in the K-G_1 gap as a result of shrinkage of the oxide coating (which is about 70 μm thick). The shrinkage is of the order of 10 μm, resulting in a decrease in cutoff voltage V_{sc} of approximately 3 V in a display monitor tube whose initial cutoff voltage is 120 V (Hayashida *et al.*, 1995).

For impregnated cathodes, evaporative barium compounds play a major role in the cutoff voltage drift. During operation these compounds build up on the G_1 surface facing the cathode, mainly around the aperture as well as inside it. This results in a decrease in aperture diameter and an increase in G_1 thickness, both of which reduce the cutoff voltage. This change occurs very slowly and continues over a long period of operation. To reduce this effect the initial G_1 aperture diameter and K-G_1 spacing should not be too small and as low a cathode temperature as possible should be used (Breeze, 1993).

V. Techniques for Improving High-Tension Stability

A. Introduction

As already discussed, in color picture tubes voltages as high as 30 kV or more may be applied between the final anode and cathode in order to obtain pictures with the desired brightness and resolution. Aside from generating X-rays, however, such voltages often cause electrical breakdown or arcing in the tube, which results in a sudden high-current flow to the electrodes of the gun. This may not only damage the gun components but also the circuits connected to the electrodes, especially those connected to the low-voltage electrodes.

If the discharge current reaches as far as the cathode, the emitter surface itself may be damaged, resulting in a complete loss of emission. In addition, the surge current may induce high-voltage transients or spikes across inductive components of nearby circuits such as those used for signal processing, causing catastrophic failure of these circuits. Special measures are thus required to either prevent the arcing from occurring or to minimize the disturbance and deleterious effects of the arcing if it occurs. These two approaches are discussed here in Sections V.C and V.D, respectively.

B. Arcing Phenomena in Picture Tubes

For a typical anode (G_4) voltage of 30 kV, the voltage applied to the adjacent focusing electrode (G_3) is about 8 kV. This produces a field strength of approximately 2×10^5 V/cm between the electrodes, which are about 1 mm apart, thereby creating a tendency for an electrical breakdown to occur in the space between them. This type of discharge is referred to as an interelectrode arc.

Another type of discharge that can take place occurs in the presence of a nearby insulator, primarily the neck glass. This is caused by a build-up of high potential on the inner surface of the neck in the neighborhood of the gun structure. Such a discharge is referred to in various terms, including flashover, creeping discharge, or trigger arc.

Many factors are involved in the occurrence of a high-voltage discharge in picture tubes and their mechanism is generally quite complicated. In subsections 1 and 2 here, some simplified models for the interelectrode arc and flashover will be discussed.

1. *Interelectrode Arc*

Although arcs generally occur between adjacent electrodes such as G_3 and G_4 of the main lens of a bipotential gun where the electric field is strongest, they can sometimes occur in the prefocus lens between G_3 and G_2, where several kilovolts are applied between these electrodes. In the case of a unipotential gun (see Section III.B.2) where the full anode voltage V_a is applied to G_3, the arcing problem at the prefocus lens is more serious than in bipotential guns.

In the initiation of the interelectrode arc, a low-level field emission may occur from the lower-potential focusing electrode (G_3) to the anode (G_4), due possibly to some irregularities on the surface of G_3 such as small protrusions, foreign particles, dust, or sharp edges, all of which can create a sufficiently high localized electric field.

Although the electrons released by field emission may be small in number, their acceleration by the strong field in the lens region can ionize residual gases in the tube. The positive ions thus produced are attracted back to the emission site, releasing additional electrons, resulting in an exponential current build-up or avalanche. The electrons freed in this process will impinge on the anode and release absorbed gases from it, or they may cause instantaneous vaporization of anode material or other particles from this electrode.

The rapid build-up of current results in an electrical breakdown in which a dense plasma is created that has a very low impedance, of the order of 10 Ω (see, e.g., Druyvesteyn and Penning, 1940). Since the external energy source cannot sustain the discharge it terminates as soon as the energy stored in the high-voltage capacitance of the tube funnel is dissipated, restoring the tube to its normal condition.

2. *Flashover*

The flashover process involves a complex and relatively uncontrollable build-up of potentials on the surface of a nearby insulator, either the inner surface of the glass neck or the surface of the multiform glass (bead) rods that support the gun electrodes. A cross section of the tube neck in the plane of the focusing (G_3) electrode is shown in Fig. 63. The neck glass, having a high percentage of potassium and lead, is somewhat conductive at elevated temperatures while the multiform glass (bead) is a hard glass with low electrical conductivity.

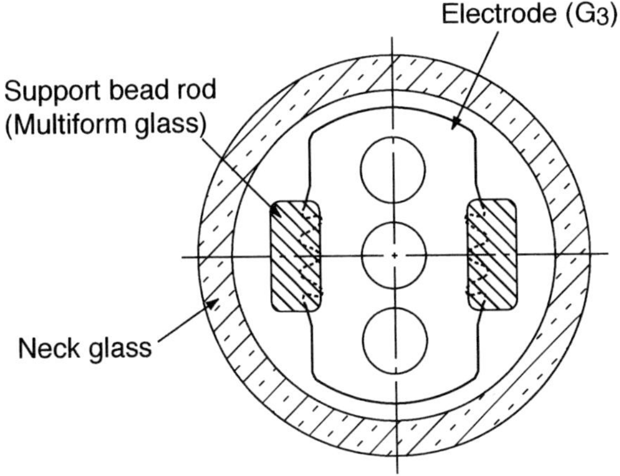

FIGURE 63. Cross section of the tube neck at the plane of the flange of focusing electrode (G_3).

According to Hernqvist (1981), the following steps occur in build-up of the flashover:

1. Due to the conductivity of the neck glass its potential as far down as the cathode rises toward the potential of the inner conductive coating of the tube funnel, which is maintained at the anode potential.
2. This causes field emission to occur from sharp points or contamination on one of the lower voltage electrodes (such as G_2, G_1 or the cathode) with the electrons from these points being drawn to the positively charged neck glass.
3. Upon striking the glass, secondary electrons are produced whose number may exceed the number of primary electrons. This causes even further positive charging of the glass, resulting in a series of electron avalanches, supported by secondary-emission, from point to point along the neck wall as shown in Fig. 64.
4. These avalanches in turn cause the release of gas from the glass surface, which then becomes ionized by the electron avalanches, resulting in the appearance of a blue glow.
5. The resulting positive ions are attracted to the field-emission sources, further augmenting the field-emission currents.
6. As a result of the regenerative process that occurs an increased gas plasma builds up and a breakdown occurs between the anode and one of the lower-voltage electrodes (such as G_2, G_1, or the cathode) along the surface of the neck glass. Also, interelectrode arcs may occur in this process.

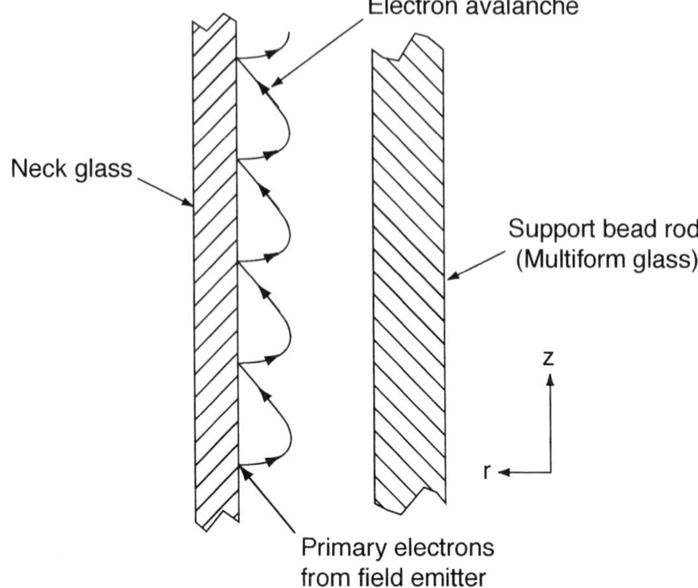

FIGURE 64. Electron avalanche formation along the surface of the neck glass in the space between the support bead and neck.

7. As soon as the charge stored in the high-voltage capacitance of the tube funnel is fully dissipated the arcs and flashovers are quenched. The residual gases are then removed by the getter of the tube or adsorbed on other tube surfaces.

Nakanishi and Imanishi (1980) report that the flashover in a bipotential gun looks like a streamer of discharge along the multiform bead rod, extending from the high-voltage funnel coating, which ends at the neck tube, down to the lower-voltage electrodes (G_2, G_1, and the cathode). Its general appearance is shown by Fig. 65. The authors also showed that in color picture tubes using typical bipotential guns the incidence of flashovers to the lower-voltage electrodes is much greater than the incidence of interelectrode arcs between the anode (G_4) and the focusing electrode (G_3).

3. *Arcing Current*

To estimate the magnitude of the discharge current, it is assumed that the capacitance between the high-voltage coating inside the tube funnel and the grounded outer shield is 2.5 nF (typical for a 25 V screen size tube) and the high-tension voltage V_a applied to the inner funnel coating is 30 kV. This results in a charge of 75×10^{-6} C and an electrical energy of about 1 J stored

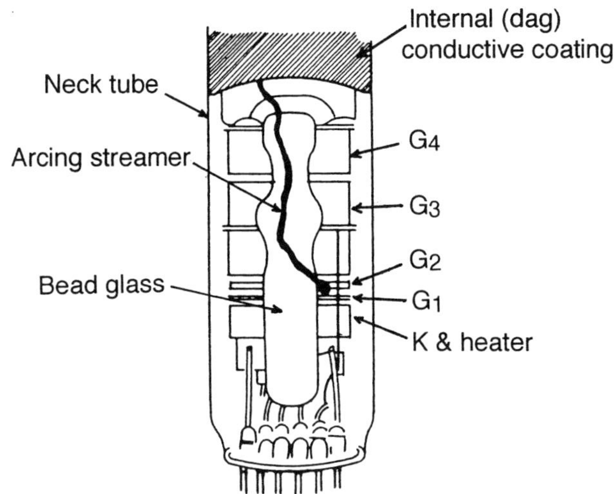

FIGURE 65. Example of arcing streamer due to flashover along the multiform bead rod, terminating on one of lower-voltage electrodes (G_2) [reproduced by permission from Nakanishi, H., and Imanishi, W. (1980). Creeping discharge in color picture tubes, *IEEE Trans. Consum. Electron.*, **CE-26**, 431–444; © 1980 IEEE].

in the capacitance. As will be discussed in section D, when an arc occurs its duration depends on the internal resistance of the discharge circuit. Assuming for simplicity a duration of 0.1 μs (corresponding to a low internal resistance of the order of 10 Ω), the average discharge current would be 750 A ($= 75 \times 10^{-6}$ C/10^{-7} s). Actually, however, the current waveform will have an initial peak that is higher than the average. In practice peak currents over 1000 A are commonly observed in tubes with a low-resistance discharge circuit.

C. Prevention of Arcs

As already mentioned, because the huge instantaneous arc currents can induce large voltage transients in low-voltage level integrated circuits, the utmost precautions are taken in the design and manufacture of color picture tubes to prevent interelectrode and flashover arcs from occurring. Clean environments must therefore be provided to prevent foreign materials and field emission sources from entering the tubes during production. Every component part and material must also be extensively washed and cleaned. Gun parts fabricated by pressing are tumbled and deburred, carefully washed, and usually fired in a hydrogen-atmosphere while assembled guns with their glass-beaded support rods are cleaned by ultrasonic washing. In addition, to evaluate the effectiveness of the various procedures employed in production, statistical analysis techniques for dealing with the arcing problems have been developed (Thierfelder and Hinnenkamp, 1979/80).

Aside from the precautions cited above, to reduce the likelihood of arc-over a "spotknocking" procedure is generally employed after the tube is sealed off from the vacuum pump. In addition, suitably located conductive coatings, metal ribbons, or wires are incorporated in all modern picture tubes to suppress flashovers to the lower-voltage electrodes. These two approaches are discussed below.

1. *Spotknocking*

The main object of spotknocking is to inactivate or destroy field emission sources such as contamination, dust, foreign particles, and minute protrusions on the surface of the focusing electrode (G_3). This is accomplished by applying repeated high-voltage pulses with a controlled amount of energy between the main lens electrodes (G_3 and G_4) to cause an arc discharge. Usually, the energy is supplied from a capacitor charged to a suitable voltage. By choosing the size of the capacitor, the energy stored and the duration of the arc can be controlled. It is important, however, that the energy delivered be no more than that required to inactivate the field emission sources that might exist on the electrode surface. Otherwise the electrode itself might be damaged. Some compromises must be made in this respect and usually no more than a few minutes are consumed in this procedure.

Other methods for generating the pulses have also been used, such as electronically generated intermittent dc voltages. The most effective method, however, involves the use of radio frequency (RF) pulses, probably due to more efficient ignition of an arc by RF voltage. As an example, an RF pulse voltage of 350 kHz and 100 kV peak-to-peak could be applied to the focusing electrode (G_3) to ignite the arc. It has been found that RF spotknocking also inactivates cold emission sources on those parts of the gun electrodes that face the neck glass and is effective in reducing flashovers to the lower-voltage electrodes (Hernqvist and Liller, 1984).

2. *Suppressing Flashovers*

As already mentioned, the flashover is triggered by the build-up of a positive potential along the inner surface of the neck glass due to conductivity of the neck glass and/or a charging up of the glass surface by secondary emission. One method for minimizing the potential rise of the glass surface at the lower end of the gun is to coat it with a highly insulating material such as a polyimide. Alternatively, by coating the neck glass with chromium-oxide, whose secondary electron emission coefficient is less than unity, potential rises on the neck surface due to secondary emission are prevented, thus preventing electron avalanches from building up (Hernqvist and Liller, 1984).

It has been found that for electron avalanches to occur a more or less specific potential distribution along the neck glass is required. Such a potential distribution can be disturbed, however, by depositing patches or bands of evaporated conductive material on the surface of the multiform bead glass facing the neck. Although the chromium-oxide coatings deposited on the neck glass mentioned here are very effective for this purpose, conductive patches or bands deposited on the multiform bead glass are more durable and easier to apply (Hernqvist, 1981). However, since these conductive areas are electrically floating they will charge up to a potential that varies with time in accordance with the high voltage conditions in the tube.

Also effective for suppressing flashovers are metal ribbons or wires, wound around the multiform bead rods, which are electrically connected to the focusing electrode (G_3) (Nakanishi and Imanishi, 1980). As in the case of conductive patches or bands, these metal ribbons serve to prevent regenerative secondary emission avalanches from building up and forming a plasma along the neck glass. The proper position for these metallic wires can be determined with the aid of the calculated potential distribution in the space between the gun electrodes of the triode section and the neck glass surface (Nakanishi, 1996).

D. Absorbing the Energy of the Arc

In spite of the utmost precautions that are taken in the design and manufacture of picture tubes, there is still a possibility that an arc will eventually occur. To prevent possible damage to low-voltage circuits during arcing, Krause (1958) proposed the use of a resistive coating R deposited between the final anode (G_4) of the gun and the internal conductive coating (dag) to which the full anode potential is applied as shown in the schematic of Fig. 66. This high-resistance coating limits the discharge current that can flow and dissipates the energy of any arc that may occur. At the same time the potential of the anode is lowered, limiting the damage from the arc. However, as mentioned already, no voltage drop occurs across this resistance during normal operation of the tube since, under these conditions, no beam current flows into the anode (G_4) of the guns customarily used in shadow-mask tubes. This method of limiting the arc current is referred to by a number of different terms such as soft-flash (Ciuciura, 1977), internal-surge-limiting (ISL) (Smithgall, 1978) and soft-arc (Thierfelder and Hinnenkamp, 1980). Although the particular form of the high-resistance element may vary from manufacturer to manufacturer, such elements are commonly used in almost all present-day color picture tubes.

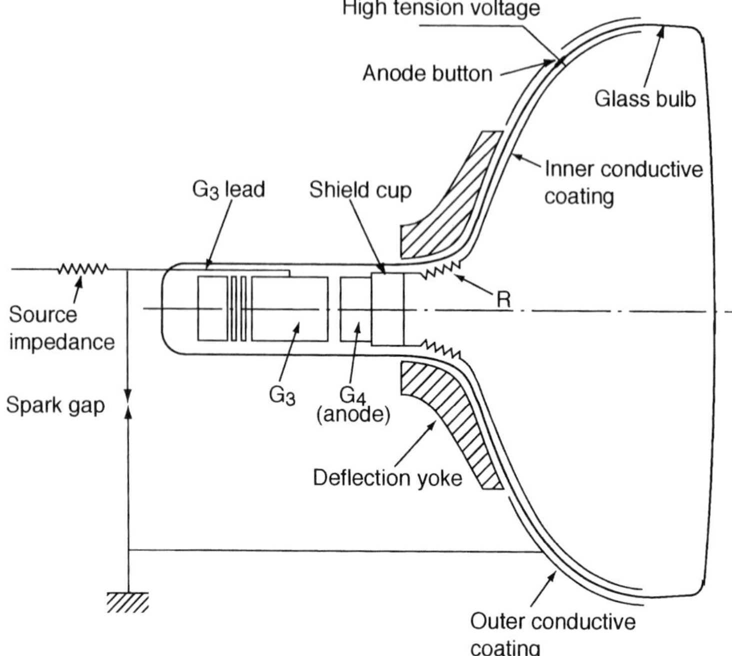

FIGURE 66. Schematic of picture tube with R representing the resistance of internal conductive (dag) coating on the tube funnel.

To optimize the effectiveness of the resistive element its value must be properly chosen. Referring again to the schematic of Fig. 66, it is assumed that an arc occurs at the main lens, thus resulting in a short-circuit in the vacuum gap between G_3 and G_4. (External to the tube a spark gap is also provided between the G_3 lead wire and ground to prevent an excessive voltage rise of the G_3 electrode during arcing.) Although the discharge path consists of a series circuit containing distributed elements it can be simulated using the discrete components shown in Fig. 67, where R represents the resistance added between the conductive coating of the funnel and G_4, C represents the capacitance between the inner and outer coatings of the tube funnel, and L represents the distributed inductance of the conductors in the discharge circuit. Closure of switch S corresponds to the initiation of a discharge.

With a low value of R such as 20 Ω, the calculated waveform of the discharge current appears as shown in Fig. 68a. The values of C and L are assumed to be 2.5 nF and 1 μH, respectively. As indicated, the current rises rapidly to a peak of about 800 A, then reverses direction and, by damped

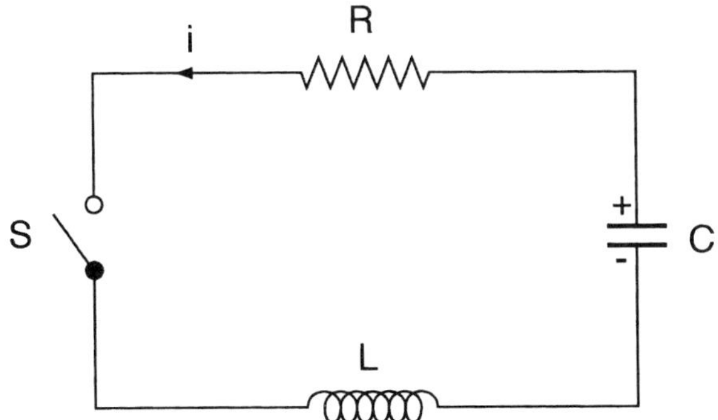

FIGURE 67. Simplified discharge circuit with discrete components, L, C, and R, initiated by closure of switch S.

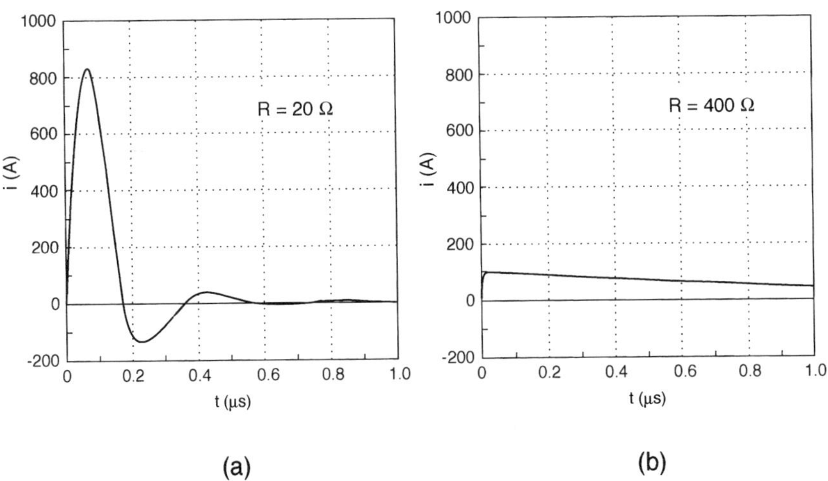

FIGURE 68. Discharge current waveforms in simplified circuits; (a) R = 20 Ω and (b) R = 400 Ω; for both (a) and (b) L = 1 μH and C = 2.5 nF charged up to 30 kV.

oscillation, falls to zero. However, with an increased resistance of R = 400 Ω, the waveform is completely changed as shown in Fig. 68b, rising to a peak current level about one-tenth the previous value, then decaying relatively slowly to zero. A detailed analysis, assuming discrete components, shows that the rate of rise in current (di/dt) is reduced by approximately one-third in the latter case (Ciuciura, 1977). Thus, by increasing the resistance R, the magnitude of the voltage spikes induced in the external circuits is greatly reduced.

In practice a value of about 400 Ω is considered optimum for the resistor. The use of a higher resistance might reduce the effectiveness of the spotknocking procedure and also prevent the spontaneous arcs that occur during normal tube operation from improving the electrode surfaces (Gerritsen, 1978).

A separate problem when a high-resistive film is deposited between G_4 and the conductive coating of the funnel is the position of the getter that is mounted in the tube. Before the use of a high-resistive coating, the so-called "antenna getter" was commonly used. In this arrangement the getter was mounted at the end of a long ribbon-type spring whose other end was attached to the shield cup (or convergence cup) of the gun anode (G_4). However, such an arrangement would defeat the purpose of the resistive coating because of a direct path along the spring from the end of the funnel coating to G_4. Even if the getter were electrically isolated from the funnel surface by a ceramic insulator, a discharge would probably occur across the insulator. In addition, the conductive barium evaporated from the getter, which would be deposited in the vicinity of the resistive coating, would short-circuit it.

To avoid this the getter unit is generally attached to the vacuum side of the anode button of the tube funnel, using a special tool to introduce it through the neck and fix it in place (Gerritsen, 1978). In another arrangement the getter is mounted on the shadow-mask frame (Smithgall, 1978). In both cases the evaporated deposit from the getter falls on the wall of the tube funnel some distance from the resistive film.

It should be noted that in place of the resistive coating, a discrete resistor can also be used (Schwartz and Fogelson, 1979) made of a microporous resistive coating (with the aid of a vitreous glass frit) whose surface is deeply roughened. As its resistance is less affected by barium deposits, it allows the getter to be mounted in its more normal position closer to gun.

In the Trinitron gun (see Section III.D.3) with its unipotential structure, a different type of arc-suppression system has been introduced (Kobori *et al.*, 1980) in which two discrete ceramic resistors are used. One is inserted in series with the wire interconnecting the G_3 and G_6 electrodes, both of which have the full anode voltage V_a applied. The other resistor is inserted in series with the lead wire, which connects the G_4 electrode to its pin at the tube base (to which the low-focus voltage is supplied). The discharge path in this case is through the high-potential electrodes of G_3 and G_6 to the G_4 electrode. The value of each resistor is in the range of 2 kΩ to 20 kΩ. Because of the low capacitance and inductance of the resistors the peak flashover current is largely determined by Ohm's law. For example, with 15 kΩ ceramic resistors and an anode voltage of 30 kV the peak current will be 2 A. The addition of these resistors still allows a standard gun design to be used. Since the resistors are located in a region of the gun away from the getter they are unaffected by the deposit of getter materials.

ACKNOWLEDGMENT

The author wishes to express his sincere appreciation to Dr. B. Kazan for his generous and continuous support and assistance in preparing the manuscript. He would also like to thank Mr. M. Fukushima who initially encouraged him to write the article and provided useful information. Finally he would like to express his appreciation to the engineers at Matsushita Electronics Corporation for their help at the start of work on this article.

REFERENCES

Aalders, A. F., Van Engelshoven, J., and Gerritsen, J. (1989). Modeling and simulation of electron gun with complex geometries, *SID Intern. Symp. Digest*, **XX**, 38–41.

Adachi, O., Wakasono, H., Kitagawa, O., and Konosu, O. (1991). Super-flat-face large-size-screen color CRT, *SID Intern. Symp. Digest*, **XXII**, 37–40.

Alig, R. C. (1980). Kinescope electron gun design, *RCA Rev.*, **41**, 517–536.

Alig, R. C., and Hughes, R. H. (1983). Expanded field lens design for inline picture tubes, *SID Symp. Digest*, **XIV**, 70–71.

Alig, R. C. (1993). Horizontal deflection defocusing in CRTs, *J. SID*, **1**, 371–374.

Alig, R. C., and Trinchero, O. (1995). A CAD/CAM tolerance analysis of color CRT electron guns, *SID Intern. Symp. Digest*, **XXVI**, 107–110.

Alig, R. C., and Fields, J. R. (1997). Computer-aided design of electron guns and deflection yokes: A review (Part 1), *J. SID*, **5**(3), 203–215.

Almer, F. H. R., and Kuiper, A. (1961/62). New developments in oxide-coated cathodes: II. An oxide-coated cathode with a half-watt heater for cathode-ray tubes, *Philips Tech. Rev.*, **23**(1), 23–27.

Amboss, K. (1993). A review of the theory of space charge limited operation, *Proc. Conf. Electron Beam Melting Refin. State Art*, 117–126.

Ando, K., Osawa, M., Shimizu, T., Maruyama, T., and Fukushima, M. (1985). A flicker-free 2448×2048 dot color CRT display, *Proc. SID*, **26**(4), 285–291.

Ashizaki, S., and Suzuki, Y. (1984). 40″ CRT display, *Denshi Tokyo (J. IEEE Tokyo Sect.)*, **23**, 108–112.

Ashizaki, S., Suzuki, H., Sugawara, K., Natsuhara, M., and Muranishi, H. (1986). In-line gun with dynamic astigmatism and focus correction, *Proc. 6th Intern. Disp. Res. Conf. (Japan Disp. '86)*, 44–47.

Ashizaki, S., Suzuki, Y., Konosu, O., and Adachi, O. (1988). 43-in. direct-view color CRT, *Proc. SID*, **29**(1), 47–51.

Barbin, R. L., and Hughes, R. H. (1972). A new color picture tube system for portable TV receivers, *IEEE Trans. Broadcast. Telev. Receivers*, **BTR-18**, 193–200.

Barbin, R. L., Simpson, T. F., and Marks, B. G. (1982). A color-data-display CRT: A product whose time has come, *RCA Engineer*, **27**(4), 23–32.

Barbin, R. L., Canevazzi, G., Cosma, P., Maresca, A., and Spina, P. (1990). A 16/9 aspect-ratio higher definition consumer color CRT family, *SID Intern. Symp. Digest*, **XXI**, 552–555.

Barkow, W. H., and Gross, J. (1974). The RCA large screen 110° precision in-line system, 2nd Annual Convention Fernseh und Kinotechnische Gesellschaft, Munich, Germany, Oct. 15.

Barten, P. G. J. (1974). The 20AX system and picture tube, *IEEE Trans. Broadcast Telev. Receivers*, **20**(4), 286–292.

Barten, P. G. J., and Kaashoek, J. (1978). 30AX self-aligning 110° in-line color TV display, *IEEE Trans. Consum. Electron.*, **CE-24**(3), 481–487.

Barten, P. G. J. (1982). Astigmatic electron lens for a cathode-ray tube, U. S. Patent 4,366,419.

Barten, P. G. J. (1984). Spot size and current density distribution of CRTs, *Proc. SID*, **25**(3), 155–159.

Barten, P. G. J. (1989). CRT; present and future, *SID Seminar Note*, S-3.

Bechis, D. J., Fields, J. R., New, D. A., Paul, W. B., Sverdlov, E. R., Winarsky, N. D., and Lausman, T. C. (1989). Applications of a fully three-dimensional electron-optics computer program to high-resolution CRT design, *Proc. SID*, **30**(3), 229–239.

Bessho, J. (1963). A study on triode electron guns, *NHK (Jpn. Broadcast. Corp.) Tech. Rep.*, **15**, 233–254.

Beynar, K. S., and Nikonov, B. P. (1964). The measurement of the work function of oxide cathodes using the method of contact potential differences, *Radio Eng. Electron.*, **9**, 1518–1524.

Bijmer, J. (1981). Experimental analysis of electron beams in a display tube, *Proc. 1st Intern. Disp. Res. Conf. (Euro. Disp. '81)*, 166–169.

Blacker, A. P., Wilson, I. M., and Schwartz, J. W. (1976). A new form of extended field lens for use in color television picture tube guns, *IEEE Chicago Spring Conf. Consum. Electron.*

Bloom, S., and Hockings, E. F. (1989). Integral grids for dynamic astigmatism control in picture-tube guns, *IEEE Trans. Electron Devices*, **ED-36**, 777–784.

Boekhorst, A., and Stolk, J. (1962). *Television Deflection Systems*, Philips Tech. Library, Eindhoven, The Netherlands, pp. 51–56.

Branton, T. W., Godfrey, R. H., Masterton, W. D., Morrel, A. M., Trim, M. E., and Weber, D. L. (1979). Recent trends in color picture tube design, *RCA Engineer*, **25**(2), 4–11.

Braun, F. (1897). Über ein Verfahren zur Demonstration und zum Studium des zeitlichen Verlaufes variabler Ströme, *Annalen der Physik*, **60**, 552–559.

Breeze, G. S. (1993). Cutoff variations due to barium evaporation from dispenser cathodes, *SID Intern. Symp. Digest*, **XXIV**, 423–426.

Buchsbaum, W. H. (1966). GE 11-inch color TV: the new look in color receivers, *Electron. World*, **75**(3), 39–41.

Bush, H. (1926). Berechnung der Bahn von Kathodenstrahlen im axialsymmetrischen elektromagnetishchen Felde, *Annalen der Physik*, **81**, 974–993.

Chen, H. Y., and Hughes, R. H. (1980). A high performance color CRT gun with an asymmetrical beam forming region, *IEEE Trans. Consum. Electron.*, **CE-26**, 459–465.

Chen, H. Y. (1982). An in-line gun for high-resolution color display, *Proc. SID*, **23**(3), 123–127.

Chen, H. Y. (1985). High-resolution electron gun designed for a new generation of color data display tubes, *Proc. SID*, **26**(4), 267–271.

Chen, H. Y., and Gorsky, R. M. (1989). A beam forming region dynamic quadrupole inline gun design, *Proc. 9th Intern. Disp. Res. Conf. (Japan Disp. '89)*, 466–469.

Chen, H. Y., and Gorsky, R. M. (1990). A new dynamic quadrupole gun optimized for high-resolution color CRT, *SID Intern. Symp. Digest*, **XXI**, 548–551.

Chen, H. Y., and Tsai, S. S. (1993). A double-hollow chain-link color CRT electron gun, *SID Intern. Symp. Digest*, **XXIV**, 411–414.

Child, D. C. (1911). Discharges from hot CaO, *Phys. Rev.*, **32**, 492–511.

Ciuciura, A. (1977). Soft-flash: a new development in color picture tube technology, *Mullard Tech. Comm.*, **14**(135), 201–208.

Corson, B. (1974). Some reliability characteristics of CRT cathode assemblies, *IEEE Trans. Reliability*, **R-23**, 226–230.

Cronin, J. L. (1981). Modern dispenser cathodes, *IEE Proc.*, **128**, 19–32.

Dallos, A., Buckbee, J. A., and Spencer, G. R. (1981). CRT cathode temperature control for high current densities, *Proc. SID*, **22**(1), 11–14.

Dasgupta, B. B. (1992). Designing self-converging CRT deflection yokes, *Inform. Disp.*, **8**(1), 15–19.

Davis, C. A., and Say, D. L. (1979). High-performance electron guns for color TV—a comparison of recent designs, *IEEE Trans. Consum. Electron.*, **CE-25**, 475–480.

Druyvesteyn, M. J., and Penning, F. M. (1940). The mechanism of electrical discharges in gases of low pressure, *Rev. Modern Phys.*, **12**(2), 140–155.

Eccles, D., Romans, G., and Held, J. (1993). HDTV:good enough for data? *Inform. Disp.*, **9**(1), 16–19.

El-Kareh, A., and El-Kareh J. (1970a). *Electron Beams, lenses, and Optics*, **1**, Academic Press, New York, pp. 54–80.

El-Kareh, A., and El-Kareh, J. (1970b). *Electron Beams, lenses, and Optics*, **2**, Academic Press, New York, pp. 50–88.

Endo, N., Mizuki, M., Ohshige, Y., Takanami, C., Amano, Y., and Ichida, K. (1997). An electron gun with extended-field elliptical aperture lens, *SID Intern. Symp. Digest*, **XXVIII**, 347–350.

Endo, S., Sasaki, S., and Nakagawa, S. (1990). Impregnated cathode for CRT use, *Tech. Group Inform. Disp. Inst. Electron. Inform. Commun. Eng. Jpn.*, **EID89-75**, 7–12.

Falce, L. R. (1983). Dispenser cathodes: the current state of the technology, *Proc. Intern. Electron. Devices Meeting*, Washington D. C., 448–451.

Falce, L. R., and Breeze, G. S. (1991). Controlled-porosity dispenser (CPD) cathodes for high-resolution CRTs, *SID Intern. Symp. Digest*, **XXII**, 703–706.

Falce, L. R. (1992). Dispenser cathodes for CRTs, *Inform. Disp.*, **8**(1), 11–14.

Farina, J. M., Maropis, N., and Werner, K. (1992). Manufacturing HDTV cathodes, *Inform. Disp.*, **8**(6), 16–18.

Francken, J. C. (1959/60). The resistance network, a simple and accurate aid to the solution of potential problems, *Philips Tech. Rev.*, **21**, 10–23.

Friedman, P. S., Stoller, R. A., and Wedding, D. K. (1991). An analysis of large-area HDTV display technology:CRT, LCD, and PDP, *Proc. SID*, **32**(2), 99–104.

Fujio, T. (1985). High-definition television systems, *Proc. IEEE*, **73**(4), 646–655.

Fukushima, M., Fukuzawa, K., Ando, K., and Yamaguchi, A. (1977). Development of color picture tube with Hi-UPF electron gun, *Tech. Group Electron. Devices Inst. Electron. Inform. Commun. Eng. Jpn.*, **ED77-71**, 1–8.

Garnier, J. P., Trinchero, O., New, D. A., and Paul, W. B. (1994). An electron gun with double-slot arrangement for improved spot uniformity, *SID Intern. Symp. Digest*, **XXV**, 659–662.

Gerritsen, J. (1978). Soft-flash picture tubes, *IEEE Trans. Consum. Electron.*, **CE-24**(4), 560–565.

Gerritsen, J., and Barten, P. G. J. (1987). An electron gun design for flat square 110° color picture tubes, *Proc. SID*, **28**(1), 15–19.

Gerritsen, J., and Himmelbauer, E. (1989). The Polygon gun . . . for a cleaner, sharper electron spot, *Electron. Components Applications*, **9**(1), 31–34.

Gerritsen, J., and Sluyterman, A. A. S. (1989). A new picture tube system with homogeneous spot performance, *Proc. 9th Intern. Disp. Res. Conf. (Japan Disp. '89)*, 458–461.

Gibson, J. W., Haas, G. A., and Thomas, R. E. (1989). Investigation of scandate cathodes: emission, fabrication, and processes, *IEEE Trans. Electron. Devices*, **ED-36**(1), 209–213.

Gold, R. D., and Schwartz, J. W. (1958). Drive factor and gamma of conventional kinescope guns, *RCA Rev.*, **24**, 564–583.

Gorog, I. (1994). Displays for HDTV:Direct-view CRTs and projection systems, *Proc. IEEE*, **82**(4), 520–536.

Grivet, P. (1972a). *Electron Optics*, 2nd ed., Pergamon Press, Oxford and New York, pp. 155–198.

Grivet, P. (1972b). *Electron Optics*, 2nd ed., Pergamon Press, Oxford and New York, pp. 287–301.

Haantjes, J., and Lubben, G. (1959). Errors of magnetic deflection II, *Philips Res. Reports.*, **14**, 65–97.

Hamano, E., Okada, H., Koshigoe, S., Suzuki, Y., and Seino, K. (1979). Supernarrow-neck color picture tube, *Toshiba Rev.*, **34**(10), 875–878.

Hara, Y., Ogawa, S., Takeuchi, K., and Kanna, K. (1990). A higher current density oxide-coated cathode for CRT use, *SID Intern. Symp. Digest*, **XXI**, 442–445.

Hasker, J., and Groendijk H. (1962). Measurement and calculation of the figure of merit of a cathode-ray tube, *Philips Res. Rep.*, **17**, 401–418.

Hasker, J. (1965). Transverse-velocity selection affects the Langmuir equation, *Philips Res. Rep.*, **20**, 34–47.

Hasker, J. (1966). The influence of initial velocities on the beam-current characteristic of electron guns, *Philips Res. Rep.*, **21**, 122–150.

Hasker, J. (1971). Astigmatic electron gun for the beam-indexing color television display, *IEEE Trans. Electron. Devices*, **ED-18**(9), 703–712.

Hasker, J. (1972). Beam-current characteristic and cathode loading of electron guns with rotational symmetry: Some important properties and method of calculation, *Philips Res. Rep.*, **27**, 513–538.

Hasker, J. (1973). Improved electron gun for the beam-indexing color television display, *IEEE Trans. Electron Devices*, **ED-20**(11), 1049–1052.

Hasker, J., and Stoffelen, H. J. H. (1985). Alternative Auger analysis reveals important properties of M-type and scandate cathodes, *Appl. Sur. Sci.*, **24**, 330–339.

?tpt=.1pt>Hasker, J., and van Dorst, P. A. M. (1989). Pitfalls in the evaluation of cathode properties from I-V characteristics, *IEEE Trans. Electron Devices*, **ED-36**, 201–207.

Hasker, J., Crombeen, J. E., and van Dorst, P. A. M. (1989). Comments on progress in scandate cathodes, *IEEE Trans. Electron. Devices*, **ED-36**, 215–219.

Hayashida, Y., Ozawa, T., and Sakurai, H. (1995). An analysis of cut-off voltage drift in the oxide cathode life, *Tech. Group Inform. Disp. Inst. Telev. Eng. Jpn.*, **IDY95-13**, 13–18.

Heijnemans, W. A. L., Nieuwendijk, J. A. M., and Vink, N. G. (1980). The deflection coils of the 30AX colour-picture system, *Philips Tech. Rev.*, **39**(6/7), 154–171.

Hellings, G. J. A., and Baalbergen, J. J. (1997). Quality assurance systems in the manufacturing of electron guns and electron gun components, *Proc. 4th Intern. Disp. Workshop (IDW '96)*, 477–480.

Hernqvist, K. G. (1981). Studies of flashovers and preventive measures for Kinescope guns, *IEEE Trans. Consum. Electron.*, **CE-27**, 117–128.

Hernqvist, K. G., and Liller, P. R. (1984). High voltage processing and arc suppression for color picture tubes, *Proc. XIth Intern. Symp. on Discharge Electrical Insulation in Vacuum*, pp. 433–435.

Herrmann, G., and Wagener, P. S. (1951a). *The Oxide Coated Cathodes*, **1**, Chapman & Hall, London, 15–73.

Herrmann, G., and Wagener, P. S. (1951b). The Oxide *Coated Cathodes*, **2**, Chapman & Hall, London, pp. 1–29.

Herrmann, G., and Wagener, P. S. (1951c). *The Oxide Coated Cathodes*, **2**, Chapman & Hall, London, pp. 35–44.

Herrmann, G., and Wagener, P. S. (1951d). *The Oxide Coated Cathodes*, **2**, Chapman & Hall, London, pp. 150–174.

Hirabayashi, K., Kitagawa, O., and Natsuhara, M. (1982). In-line type high-resolution color display tube, *Natl. Tech. Rep.*, (*Matsushita Electr. Ind. Co.*), **28**(1), 94–106.

Hosokoshi, K., Ashizaki, S., and Suzuki, H. (1980). A new approach to a high performance electron gun design for color picture tubes, *IEEE Trans. Consum. Electron.*, **CE-26**, 452–458.

Hosokoshi, K., Ashizaki, S., and Suzuki, H. (1983). Improved OLF in-line gun system, *Proc. 3rd Intern. Disp. Res. Conf. (Japan Disp. '83)*, 272–275.

Hughes, R. H., and Chen, H. Y. (1979). A novel high-voltage bipotential CRT gun design, *IEEE Trans. Consum. Electron.*, **CE-25**(2), 185–191.

Ichida, K., Nakayama, Y., and Inoue, H. (1987). A complex lens Trinitron gun for a high-resolution color tube, *Proc. 7th Intern. Disp. Res. Conf. (Euro Disp. '87)*, 204–207.

Ichida, K., Watanabe, Y., and Inoue, H. (1988). A Trinitron gun with dynamic quadrupole lens providing focus control and astigmatism correction, *Proc. 8th Intern. Disp. Res. Conf.*, pp. 13–16.

Iguchi, Y., Hasegawa, K., Hayashi, M., and Nakayama, Y. (1992). A new method of designing a DQL Trinitron gun to reduce the dynamic focus voltage, *SID Intern. Symp. Digest*, **XXIII**, 881–884.

Ikegami, K., Okuda, S., Sano, K., and Nosaka, E. (1986). Measurement of emittance diagram of electron gun asymmetrical beam-forming region, *IEEE Trans. Electron. Devices*, **ED-33**(8), 1145–1148.

Imai, T. (1962). Thermionic cathodes for electron tubes: A survey, *Oyo Butsuri (Jpn.)*, **31**(12), 963–983.

Inoue, T., Nakamura, M., Ohmura, N., Ohta, Y., and Takamura, T. (1992). A super-flat high-resolution Trinitron, *SID Intern. Symp. Digest*, **XXIII**, 877–880.

Ishida, R., Yamane, H., Toshiyasu, M., Okuda, S., and Nakanishi, H. (1986). Dynamic beam shaping of in-line color CRTs, *SID Intern. Symp. Digest*, **XVII**, 327–329.

Iwasaki, K., and Konosu, O. (1987). Coma correction free deflection yoke, Tech. *Group Electron. Devices Inst. Telev. Eng. Jpn.*, **ED87-55**, 19–24.

Jenkins, R. O. (1969). A review of thermionic cathodes, *Vacuum*, **19**, 353–359.

Johnson, J. B. (1922). A low voltage cathode ray oscilloscope, *J. Opt. Soc. Am.*, **6**, 701–702.

Kaashoek, J. (1974). Deflection in the 20AX system, *IEEE Trans. Broadcast Telev. Receivers*, **20**(4), 293–298.

Katsuma, T., Hamano, E., Shimaohgi, T., and Umezu, N. (1988). Dynamic astigmatism control quadra potential focus gun for 21-in. flat square color display tube, *SID Intern. Symp. Digest*, **XIX**, 136–139.

Kawakami, H., Ishigaki, I., Nishida, J., and Maeda, H. (1975). Low-blooming electron gun for color picture tube, *J. Inst. Telev. Eng. Jpn.*, **29**(12), 1000–1006.

Kikuchi, N., Natori, M., and Amano, Y. (1996). A new electron gun with multi-astigmatism lens system, *Proc. 3rd Intern. Disp. Workshop (IDW '96)*, 213–216.

Kimura, S., Kobayashi, K., Higuchi, T., Yakabe, T., Matsumoto, S., Yamamoto, E., Miyazaki, D., Ohno, C., and Hara, A. (1989). Dispenser cathodes for electron tubes, *Tech. Group Electron. Devices Inst. Electron. Inform. Commun. Jpn.*, **ED89-120**, 15–21.

Kimura, S., Yatabe, T., Matsumoto, S., Miyazaki, D., Yoshii, T., Fujiwara, M., and Koshigoe, S. (1990). Ir-coated dispenser cathode for CRT, *IEEE Trans. Electron. Devices*, **ED-37**(12), 2564–2567.

Kitagawa, O., Iwasaki, K., Adachi, O., Fujisawa, H., and Kuwabara, Y. (1987). Hyperbolic flat ART color picture tube series, *Natl. Tech. Rep. (Matsushita Electr. Ind. Co.)*, **33**(2). 21–35.

Klemperer, O., and Barnett, M. E. (1971a). *Electron Optics*, 3rd ed., Cambridge Univ. Press, London, pp. 229–232.

Klemperer, O., and Barnett, M. E. (1971b). *Electron Optics*, 3rd ed., Cambridge Univ. Press, London, pp. 38–40.

Klemperer, O., and Barnett, M. E. (1971c). *Electron Optics*, 3rd ed., Cambridge Univ. Press, London, pp. 22–25.

Klemperer, O., and Barnett, M. E. (1971d). *Electron Optics*, 3rd ed., Cambridge Univ. Press, London, pp. 169–175.

Klemperer, O., and Barnett, M. E. (1971e). *Electron Optics*, 3rd ed., Cambridge Univ. Press, London, pp. 390–392.

Klemperer, O., and Barnett, M. E. (1971f). *Electron Optics*, 3rd ed., Cambridge, Univ. Press, London, pp. 81–92.

Klemperer, O., and Barnett, M. E. (1971g). *Electron Optics*, 3rd ed., Cambridge, Univ. Press, London, pp. 33–37, 100–106.

Kobori, Y., Katagiri, Y., Kaji, T., and Hasegawa, N. (1980). A novel arc-suppression technique for cathode ray tubes, *IEEE Trans. Consum. Electron.*, **CE-26**, 446–451.

Koganezawa, N., and Yamamoto, S. (1991). Characteristics of an impregnated cathode for HD-CRTs, *SID Intern. Symp. Digest*, **XXII**, 707–710.

Konosu, O. (1995). The role of electron beam in display devices, *Nuclear Inst. Meth. Phys. Res.*, Sect. A, **363**, 330–336.

Krause, A. V. D. V. (1958). Cathode-ray tubes, U. S. Patent No. 2, 829, 292.

Kubo, T. (1982). Development of high-definition TV displays, *IEEE Trans. Broadcast.*, **BC-28**, 51--64.

Kuramoto, T., Honda, M., Yoshida, T., and Konosu, O. (1989). The SSC deflection yoke for in-line color CRTs, *Proc. SID*, **30**(1), 29–32.

Langmuir, D. B. (1937). Theoretical limitation of cathode ray tubes, *Proc. IRE*, **25**, 977–991.

Langmuir, I. (1923). The effect of space charge and initial velocities on the potential distribution and thermionic current between parallel plane electrodes, *Phys. Rev.*, **21**, 419–435.

Law, R. R. (1937). High current electron gun for projection kinescopes, *Proc. IRE*, **25**, 954–976.

Law, R. R. (1942). Factors governing performance of electron guns in television cathode-ray tubes, *Proc. IRE*, **30**, 103–105.

Lechner, B. (1985). High definition TV, *SID Intern. Symp. Digest*, **XVI**, 14–19.

Lee, S. Y., Nakanishi, H., and Lee, M. Y. (1994). CRT grid emission and its prevention when using dispenser cathodes, *SID Intern. Symp. Digest*, **XXV**, 516–519.

Lemmens, H. J., Tnasen, M. J., and Loosjes, R. (1950). A new thermionic cathode for heavy loads, *Philips Tech. Rev.*, **11**, 341–350.

Lemmens, H. J., and Zalm, P. (1961/62). New developments in oxide-coated cathodes: I. Oxide-coated cathodes for loads of 1 to 2 A/cm^2, *Philips Tech. Rev.*, **23**(1), 19–23.

Levi, R. (1955). U.S. Patent No. 2, 700, 000.

Levi, R. (1957/58). The impregnated cathode, *Philips Tech. Rev.*, **19**(6), 186–190.

Levi, R. (1965). U.S. Patent No. 3, 201, 639.

Loosjes, R., and Vink, H. J. (1949). The conduction mechanism in oxide-coated cathodes, *Philips Res. Rep.*, **4**, 449–475.

Loosjes, R., and Vink, H. J. (1950). Conduction processes in the oxide-coated cathode, *Philips Tech. Rev.*, **11**(9), 271–278.

Loty, C. (1984). Simulation of cathode-region electron beams, *SID Intern. Symp. Digest*, **XV**, 262–263.

Lucchesi, B. F., and Carpenter, M. E. (1979). "Pictures" of deflected electron spot from a computer, *IEEE Trans. Consum. Electron.*, **CE-25**, 468–474.

MacGregor, D. M. (1983). Computer-aided design of color picture tubes with a three-dimensional model, *IEEE Trans. Consum. Electron.*, **CE-29**(3), 318–325.

MacGregor, D. M. (1986). Three-dimensional analysis of cathode ray tubes for color television, *IEEE Trans. Consum. Electron.*, **CE-33**(8), 1098–1106.

Medicus, G. (1979). Thermionic constants determined from retarding-field data, *J. Appl. Phys.*, **50**(5), 3666–3673.

Mitsuda, K., Sugawara, K., Konosu, O., and Ashizaki, S. (1991). 107 cm diagonal 16:9 color CRT for HDTV displays, *Proc. 11th Intern. Disp. Res. Conf.*, 35–38.

Miyazaki, M., Izumida, Y., Yamazaki, E., and Shirai, S. (1986). An elliptical aperture, concaved surface main lens for in-line color CRT, *Proc. 6th Intern. Disp. Res. Conf. (Japan Disp. '86)*, 48–51.

Morrell, A. M., Law, H. B., Ramberg, E. G., and Herold, E. W. (1974a). *Color Television Picture Tubes* (B. Kazan, Ed.), Academic Press, New York, pp. 91–99.

Morrell, A. M., Law, H. B., Ramberg, E. G., and Herold, E. W. (1974b). *Color Television Picture Tubes.*, Academic Press, New York, pp. 129–133.

Morrell, A. M. (1974). Design principles of the RCA large-screen 110° precision in-line color picture tube (B. Kazan, Ed.), 2nd Annual Convention Fernseh und Kinotechnische Gesellschaft, Munich, Germany, Oct. 15.

Morrell, A. M. (1982). An overview of the COTY-29 tubes system: an improved generation of color picture tubes, *IEEE Trans. Consum. Electron.*, **CE-28**(3), 290–296.

Morrell, A. M. (1983). The RCA COTY-29 tube system, *RCA Engineer*, **28**(4), 28–32.

Moss, H. (1946). The electron gun of the cathode-ray tube, Pt. II, *J. Brit. IRE*, **6**, 99–129.

Moss, H. (1961). Measurements of the limiting image current density produced by electron guns of rotational symmetry, *J. Electron. Control*, **10**, 341–364.

Moss, H. (1968a). *Narrow Angle Electron Guns and Cathode Ray Tubes*, Academic Press, New York, pp. 81–95.

Moss, H. (1968b). *Narrow Angle Electron Guns and Cathode Ray Tubes*, Academic Press, New York, pp. 15–26.

Moss, H. (1968c). *Narrow Angle Electron Guns and Cathode Ray Tubes*, Academic Press, New York, pp. 53–77.

Moss, H. (1968d). *Narrow Angle Electron Guns and Cathode Ray Tubes*, Academic Press, New York, pp. 156–158.

Mueller, W. (1989). Electronic structure of BaO/W cathode surfaces, *IEEE Trans. Electron. Devices*, **ED-36**, 180–187.

Mulvey, T. (1967). Electron microprobes, *in Focusing of Charged Particles* (A. Septier, Ed.), **1**, Academic Press, New York, pp. 475–476.

Muranishi, H., and Okamoto, T. (1977). An improvement of electron gun for color CRT, *Tech. Group Electron. Devices Inst. Electron. Inform. Commun. Jpn.*, **ED77-72**, 9–16.

Nakanishi, H., and Imanishi, W. (1980). Creeping discharge in color picture tubes, *IEEE Trans. Consum. Electron.*, **CE-26**, 431–444.

Nakanishi, H., Okuda, S., Yoshida, T., and Sugahara, T. (1986). A high-resolution color CRT for CAD/CAM use, *IEEE Trans. Electron. Devices*, **ED-33**(8), 1141–1144.

Nakanishi, H. (1987). Thermionic cathodes for CRTs, *Oyo Butsuri (Jpn.)*, **56**(11), 1423–1432.

Nakanishi, H., and Sano, K. (1989). High current density cathodes for CRT use, *Tech. Group Inform. Disp. Inst. Telev. Eng. Jpn.*, **ID89-23**, 1–8.

Nakanishi, H., Sano, K., Suzuki, R., and Saito, M. (1990). An oxide-coated cathode for CRT use at high current density, *Proc. SID*, **31**, 165–169.

Nakanishi, H. (1994). Cathodes for CRTs, *Display and Imaging (Jpn.)*, **3**, 99–106.

Nakanishi, H., Takechi, K., Furuya, K., and Tawa, Y. (1994). A new type dispenser cathode containing Y_2O_3 for CRTs, *Tech. Group Electron. Inform. Display Inst. Electron. Inform. Commun. Eng. Jpn.*, **EID94-23**, 49–54.

Nakanishi, H. (1996). CRT stray emission study with 2D-FEM, *Proc. 16th Intern. Disp. Res. Conf. (Euro Disp. '96)*, 189–192.

Natori, M., Kikuchi, N., and Amano, Y. (1997). A new method for correcting delta deflection defocusing, *Proc. 4th Intern. Disp. Workshop (IDW '96)*, 473–476.

Nikaido, S., Ouchi, Y., Kimura, S., Yamamoto, E., Kobayashi, K., and Onoe, A. (1982). Some considerations of operation mechanism of oxide cathodes, *Tech. Group Electron. Devices. Inst. Electron. Commun. Eng. Jpn.*, **ED82-90**, 49–56.

Niklas, W. F., Szegho, C. S., and Wimpffen, J. (1957). A television picture tube with increased effective perveance for cathode modulation, *J. Telev. Soc.*, **8**, 368–375.

Ninomiya, K., Urano, T., and Okoshi, T. (1971). Digital computer analysis of electron guns for cathode-ray tubes by taking into account initial thermal velocities, *J. Inst. Electron. Inform. Commun. Eng. Jpn.*, **54-B**, 490–497.

Nose, H., Shirai, S., Ueyame, T., Maehara, M., and Sakurai, S. (1995). Development of 56 cm color display tube, *Proc. 15th Intern. Disp. Res. Conf. (Asia Disp. '95)*, 749–752.

Ogusu, C. (1969). Electron-optical study on electron beams of CRTs with initial thermal velocities, *NHK (Jpn. Broadcast. Corp.) Tech. Rep.*, **21**, 287–338.

Ohsawa, M., Sakurai, S., Ando, K., Funada, E., Kaku, M., and Shiomi, M. (1990). A 32-in. high-brightness color display for HDTV, *SID Intern. Symp. Digest*, **XXI**, 556–559.

Ohta, Y., Takenaka, S., and Ikegaki, M. (1972). RIS system 110-degree-deflection color picture tube, *Toshiba Rev.*, **27**(11), 1007–1010.

Oku, K., Tobe, A., Imazeki, M., Shirai, K., and Osakabe, K. (1992). Analysis on electron beams of CRTs with axi-asymmetric electron guns, *Proc. 12th Intern. Disp. Res. Conf. (Jpn. Disp. '92)*, 273–276.

Ouchi, Y., Nikaido, M., Kimura, S., Nakayama, S., and Onoe, A. (1982). Structure of basemetal of indirect oxide cathode under operation, *Tech. Group Electron. Devices Inst. Electron. Commun. Eng. Jpn.*, **ED82-89**, 43–48.

Pierce, J. R. (1939). Limiting current densities in electron beams, *J. Appl. Phys.*, **10**, 715–724.

Ploke, M. (1951). Elementary theory of electron beam formation with triode system I, *Z. Angew. Phys.*, **3**, 441–449.

Ploke, M. (1952). Elementary theory of electron beam formation with triode system II, *Z. Angew. Phys.*, **4**, 1–12.

Ramberg, E. G. (1942). Variation of the axial aberrations of electron lenses with lens strength, *J. Appl. Phys.*, **13**, 582–594.

Rittner, E. S. (1953). A theoretical study of the chemistry of the oxide cathode, *Philips Res. Rep.*, **8**, 184–238.

Rutter, V. E., and Schimmel, D. G. (1979). Coated powder cathodes-applications and technology, *Appl. Surf. Sci.*, **2**, 118–127.

Saito, T., Kubota, S., and Miyaoka, S. (1973). Digital computer studies of electron lenses for cathode-ray tubes, *J. Appl. Phys.*, **44**(10), 4505–4510.

Saito, T., Kikuchi, M., and Sovers, O. J. (1979). Design of complex electron lenses consisting of two double cylinders, *J. Appl. Phys.*, **50**(10), 6123–6128.

Saito, M., Ishida, M., Fukuyama, K., Watanabe, K., Kamata, T., Sano, K., and Nakanishi, H. (1986). Higher current density oxide cathode for CRT, *NTG Fachber*, **95**, 165–170.

Saito, M., Suzuki, R., Fukuyama, K., Watanabe, K., Sano, K., and Nakanishi H. (1990). Investigation of Sc_2O_3 behavior in Sc_2O_3-dispersed oxide cathodes, *IEEE Trans. Electron Devices*, **ED-37**(12), 2605–2611.

Saito, M., Ohira, T., Suzuki, R., Watanabe, K., Shinjo, T., Sano, K., and Kamata, T. (1995). Characteristics and fabrication technologies of Sc_2O_3-dispersed oxide cathode for CRT applications, *SID Digest Manufac. Tech. Conf.*, **II**, 59–60.

Sakuraya, S., Uba, T., Tamai, H., Oguchi, S., Makino, T., Sudo, M., and Ashiya, R. (1991). 36-in. 34 V high-brightness wide Trinitron CRT for HDTV, *SID Intern. Symp. Digest*, **XXII**, 41–44.

Sato, R., and Yamamoto, H. (1954). Studies on activation of oxide-coated cathodes, *Toshiba Rev.*, **9**(10), 977–983.

Say, D. L. (1978). The "high voltage bipotential" approach to enhanced color tube performance, *IEEE Trans. Consum. Electron.*, **CE-24**(1), 75–80.

Sayama, T., Maki, H., Tominaga, N., Wakasono, H., Mitsuda, K., Miyamoto, H., Kitada, K., and Harada, M. (1991). 16:9 color CRTs for HDTV, *Natl. Tech. Rep.*, (Matsushita Electr. Ind. Co.), **37**(5), 85–92.

Schwartz, J. W. (1957). Space charge limitation on the focus of electron beams, *RCA Rev.*, **18**, 1–11.

Schwartz, J. W., and Fogelson, M. (1979). Recent developments in arc suppression for picture tubes, *IEEE Trans. Consum. Electron.*, **CE-25**, 82–90.

Shafer, D., and Turnbull, J. (1981). The emission carbonate crystallite and oxide cathode performance in electron tubes, *Appl. Surf. Sci.*, **8**, 225–230.

Shirai, S., Takano, H., Fukushima, M., Yamauchi, M., and Iidaka, Y. (1983). A rotationally asymmetric electron lens with elliptical apertures in color picture tubes, *Proc. 3rd Intern. Disp. Res. Conf. (Japan Disp. '83)*, 276–279.

Shirai, S., and Fukushima, M. (1987). Quadrupole lens for dynamic focus and astigmatism control in an elliptical aperture lens gun, *SID Intern. Symp. Digest*, **XVIII**, 162–165.

Shirai, S., Noguchi, K., Miyamoto, S., and Miyazaki, M. (1989). Enhanced elliptical aperture lens gun, *Proc. 9th Intern. Disp. Res. Conf. (Japan Disp. '89)*, 470–473.

Shirai, S., and Noguchi, K. (1991). Reduction of dynamic focus voltage in color CRTs, *SID Intern. Symp. Digest*, **XXII**, 695–698.

Shirai, S., Uchida, G., Yoshiwara, Y., Ishii, K., Kinami, T., Mizukami, N., and Yamauchi, M. (1994). Small neck color display tubes with NEAT (Narrow Elliptical Aperture Technology) electron gun, *Proc. 14th Intern. Disp. Res. Conf.*, pp. 468–471.

Smithgall, H. E. (1978). Internal surge limiting for picture tubes, *IEEE Trans. Consum. Electron.*, **CE-24**(3), 488–491.

Spangenberg, K. R. (1948a). *Vacuum Tubes*, McGraw-Hill, New York, pp. 46–48.

Spangenberg, K. R. (1948b). *Vacuum Tubes*, McGraw-Hill, New York, pp. 168–173.

Spangenberg, K. R. (1948c). *Vacuum Tubes*, McGraw-Hill, New York, pp. 337–349.

Spangenberg, K. R. (1948d). *Vacuum Tubes*, McGraw-Hill, New York, pp. 349–393.

Spanjer, T. G., van Gorkum, A. A., and van Soest, T. L. (1987). A high-resolution electron gun for color projection CRTs, *SID Intern. Symp. Digest*, **XVIII**, 170–173.

Spanjer, T. G., van Gorkum, A. A., and van Alphen, W. M. (1989). Electron guns for projection television, *Philips Tech. Rev.*, **44**(11/12), 348–356.

Spanjer, T. G., Vrijssen, G. A. H. M., and van Alphen, W. M. (1991). A compact high-resolution electron gun with electrostatic focus for projection CRTs, *Proc. SID*, **32**(4), 273–278.

Sudo, M., Ashiya, R., Uba, T., Murata, A., and Amano, Y. (1986). High-resolution 20 V$''$ $\times$ 20 V$''$ Trinitron and monochrome CRT, *SID Intern. Symp. Digest*, **XVII**, 338–341.

Sugawara, S., Kimiya, J., Kamohara, E., and Fukuda, K. (1995). A new dynamic-focus electron gun for color CRTs with tri-quadrupole electron lens, *SID Intern. Sym. Digest*, **XXVI**, 103–106.

Suzuki, H., Masuda, M., and Konosu, O. (1976). Light intensity distribution of a beam spot measured on the CRT screen, *Tech. Group Electron. Devices Inst. Telev. Eng. Jpn.*, **ED76-55**, 1–10.

Suzuki, H., Konosu, O., Ashizaki, S., and Natsuhara, M. (1983). An in-line gun for high-resolution color display tubes, *Proc. 3rd Intern. Disp. Res. Conf. (Japan Disp. '83)*, 612–615.

Suzuki, H., Ashizaki, S., Sugawara, K., Natsuhara, M., and Muranishi, H. (1987). Progressive-scanned 33$''$ 110° flat-square color CRT, *Proc. SID*, **28**(4), 403–407.

Suzuki, H., Mitsuda, K., Muranishi, H., Iwasaki, K., and Ashizaki, S. (1990). A 27 V flat square high-resolution color CRT for graphic displays, *Pro. SID*, **31**(3), 221–225.

Suzuki, H., Tominaga, N., Natsuhara, M., and Sugawara, K. (1992). Improved gun design for self-converging HDTV color CRTs, *Proc. 12th Intern. Disp. Res. Conf. (Japan Disp. '92)*, 269–272.

Suzuki, H., Ueda, Y., Tominaga, N., Hiromitsu, N., and Sugawara, K. (1993). Double-quadrupole DAF gun for self-converging color CRTs, *J. SID*, **1**, 129–134.

Takano, Y. (1983). Recent advances in color picture tubes, *in Advances in Image Pickup and Display*, (B. Kazan, Ed.), **6**, Academic Press, New York, 28–32.

Takenaka, S., Hamano, E., Koshigoe, S., and Kamohara, E. (1979). New hi-fi focus electron gun for color cathode-ray tube, *Toshiba Rev.*, **121**, 30–35.

Thierfelder, C. W., and Hinnenkamp, F. J. (1979/80). Low arc picture tube development using statistically designed experiments, *RCA Engineer*, **25**(4), 27–33.

Thomas, R. E., Gibson, J. W., Haas, G. A., and Abrams, Jr., R. H. (1990). Thermionic sources for high-brightness electron beams, *IEEE Trans. Electron. Devices*, **37**(3), 850–861.

Thompson, B. J., and Headrick, L. B. (1940). Space charge limitations on the focus of electron beams, *Proc. IRE*, **28**, 318–324.

Ueda, Y., Sukeno, M., and Suzuki, H. (1995). Beam spot shift in a dynamic astigmatism correction type (DQ-DAF) electron gun, *Nuclear Instr. Meth. Phys. Res.*, Sect. A, **363**, 389–392.

Ueda, Y., Ohta, K., and Itoh, T. (1996). New main lens system with larger lens diameter for color CRTs, *Proc. 16th Intern. Disp. Res. Conf. (Euro Disp. '96)*, 493–496.

van den Broek, M. H. L. M. (1985). Simulation of 3D electron beams with a fitting technique, *Optik*, **71**(1), 27–30.

van den Broek, M. H. L. M. (1986a). Electron-optical simulation of rotationally symmetric triode electron guns, *J. Appl. Phys.*, **60**(11), 3825–3835.

van den Broek, M. H. L. M. (1986b). Calculation of the radius of the emitting area and the current in triode electron guns, *J. Phys. D: Appl. Phys.*, **19**, 1389–1399.

van der Heide, P. M. A. (1992). OsRu coated impregnated cathode for use in cathode ray tube, Conf. Record 1992 Tri-Service/NASA Cathode Workshop, Greenbelt, Maryland, 155–159.

van Gorkum, A. A. (1985). The cup model for the cathode lens in triode electron guns, *Optik*, **71**(3), 93–104.

van Gorkum, A. A., and van den Broek, M. H. L. M. (1985). Spot reduction in electron guns using a selective prefocusing lens, *J. Appl. Phys.*, **58**(8), 2902–2908.

van Gorkum, A. A., and Spanjer, T. G. (1986). A generalized comparison of spherical aberration of magnetic and electrostatic electron lenses, *Optik*, **72**(4), 134–136.

van Oostrum, K. J. (1985). CAD in light optics and electron optics, *Philips Tech. Rev.*, **42**(3), 69–84.

van Raalte, J. A. (1992). CRTs for high-definition television, *Inform. Disp.*, **8**(1), 6–10.

Washino, S., Ueyama, Y., and Takenobu, S. (1979). Development of a new electron gun for color picture tubes, *IEEE Trans. Consum. Electron.*, **CE-25**, 481–490.

Watanabe, Y., Nakayama, Y., and Saito, T. (1985). Improved uniform focus character display with CFD, *SID Intern. Symp. Digest*, **XVI**, 187–189.

Weber, C. (1964). The electron beam in a cathode-ray tube, *Proc. IEEE*, **52**, 996–1001.

Weber, C. (1967a). Analogue and digital methods for investigating electron-optical systems, *Philips Res. Rep. Suppl.*, No. 6.

Weber, C. (1967b). Numerical solution of Laplace's and Poisson's equations and the calculation of electron trajectories and electron beams, *in Focusing of Charged Particles* (A. Septier, Ed.) **1**, Academic Press, New York, pp. 45–99.

Wehnelt, A. (1904). Uber den Austritt negativer Ionen aus gluhenden Metallverbindungen und damit zusammenhangende Erscheinungen, *Ann. der Phys.*, **14**(8), 425–468.

Wilson, A. H. (1931). The theory of electronic semiconductors, *Proc. Roy. Soc. London*, **133**, 458–491.

Wilson, I. M. (1975). Theoretical and practical aspects of electron gun design for color picture tubes, *IEEE Trans. Consum. Electron.*, **CE-21**, 32–38.

Yamada, Y., Ashiya, R., Kikuchi, M., and Saita, K. (1980). A new high-resolution Trinitron picture tube for display application, *IEEE Trans. Consum. Electron.*, **CE-26**, 466–472.

Yamaguchi, A., Yamauchi, M., and Fukushima, M. (1979). Development of B-U type electron gun, *Tech. Group Electron Devices, Inst. Telev. Eng. Jpn.*, **ED-459**, 17–20.

Yamamoto, S. Taguchi, S., Watanabe, I., and Kawase, S. (1986). Electron emission properties and surface atom behavior of an impregnated cathode coated with tungsten film containing Sc_2O_3, *Jpn. J. Appl. Phys.*, **25**(7), 971–975.

Yamamoto, S. (1988). Monoatomic layer on electron emissive materials, *Oyo Butsuri (Jpn.)*, **57**(11), 1734–1741.

Yamamoto, S., Watanabe, I., Sasaki, S., Yaguchi, T., and Taguchi, S. (1989). Impregnated cathode operable at low temperatures, *Tech. Group Inform. Disp. Inst. Telev. Eng. Jpn.*, **ID89-24**, 9–16.

Yamane, H., Ishida, R., Ikegami, K., Okuda, S., and Nakanishi, H. (1986). An in-line color CRT with dynamic beam shaping for data display, *Proc. 6th Intern. Disp. Res. Conf. (Japan Disp. '86)*, 160–163.

Yamazaki, E. (1988). Future requirements for high-resolution full-color display, *Trans. Inst. Electron. Inform. Commun. Eng. Jpn.*, **E71**(11), 1047–1049.

Yamazaki, H., Nakamura, M., and Takahashi, T. (1987). 32-in. full-square color picture tube, *Toshiba Rev.*, **42**(8), 605–608.

Yoshida, S., Ohkoshi, A., and Miyaoka, S. (1968). The "TRINITRON"— a new color tube, *IEEE Trans. Broadcast. Telev. Receivers*, **BTR-14**, 19–27.

Yoshida, S., Ohkoshi, A., and Miyaoka, S. (1973). A wide-deflection angle (114°) Trinitron color picture tube, *IEEE Trans. Broadcast. Telev. Receivers*, **ED-19**(4), 231–238.

Yoshida, S., Ohkoshi, A., and Miyaoka, S. (1974). 25 V inch 114 degree Trinitron color picture tube and associated new developments, *IEEE Chicago Spring Conference on BTR.*

Yoshida, S. (1977). A new 30 V-inch screen size color picture tube, *IEEE Trans. Consum. Electron.*, February 1–7.

Yun, N. Y., Kim, H. K., Kim, H. C., and Kim, I. T. (1994). An in-line gun with a double-slot beam-forming region developed for flat square 110° CRTs, *SID Intern. Symp. Digest*, **XXV**, 655–658.

Zalm, P., and van Stratum, A. J. A. (1966). Osmium dispenser cathodes, *Philips Tech. Rev.*, **27**(3/4), 69–75.

Zmuda, E. I., Say, D. L., and Lucchesi, B. F. (1983). The conical field focus (CFF) electron lens—a new dimension in color electron optics, *Proc. 3rd Intern. Disp. Res. Conf. (Japan Disp. '83)*, 268–271.

Zworykin, V. K. (1929). Television with cathode-ray tube for receiver, *Radio Engineering*, **9**(Dec.), 38–41.

Zworykin, V. K., Morton, G. A., Ramberg, E. G., Hillier, J., and Vance, A. W. (1945a). *Electron Optics and the Electron Microscope*, John Wiley and Sons, New York, pp. 423–426.

Zworykin, V. K., Morton, G. A., Ramberg, E. G., Hillier, J., and Vance, A. W. (1945b). *Electron Optics and the Electron Microscope*, John Wiley and Sons, New York, pp. 437–456.

Zworykin, V. K., and Morton, G. A. (1954). *Television*, 2nd ed., John Wiley and Sons, New York.

INDEX

L

M

N

O

P

T

ISBN 0-12-014747-5